Advances in Experimental Medicine and Biology

Cell Biology and Translational Medicine

Volume 1347

Cell Biology and Translational Medicine aims to publish articles that integrate the current advances in Cell Biology research with the latest developments in Translational Medicine. It is the latest subseries in the highly successful Advances in Experimental Medicine and Biology book series and provides a publication vehicle for articles focusing on new developments, methods and research, as well as opinions and principles. The Series will cover both basic and applied research of the cell and its organelles' structural and functional roles, physiology, signalling, cell stress, cell-cell communications, and its applications to the diagnosis and therapy of disease.

Individual volumes may include topics covering any aspect of life sciences and biomedicine e.g. cell biology, translational medicine, stem cell research, biochemistry, biophysics, regenerative medicine, immunology, molecular biology, and genetics. However, manuscripts will be selected on the basis of their contribution and advancement of our understanding of cell biology and its advancement in translational medicine. Each volume will focus on a specific topic as selected by the Editor. All submitted manuscripts shall be reviewed by the Editor provided they are related to the theme of the volume. Accepted articles will be published online no later than two months following acceptance.

The Cell Biology and Translational Medicine series is indexed in SCOPUS, Medline (PubMed), Journal Citation Reports/Science Edition, Science Citation Index Expanded (SciSearch, Web of Science), EMBASE, BIOSIS, Reaxys, EMBiology, the Chemical Abstracts Service (CAS), and Pathway Studio.

More information about this series at https://link.springer.com/bookseries/5584

Kursad Turksen
Editor

Cell Biology and Translational Medicine, Volume 14

Stem Cells in Lineage Specific Differentiation and Disease

Editor
Kursad Turksen (emeritus)
Ottawa Hospital Research Institute
Ottawa, ON, Canada

ISSN 0065-2598 ISSN 2214-8019 (electronic)
Advances in Experimental Medicine and Biology
ISSN 2522-090X ISSN 2522-0918 (electronic)
Cell Biology and Translational Medicine
ISBN 978-3-030-80494-7 ISBN 978-3-030-80492-3 (eBook)
https://doi.org/10.1007/978-3-030-80492-3

This Springer imprint is published by the registered company Springer Nature Switzerland AG
The registered company address is: Gewerbestrasse 11, 6330 Cham, Switzerland

Preface

In this next volume in the Cell Biology and Translational Medicine series, we continue to explore the potential utility of stem cells in regenerative medicine. In this volume, the authors have addressed such topics as the use of various types or sources of stem cells for generation of tissue-specific differentiated cells, the potential utility of specifically-designed biomaterials such as bilayer scaffolds at tissue interfaces and nanoparticles for drug delivery, the use of organoids in transplantation and, of course, stem cells as therapeutic options in various disease states.

I remain very grateful to Gonzalo Cordova, the Associate Editor of the series and wish to acknowledge his continued support.

I would also like to acknowledge and thank Mariska van der Stigchel, Assistant Editor, for her outstanding efforts in helping to bring this volume to the production stages.

A special thank you goes to Shanthi Ramamoorthy and Rathika Ramkumar for their outstanding efforts in the production of this volume.

Finally, sincere thanks to the contributors not only for their support of the series, but also for their willingness to share their insights and all their efforts to capture both the advances and the remaining obstacles in their areas of research. I trust readers will find their contributions as interesting and helpful as I have.

Ottawa, ON, Canada — Kursad Turksen

Contents

Adv Exp Med Biol - Cell Biology and Translational Medicine (2021) 14: 1–27
https://doi.org/10.1007/5584_2021_653

Published online: 24 August 2021

Recent Advances in the Generation of β-Cells from Induced Pluripotent Stem Cells as a Potential Cure for Diabetes Mellitus

Akriti Agrawal, Gloria Narayan, Ranadeep Gogoi, and Rajkumar P. Thummer

Abstract

Diabetes mellitus (DM) is a group of metabolic disorders characterized by high blood glucose levels due to insufficient insulin secretion, insulin action, or both. The present-day solution to diabetes mellitus includes regular administration of insulin, which brings about many medical complications in diabetic patients. Although islet transplantation from cadaveric subjects was proposed to be a permanent cure, the increased risk of infections, the need for immunosuppressive drugs, and their unavailability had restricted its use. To overcome this, the generation of renewable and transplantable β-cells derived from autologous induced pluripotent stem cells (iPSCs) has gained enormous interest as a potential therapeutic strategy to treat diabetes mellitus permanently. To date, extensive research has been undertaken to derive transplantable insulin-producing β-cells (iβ-cells) from iPSCs in vitro by recapitulating the in vivo developmental process of the pancreas. This in vivo developmental process relies on transcription factors, signaling molecules, growth factors, and culture microenvironment. This review highlights the various factors facilitating the generation of mature β-cells from iPSCs. Moreover, this review also describes the generation of pancreatic progenitors and β-cells from diabetic patient–specific iPSCs, exploring the potential of the diabetes disease model and drug discovery. In addition, the applications of genome editing strategies have also been discussed to achieve patient-specific diabetes cell therapy. Last, we have discussed the current challenges and prospects of iPSC-derived β-cells to improve the relative efficacy of the available treatment of diabetes mellitus.

Keywords

Cell reprogramming · Diabetes mellitus · Disease modeling · Genome editing · Growth factors · Induced pluripotent stem cells · Microenvironment · Pancreatic progenitors ·

Authors Akriti Agrawal and Gloria Narayan have equally contributed to this chapter.

A. Agrawal, G. Narayan, and R. P. Thummer (✉)
Laboratory for Stem Cell Engineering and Regenerative Medicine, Department of Biosciences and Bioengineering, Indian Institute of Technology Guwahati, Guwahati, Assam, India
e-mail: a.akriti@iitg.ac.in; gloria@iitg.ac.in; rthu@iitg.ac.in

R. Gogoi
Department of Biotechnology, National Institute of Pharmaceutical Education and Research Guwahati, Changsari, Guwahati, Assam, India
e-mail: gogoiranadeep@yahoo.co.in

Small molecules · Transcription factors · β-cells

Abbreviations

4PBA	4-phenyl butyric acid
ALK	Activin receptor-like kinase
BMP	Bone morphogenetic protein
CYC	KAAD-cyclopamine
DM	Diabetes mellitus
EGF	Epidermal growth factor
ER	Endoplasmic reticulum
FGF	Fibroblast growth factor
GLP1	Glucagon-like peptide-1
GLUT2	Glucose transporter 2
GSIS	Glucose-stimulated insulin secretion
GSK3β	Glycogen synthase kinase 3β
HDAC	Histone deacetylase
HGF	Hepatocyte growth factor
HIF	Hypoxia-inducible factor
IBMX	3-isobutyl-1-methylxanthine
IGF-II	Insulin-like growth factor II
ILV	Indolactam V
iPSCs	Induced pluripotent stem cells
iβ-cells	Insulin-producing β-cells
KGF	Keratinocyte growth factor
MAPK	Mitogen-activated protein kinase
miR	MicroRNA
MODY	Maturity-onset diabetes of the young
OSK	Oct4, Sox2, and Klf4
OSKM	Oct4, Sox2, Klf4, and c-Myc
PdBU	Phorbol 12,13-dibutyrate
PI3K	Phosphoinositide 3-kinase
PLLA/PVA	Poly-L-lactic acid/polyvinyl alcohol
RA	Retinoic acid
ROCK	Rho-associated protein kinase
SHH	Sonic hedgehog
T1DM	Type 1 Diabetes mellitus
T2DM	Type 2 Diabetes mellitus
T3	Triiodothyronine
TGF-β	Transforming growth factor-β
WFS	Wolfram syndrome
WNT3a	Wingless-related integration site-3a

1 Introduction

The prevalence of diabetes mellitus (DM) is increasing day by day, and diabetic patients suffer from impaired blood glucose levels, carbohydrate, fat, and protein metabolism resulting in multiple health complications like retinopathy, neuropathy, and nephropathy (Fong et al. 2004; Pop-Busui et al. 2017; Molitch et al. 2004). Moreover, diabetic patients are prone to atherosclerotic cardiovascular and cerebrovascular diseases (Tzoulaki et al. 2009; Quinn et al. 2011; Holman et al. 2014; Leon and Maddox 2015). The present number of affected lives accounts for more than 463 million (around 9.3%), and this number is further expected to rise to 700 million by 2045 (Saeedi et al. 2019). Diabetes-associated cardiovascular complications are the principal causes of most deaths occurring in DM, partly due to damage to the renal microvasculature (Carmines 2010).

DM is broadly classified into two major etiopathogenetic categories: First, type 1 DM (T1DM), previously known as juvenile-onset diabetes, accounts for around 5–10% of the total diabetic population. T1DM is signified by a varying rate of destruction of β-cells, from fast destruction (mainly in infants) to slow destruction (in the case of adults) and leads to no or reduced insulin secretion affecting children aged below 14 years. Individuals at an increased risk of developing this type of DM can often be identified by the occurrence of an autoimmune pathological process in the islets and also by genetic markers (Katsarou et al. 2017). Second, type 2 DM (T2DM), previously referred to as non-insulin-dependent diabetes, is the most common form accounting for 90–95% of the cases encompassing individuals with insulin resistance or lower insulin secretion compared with normal values (Lorenzo et al. 2017). In T2DM, insulin resistance by peripheral tissues is the primary abnormality. At this stage, the majority of the patients exhibit normoinsulinemia. As the disease advances, it is characterized by intact but progressively dysfunctional β-cells (Kahn 2003; Elsayed et al. 2021). The third category is gestational DM and considered as one of the most ubiquitous conditions complicating pregnancy that has a prevalence of 1–28% worldwide (Saeedi et al. 2021).

The present-day solutions to DM include antidiabetic oral drugs, regular administration of insulin shots, bariatric surgery (cases where the overweight is the prime cause), islet, and complete organ transplantation. Moreover, the bioartificial pancreas can be a possible promising solution in the near future. However, most of these treatments are temporary or are still under development facing different challenges or cause various medical complications. Among these, islet or complete organ transplantation emerged as a promising alternative for replacing the lost β-cells, but ethical issues, immune rejection, transmission of the virus from porcine to human (in case of porcine islet transplantation), and scarcity of cadaveric donors make it less appealing (Merani and Shapiro 2006; Pellegrini et al. 2016). When immunosuppressive drugs were used to suppress the immune rejection, serious opportunistic infections became another barrier in this path. Apart from insulin therapy, many drugs of synthetic or semisynthetic origin are playing a pivotal role in the management of diabetes. However, it is becoming more evident that the use of these drugs is associated with an increased risk of cardiovascular diseases. Likewise, insulin therapy requires multiple injections of insulin (after every carbohydrate diet) in a day. This makes insulin therapy extremely cumbersome and painful. Furthermore, weight gain during long-term insulin treatment significantly increases cardiovascular risks (Cichosz et al. 2017). The most feasible approaches for the generation of β-cells and thereby available treatment options for diabetes are shown in Fig. 1.

The generation of induced pluripotent stem cells (iPSCs) offers a great promise for the treatment of degenerative metabolic diseases or disorders such as DM. Any somatic cell type can be reprogrammed to a pluripotent state by ectopic expression of defined factors to generate these cells. Eventually, these pluripotent stem cells will give rise to any desired cell type. Similarly, somatic cells from a diabetic patient can be reprogrammed to iPSCs followed by differentiation toward insulin-producing β-cells (i-β-cells). These cells can then be transplanted back to the same patient to replace the destroyed (in case of T1DM) and dysfunctional (in case of T2DM) pancreatic β-cells. Further, these differentiated β-cells can be used for cell therapy, biobanking, drug screening, disease modeling, and tissue engineering (Fig. 2).

2 iPSCs

iPSCs are generated by reprogramming somatic cells into an early embryonic-like state by introducing a combination of transcription factors. In 2006, a groundbreaking study reported iPSC generation by reprogramming the mouse embryonic fibroblasts using a cocktail of stem cell–specific reprogramming factors (Oct4, Sox2, Klf4, and c-Myc [OSKM]) (Takahashi and Yamanaka 2006). These iPSCs were found to be very similar to embryonic stem cells with respect to genetic, epigenetic, developmental, and functional features (Takahashi et al. 2007; Yu et al. 2007). Moreover, they surpass the ethical and immunological concerns associated with embryonic stem cells (Takahashi et al. 2007; Yu et al. 2007; Hossain et al. 2016; Saha et al. 2018). The classical reprogramming techniques utilized viral-based integrative approaches such as γ-retro- and lentiviral vectors to derive iPSCs from somatic cells (Hu 2014). These approaches are highly efficient and robust, but they carry a risk of insertional mutagenesis and tumorigenesis, thereby restricting their clinical applicability (Hu 2014; Borgohain et al. 2019; Saha et al. 2018; Okita et al. 2007). To overcome these concerns, safer nonintegrative reprogramming approaches with minimal or no transgene integrations are developed and employed to derive iPSCs (Hu 2014; Dey et al. 2016; Saha et al. 2018; Haridhasapavalan et al. 2019; Borgohain et al. 2019). The nonintegrative reprogramming approaches can be further categorized into integration-free viral (including

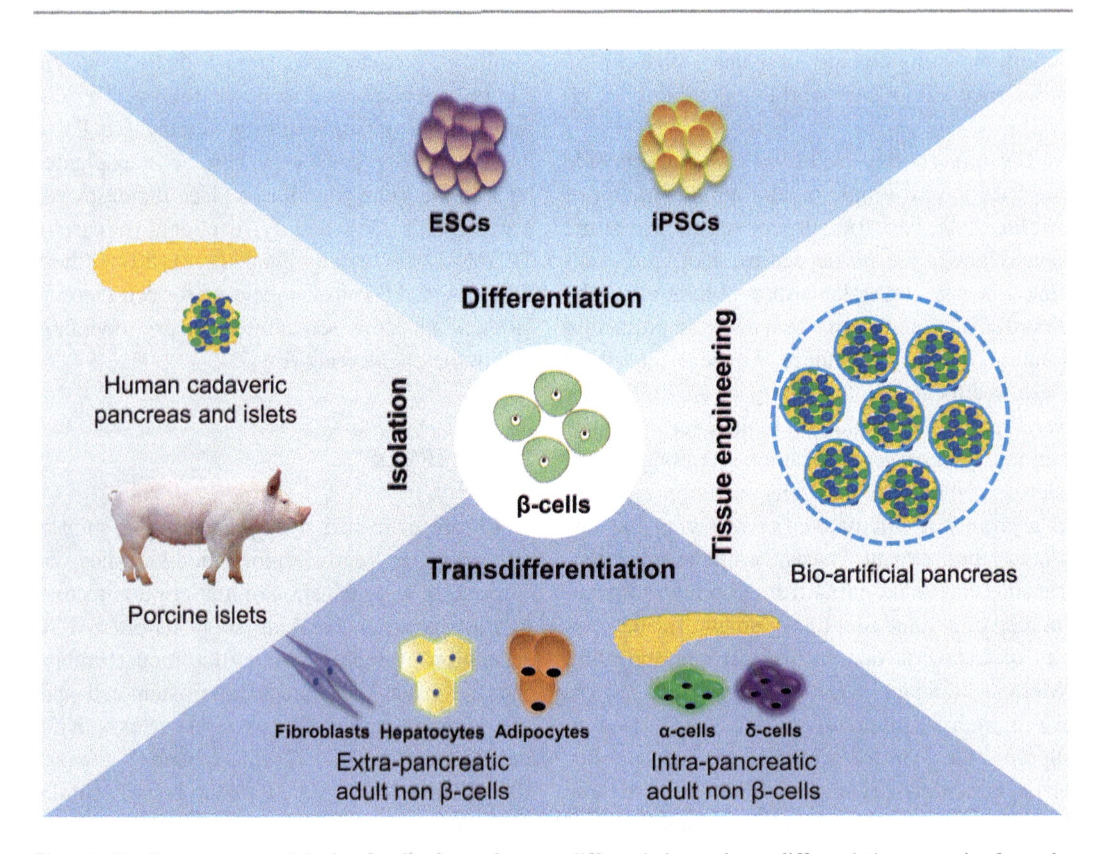

Fig. 1 Various sources rich in β-cells have been identified to replenish the lost or dysfunctional β-cells for the treatment of DM. Different strategies have been developed, which offer an unlimited source of β-cells for the treatment of DM. β-cells can be isolated from sources like cadaveric human pancreas or islets and porcine islets. In addition, β-cells can be obtained via advanced differentiation and transdifferentiation strategies from pluripotent stem cells and adult somatic cells, respectively. Moreover, the tissue-engineered bioartificial pancreas is also considered a potentially promising source rich in functional β-cells. These strategies can prove beneficial for the treatment of DM

adenoviral and Sendai viral), nonviral, deoxyribonucleic acid (DNA)-based (including plasmid, minicircle, episomal, and transposon) (Haridhasapavalan et al. 2019; Dey et al. 2021), and DNA-free (including recombinant proteins, micro-ribonucleic acids (RNAs) (miRs), small molecules, and synthetic messenger RNA) (Borgohain et al. 2019; Dey et al. 2016, 2021) methods. These approaches allow no genomic alterations; moreover, some small molecules also promote and maintain pluripotency by enhancing reprogramming and their subsequent differentiation (Hossain et al. 2016; Borgohain et al. 2019). However, collectively, these nonintegrative approaches are reported to be less efficient with slower kinetics (Borgohain et al. 2019; Dey et al. 2021), due to the presence of various reprogramming barriers (Haridhasapavalan et al. 2020; Saha et al. 2018; Ebrahimi 2015).

In essence, iPSCs are noncontroversial cell types, providing an unlimited source of pluripotent stem cells allowing their wider applications in personalized medicine (Ferreira and Mostajo-Radji 2013). Research findings have suggested the role of iPSCs in disease modeling (e.g., diabetes mellitus) (Millman et al. 2016), drug development and screening (Young 2012), several cell-based therapeutics (Singh et al. 2015), tissue engineering (Loskill and Huebsch 2019), and so forth. The clinical applications of iPSC-derived differentiated cell type(s) are still under investigation and require further improvements concerning

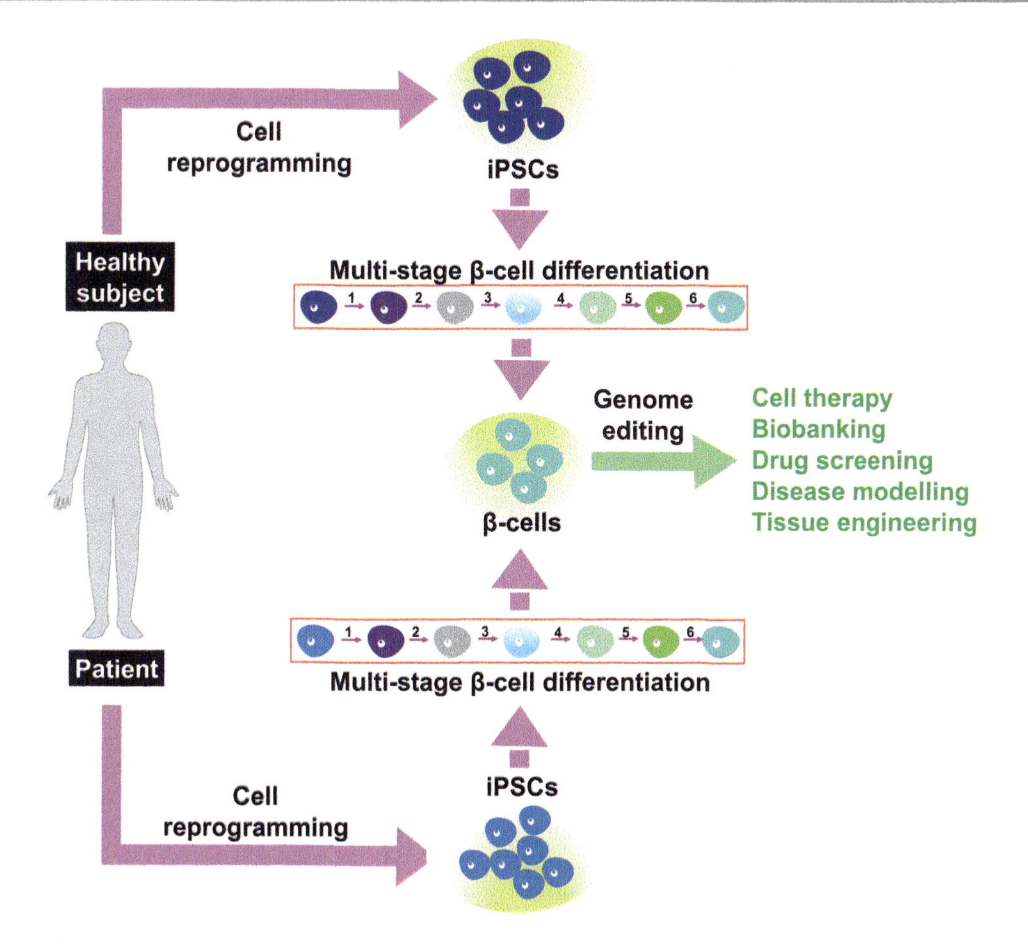

Fig. 2 iPSCs as a promising cell source giving rise to insulin-producing β-cells to treat DM and its other diverse applications. Autologous somatic cells from diabetic patient/healthy subjects can be reprogrammed to iPSCs and subsequently differentiated to β-cells through established multistage differentiation protocols. These personalized β-cells can also be genome-edited, if required, and used for various biomedical applications. In addition, iPSCs and the β-cells generated from them can be biobanked for future research and therapy

several bottlenecks associated with genomic stability and safety.

3 Differentiation of iPSCs to iβ-cells

Pancreas organogenesis in humans and mice demonstrates a series of stages marked by the endogenous expression of several transcription factors, which varies from stage to stage (Fig. 3) (Jennings et al. 2013; Pan and Wright 2011). To date, numerous studies have reported the differentiation of iPSCs into insulin-producing cells or β-cells in a similar fashion that mimics the embryonic pancreas organogenesis stages (Tateishi et al. 2008; Zhang et al. 2009; Zhu et al. 2011; Jeon et al. 2012; Shahjalal et al. 2014; Pellegrini et al. 2015; Millman et al. 2016; Enderami et al. 2018; Yabe et al. 2019). This differentiation process generally includes the sequential treatment of iPSCs with several small molecules and growth factors at multiple differentiation stages (Table 1), which successfully induces the expression of key pancreatic-specific transcription factors (Tateishi et al. 2008; Maehr et al. 2009; Zhang et al. 2009; Kunisada et al. 2011; Pellegrini et al. 2015; Yabe et al. 2019). Most iPSCs to β-cell differentiation

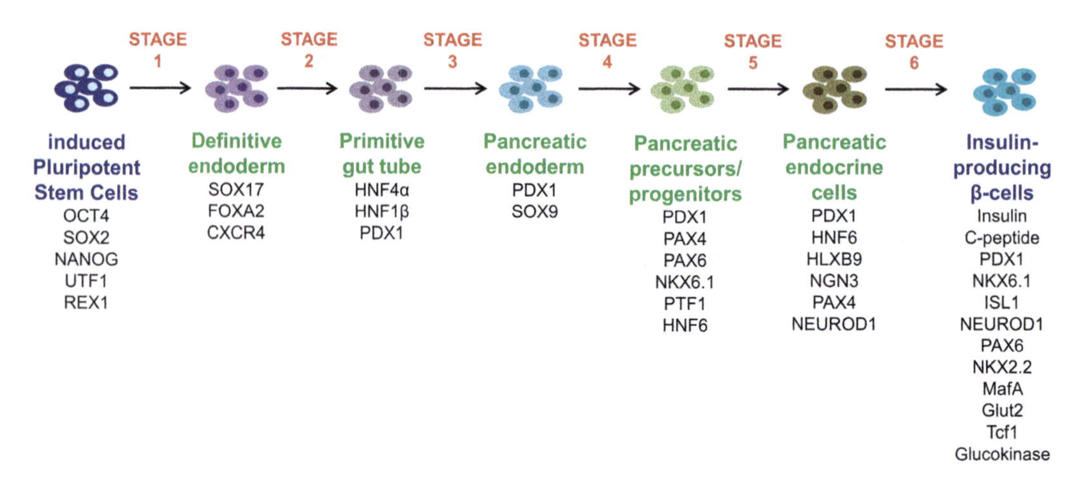

Fig. 3 Differentiation of iPSCs to β-cells in a stage-wise manner. Differentiation of iPSCs to β-cells progresses through a sequence of stages, mimicking the in vivo development (Stages 1–6). The corresponding genetic markers are also mentioned below each development stage that identifies each of these stages

protocols follow sequential pancreas developmental stages (Fig. 3), which includes the formation of (1) definitive endoderm, (2) primitive gut tube, (3) pancreatic endoderm, (4) pancreatic progenitors/precursors, (5) pancreatic endocrine cells, and (6) β-cells (Maehr et al. 2009; Thatava et al. 2010; Zhu et al. 2011; Kaitsuka et al. 2014; Hakim et al. 2014; Shahjalal et al. 2014; Pellegrini et al. 2015; Millman et al. 2016; Pellegrini et al. 2018; Wang et al. 2019; Haller et al. 2019; Yabe et al. 2019).

During gastrulation, definitive endoderm is formed, where it contributes to the development of the digestive tract, including the stomach, intestine, liver, and pancreas (Walczak et al. 2016). Moreover, definitive endoderm also participates in the morphogenesis of lungs, thymus, and thyroid (Walczak et al. 2016). The formation of definitive endoderm is marked by the expression of FOXA2, SOX17, and CXCR4 (Tateishi et al. 2008; Haller et al. 2019). A commonly used inducer of definitive endoderm is activin A (a transforming growth factor-β [TGF-β] family member, NODAL activator) (Hossain et al. 2016) along with wingless-related integration site-3a (WNT3a) (β-catenin activator) or CHIR99021 (glycogen synthase kinase 3β [GSK3β] inhibitor) (Kunisada et al. 2011). Moreover, the combination of activin A and wortmannin (phosphoinositide 3-kinase [PI3K] signaling inhibitor) (Zhang et al. 2009), as well as activin A and sodium butyrate (histone deacetylase [HDAC] inhibitor) (Tateishi et al. 2008), has also been reported to promote definitive endoderm formation (Zhang et al. 2009; Tateishi et al. 2008). Concurrently, efficient definitive endoderm induction is also reported by the combination of activin A with fibroblast growth factor (FGF)-2, bone morphogenetic protein (BMP)-4 (BMP4, a transforming growth factor [TGF]-β family member), and CHIR99021 (Yabe et al. 2019). The definitive endoderm further specializes and forms a primitive gut tube from where the dorsal and ventral pancreatic buds appear (Jennings et al. 2013). At this early stage, the formation of pancreatic anlage depends on the addition of FGFs, retinoid acid (RA), and sonic hedgehog (SHH) signaling inhibitor (Haller et al. 2019; Shahjalal et al. 2014). Treatment of definitive endoderm with FGF family member at lower concentrations, essentially FGF10 or keratinocyte growth factor (KGF), and KAAD-cyclopamine (CYC) (SHH inhibitors), prime definitive endoderm toward primitive gut tube development (Thatava et al. 2010; Shahjalal et al. 2014; Haller et al. 2019; Gheibi et al. 2020b). The formation of the primitive gut tube is marked by the expression of HNF4α, HNF1β,

Table 1 Small molecules and growth factors (with their respective functions) involved at different stages in the iPSC differentiation process eventually giving rise to insulin-producing cells

Stage No.	Stages	Genes expressed	Essential small molecules and growth factors	Major function	References
1	iPSCs → Definitive endoderm	FOXA2, SOX17, CXCR4	Activin A	TGF-β family member, NODAL growth differentiation activator	Tateishi et al. (2008), Hosokawa et al. (2018), Hossain et al. (2016), Jeon et al. (2012), Kunisada et al. (2011), Maehr et al. (2009), Millman et al. (2016), Shahjalal et al. (2014), Walczak et al. (2016), Wang et al. (2019), Yabe et al. (2019), Zhang et al. (2009), Zhu et al. (2011)
			WNT3a	β-catenin activator	Maehr et al. (2009), Kunisada et al. (2011), Zhu et al. (2011)
			CHIR99021	GSK3β inhibitor, induces endoderm	Hosokawa et al. (2018), Kunisada et al. (2011), Millman et al. (2016), Shahjalal et al. (2014), Walczak et al. (2016), Wang et al. (2019)
			Wortmannin	PI3K signaling inhibitor	Jeon et al. (2012), Powis et al. (1994), Zhang et al. 2009)
			Sodium butyrate	HDAC inhibitor	Tateishi et al. (2008), Liu et al. (2018)
			FGF2	Induces MAPK signaling pathway and thereby initiates PDX1 expression	Ameri et al. (2010), Yabe et al. (2019)
			BMP4	TGF-β family member	Herrera & Inman (2009), Yabe et al. (2019)
			Y27632	ROCK inhibitor, promotes cluster integrity and enhances progenitor formation	Hosokawa et al. (2018), Gheibi et al. (2020b)
			PI-103	Antagonist of PI3K signaling	Park et al. (2008), Wang et al. (2019)
2	Definitive endoderm → Primitive gut tube	HNF4α, HNF1β, PDX1	CYC	SHH signaling inhibitors	Maehr et al. (2009), Shahjalal et al. (2014)
			FGF10	Promotes pancreatic epithelium proliferation	Amour et al. (2006), Maehr et al. (2009)
			KGF	FGF family member, an inducer of PDX1 expression and subsequently promotes NKX6.1 expression	Millman et al. (2016), Russ et al. (2015), Wang et al. (2019)
			FGF7	Induces the expression of PDX1, promotes 3D cluster formation	Gheibi et al. (2020b), Yabe et al. (2019)
3	Primitive gut tube → Pancreatic endoderm	PDX1, SOX9	RA	Inducer of PDX1 expression	Johannesson et al. (2009), Kaitsuka et al. (2014), Millman et al. (2016), Pellegrini et al. (2018)
			CYC	SHH inhibitor	Kaitsuka et al. (2014), Pellegrini et al. (2018), Shahjalal et al. (2014)
			FGF10	Promotes pancreatic epithelium proliferation	Amour et al. (2006), Kaitsuka et al. (2014)

(continued)

Table 1 (continued)

Stage No.	Stages	Genes expressed	Essential small molecules and growth factors	Major function	References
			SANT1	SHH antagonist	Millman et al. (2016), Pagliuca et al. (2014)
			LDN	BMP type I receptor inhibitor	Millman et al. (2016), Pagliuca et al. (2014)
			Phorbol 12,13-dibutyrate (PdBU)	Protein kinase C activator	Millman et al. (2016), Pagliuca et al. (2014)
			KGF	**FGF family member, an inducer of PDX1 expression and subsequently promotes NKX6.1 expression**	**Millman et al. (2016), Pellegrini et al. (2018), Russ et al. (2015)**
4	**Pancreatic** endoderm → Pancreatic progenitors /precursors	**PDX1, PAX4,** PAX6, NKX6.1, PTF1, HNF6	**RA**	**Inducer of PDX1 expression**	**Hosokawa et al. (2018), Jeon et al. (2012), Johannesson et al.** (2009), Kunisada et al. (2011), Maehr et al. (2009), Millman et al. (2016), Shahjalal et al. (2014), Wang et al. (2019), Zhang et al. (2009), Zhu et al. (2011)
			CYC	SHH inhibitor	Maehr et al. (2009), Shahjalal et al. (2014)
			Epidermal growth factor	Pancreatic progenitor cell expansion and maturation; induces PDX1 expression	Tateishi et al. (2008), Zhang et al. (2009), Zhu et al. (2011)
			Noggin	BMP signaling inhibitors, induces PDX1 expression	Tateishi et al. (2008), Jeon et al. (2012), Kunisada et al. (2011), Shahjalal et al. (2014), Zhang et al. (2009)
			FGF10	Promotes pancreatic epithelium proliferation	Maehr et al. (2009), Yabe et al. (2019)
			KGF	FGF family member, an inducer of PDX1 expression and subsequently promotes NKX6.1 expression	Millman et al. (2016), Russ et al. (2015), Wang et al. (2019), Zhu et al. (2011)
			SANT1	SHH antagonist	Millman et al. (2016), Pagliuca et al. (2014), Wang et al. (2019), Yabe et al. (2019)
			EC23	Agonist of retinoic acid receptor	Gheibi et al. (2020b), Yabe et al. (2019)
			LDN	BMP type I receptor inhibitor	Wang et al. (2019), Yabe et al. (2019)
			Indolactam V	Protein kinase C activator	Maehr et al. (2009), Yabe et al. (2019)
			Phorbol 12,13-dibutyrate	Protein kinase C activator	Pagliuca et al. (2014), Wang et al. (2019)
			FGF7	Induces the expression of PDX1, promotes 3D cluster formation	Jeon et al. (2012), Zhang et al. (2009)

			SB431542	TGF-β type I receptor kinase inhibitor VI	Hosokawa et al. (2018), Shahjalal et al. (2014)
			Dorsomorphin	Inhibitor of BMP signaling, inhibits BMP type I receptors Induces PDX1 expression when used with RA	Kunisada et al. (2011)
5	Pancreatic progenitors/ precursors → Pancreatic endocrine cells	PDX1, HNF6, HLXB9, NGN3, PAX4, NEUROD1	Epidermal growth factor	Pancreatic progenitor cell expansion and maturation; induces PDX1 expression	Tateishi et al. (2008), Zhang et al. (2009), Zhu et al. (2011)
			FGF10	Promotes pancreatic epithelium proliferation	Hakim et al. (2014)
			RA	Inducer of PDX1 expression	Hakim et al. (2014), Millman et al. (2016), Walczak et al. (2016), Wang et al. (2019)
			Indolactam V	Protein kinase C activator	Shahjalal et al. (2014)
			Noggin	BMP signaling inhibitors, induces PDX1 expression	Tateishi et al. (2008), Kunisada et al. (2011)
			SB431542	TGF-β type I receptor inhibitor, induces NGN3, NEUROD1, PDX1, SOX9, and HLXB9 expression when combined with dorsomorphin and RA	Kunisada et al. (2011), Walczak et al. (2016), Zhu et al. (2011)
			Dorsomorphin	Inhibitor of BMP signaling, inhibits BMP type I receptors	Walczak et al. (2016)
			A-83-01, SB431542	TGF-β type I receptor inhibitor	Kunisada et al. (2011), Wang et al. (2019)
			SANT1	SHH antagonist	Millman et al. (2016), Wang et al. (2019)
			T3	Induces β-cell maturation marker expression	Millman et al. (2016), Wang et al. (2019), Gheibi et al. (2020b)
			XXI	γ-secretase inhibitors	Millman et al. (2016), Pellegrini et al. (2018), Wang et al. (2019)
			Betacellulin	Maintains PDX1 and NKX6.1 expression	Millman et al. (2016), Wang et al. (2019), Gheibi et al. (2020b)
			Y27632	ROCK inhibitor, promotes cluster integrity and enhances progenitor formation	Yabe et al. (2019)
			Exendin 4	Enhances NEUROD1 expression	Kaitsuka et al. (2014), Gheibi et al. (2020b)
			DAPT	γ-secretase inhibitors	Kaitsuka et al. (2014)
			Alk5 inhibitor	TGF-β type I receptor kinase inhibitor type II	Millman et al. (2016), Shahjalal et al. (2014)
6	Pancreatic endocrine cells → β-cells	C-peptide, PDX1, NKX6.1, ISL1, NEUROD1, PAX6, NKX2.2, MafA, Glut2,	TGF-β type I receptor inhibitor VIII or SB431542 or A-83-01	Inhibitors of TGF-β type I receptor, they induce INS expression	Kunisada et al. (2011), Pellegrini et al. (2018), Wang et al. (2019)

(continued)

Table 1 (continued)

Stage No.	Stages	Genes expressed	Essential small molecules and growth factors	Major function	References
		Tcf1, glucokinase, urocortin-3, islet amyloid polypeptide, ZnT8 (SLC30A8)	IBMX	Phosphodiesterase	Pellegrini et al. (2015), Shahjalal et al. (2014)
			DAPT	γ-secretase inhibitors	Jiang (2011), Maehr et al. (2009)
			XXI	γ-secretase inhibitors	Pellegrini et al. (2018), Wang et al. (2019)
			Forskolin	Activation of adenylate cyclase, increase cAMP level	Hosokawa et al. (2018), Kunisada et al. (2011), Walczak et al. (2016), Yabe et al. (2019)
			Exendin 4	Peptide agonist of glucagon-like peptide-1 receptor	Kaitsuka et al. (2014), Maehr et al. (2009), Shahjalal et al. (2014), Yabe et al. (2019), Zhang et al. (2009), Zhu et al. (2011)
			Nicotinamide	Induces β-cell differentiation, ADP-ribose synthetase inhibitor, promotes PDX1 expression	Gheibi et al. (2020b), Hakim et al. (2014), Hosokawa et al. (2018), Kunisada et al. (2011), Shahjalal et al. (2014), Tateishi et al. (2008), Walczak et al. (2016), Yabe et al. (2019), Zhang et al. (2009), Zhu et al. (2011)
			Insulin growth factor I	Promotes β-cell differentiation and maturation	Kaitsuka et al. (2014), Maehr et al. (2009), Zhu et al. (2011)
			Insulin growth factor II	Activator of β-cell insulin-like growth factor 1 receptor	Modi et al. (2015), Tateishi et al. (2008)
			GLP1	Glucagon-like peptide-1 receptor analog	Hakim et al. (2014)
			Hepatocyte growth factor	Promotes β-cell maturation	Kaitsuka et al. (2014), Maehr et al. (2009), Yabe et al. (2019), Zhu et al. (2011), Gheibi et al. (2020b)
			BMP4	TGF-β family member	Yabe et al. (2019), Zhang et al. (2009)
			Betacellulin	Induces MAFA expression	Wang et al. (2019), Gheibi et al. (2020b)
			Dexamethasone	Promotes β-cell differentiation and proliferation	Hosokawa et al. (2018), Kunisada et al. (2011), Walczak et al. (2016), Gheibi et al. (2020b)
			T3	Promotes MafA expression and mono-hormonal Insulin$^+$ cells	Millman et al. (2016), Pelleg`ni et al. (2018), Wang et al. (2019), Gheibi et al. (2020b)
			Heparin	Enhances maturation of β-cells from progenitor cells	Yabe et al. (2019)
			RA	PDX1 inducer	Wang et al. (2019)
			Alk5i II	TGF-β type I receptor inhibitor type I	Hosokawa et al. (2018), Kunisada et al. (2011), Millman et al. (2016), Pellegrini et al. (2018), Walczak et al. (2016)

iPSCs induced pluripotent stem cells, *TGF-β* transforming growth factor-β, *WNT3a* wingless-related integration site-3a, *GSK3β* glycogen synthase kinase 3β, *PI3K* phosphoinositide 3-kinase; *HDAC* histone deacetylase, *FGF* fibroblast growth factor, *MAPK* mitogen-activated protein kinase, *BMP* bone morphogenetic protein, *ROCK* Rho-associated protein kinase, *CYC* KAAD-cyclopamine, *SHH* sonic hedgehog, *KGF* keratinocyte growth factor, *GLP1* glucagon-like peptide-1, *T3* triiodothyronine, *RA* retinoic acid, *ALK* activin receptor-like kinase, *IBMX* 3-isobutyl-1-methylxanthine

and PDX1 (Maehr et al. 2009; Thatava et al. 2010; Pellegrini et al. 2018). At this stage, the expression of PDX1 is observed (Jennings et al. 2013), which is induced by FGF family members (Gheibi et al. 2020b). Moreover, the supplementation of primitive gut tube cells with RA, CYC, and FGF members at higher concentrations promotes pancreatic endoderm specification (Pellegrini et al. 2018; Gheibi et al. 2020b). The pancreatic endoderm specification is characterized by the expression of transcription factors, namely PDX1 and SOX9 (Haller et al. 2019). The next phase of pancreatic development is the formation of multipotent pancreatic progenitors/precursors via further development of pancreatic endoderm. These pancreatic precursor cells give rise to the endocrine, exocrine, and ductal cells of the pancreas (Amour et al. 2006). The combination of RA, CYC, FGF, epidermal growth factor (EGF), and inhibitors of various signaling pathways such as BMP, TGF-β type I receptor, and so forth induce differentiation of pancreatic precursors (Tateishi et al. 2008; Shahjalal et al. 2014; Hosokawa et al. 2018). The pancreatic precursor cells expresses PDX1, NKX6.1, PTF1, and HNF6 (Tateishi et al. 2008; Shahjalal et al. 2014). Following pancreatic precursors specification, the cells were differentiated into pancreatic endocrine cells (Zhu et al. 2011; Kunisada et al. 2011; Shahjalal et al. 2014; Wang et al. 2019). This was achieved by continual inhibition of signaling pathways, namely SHH, TGF-β type I receptor, activin receptor-like kinase (Alk)-5 inhibitor II, BMP, and Notch (γ-secretase), which results in the formation of pancreatic endocrine cells. In addition, other small-molecule cocktails that comprise EGF, FGF, KGF, RA, indolactam V (ILV; protein kinase C activator), and nicotinamide have also resulted in the formation of pancreatic endocrine cells (Thatava et al. 2010; Zhu et al. 2011; Kunisada et al. 2011; Pellegrini et al. 2015; Yabe et al. 2019). These small molecules and growth factors together induce the expression of pancreatic endocrine cell markers, namely PDX1, HNF6, HLXB9, NGN3, PAX4, and NEUROD1 (Thatava et al. 2010; Zhu et al. 2011; Kunisada et al. 2011; Shahjalal et al. 2014). In a different study, the effect of O_2 concentration in the differentiation of iPSCs into pancreatic lineages is also demonstrated (Hakim et al. 2014). Early-stage treatment with high O_2 concentration increased NGN3 expression, eventually giving rise to insulin-positive cells. This effect was mediated by the repression of hypoxia-inducible factor-1α (HIF-1α) and activation of Wnt signaling (Hakim et al. 2014). Last, the iβ-cells were formed from pancreatic endocrine cells. Examination of several molecules suggested that mature β-cell markers are promoted by the inhibition of signaling pathways such as TGF-β and BMP and enzymes, namely phosphodiesterase and γ-secretase (Maehr et al. 2009; Zhang et al. 2009; Kunisada et al. 2011; Shahjalal et al. 2014; Pellegrini et al. 2018). Moreover, the activation of adenylate cyclase (using forskolin) promotes cell maturation, and when combined with small molecules, it facilitates β-cell maturation (Kunisada et al. 2011; Shahjalal et al. 2014). Also, several other compounds, namely Exendin 4 (glucagon-like peptide-1 [GLP1] agonist), nicotinamide, insulin-like growth factor II (IGF-II), glucagon-like peptide-1 (GLP1), hepatocyte growth factor (HGF), BMP4, and betacellulin, are used for β-cell maturation (Tateishi et al. 2008; Maehr et al. 2009; Zhang et al. 2009; Thatava et al. 2010; Kunisada et al. 2011; Shahjalal et al. 2014; Hakim et al. 2014; Pellegrini et al. 2018). These matured iβ-cells are distinguished by the expression of insulin, C-peptide, PDX1, NKX6.1, ISL1, NEUROD1, PAX6, NKX2.2, MafA, glucose transporter 2 (Glut2), Tcf1, glucokinase, urocortin-3, islet amyloid polypeptide, and ZnT8 (SLC30A8) (Tateishi et al. 2008; Zhang et al. 2009; Kunisada et al. 2011; Shahjalal et al. 2014; Thatava et al. 2010; Zhu et al. 2011; Pellegrini et al. 2018). The description of various small molecules and growth factors involved at different stages in the iPSC differentiation process eventually giving rise to mature iβ-cells is listed in Table 1.

4 Promotion of Differentiation of iPSCs into iβ-cells

Despite rigorous screening of chemical compounds, growth factors, and media conditions at different stages to efficiently give rise to a pure population of iβ-cells, the aforementioned studies were marred by low efficiency and heterogeneous population of cells. To improvise, various studies have not only utilized small molecules and optimal media conditions but also induced expression of key pancreatic transcription factors during the differentiation process (Kim et al. 2020; Saxena et al. 2016; Walczak et al. 2016; Wang et al. 2014). An interesting article demonstrated that the assembly of synthetic lineage control networks with sequential ON and OFF of Ngn3, Pdx1, and MafA could reprogram iPSCs to β-like cells efficiently (Saxena et al. 2016). Another study induced overexpression of PDX1 using an adenoviral vector for efficient and functional iβ-cell generation within 2 weeks (Kim et al. 2020). Likewise, the effect of combined induction of PDX1, NEUROD1, and MAFA using adenoviral vectors in iPSCs has also been reported to promote the formation of iβ-cells (Wang et al. 2014). The results suggested that the induction of these factors in generating iβ-cells was efficient but did not recapitulate the complete in vivo pancreas development process (Wang et al. 2014). The pancreas development process includes synergistic functioning of several transcription factors and small molecules. This study was limited to only three transcription factors, which were insufficient to recapitulate entire pancreatic development. Furthermore, Walczak et al. reported that the expression of PDX1 and NKX6.1 using an integrative approach enhanced the differentiation of iPSCs into iβ-cells (Walczak et al. 2016). However, the use of these viral-based approaches is limited because of genetic integration and the introduction of possible insertional mutagenesis (Dey et al. 2021).

Few studies also reported the use of protein transduction and miR overexpression in iPSCs to induce differentiation toward iβ-cells (Kaitsuka et al. 2014; Lahmy et al. 2014). Kaitsuka et al. developed an effective iPSCs to iβ-cell differentiation process using protein transduction; the latter poses no risk of genetic integration. They transduced PDX1, NEUROD, and MAFA, the key transcription factors involved in pancreatic β-cell development, in mouse and human iPSCs to induce differentiation. Further, they used laminin-5- and fibronectin-rich extracellular matrix media; these media components are known to promote the efficiency of rat β-cell survival, proliferation, and function. This approach facilitated a significant increase in the expression of the key endocrine progenitor marker, NGN3. However, insulin gene expression was observed only in mouse iPSC lines and not in human iPSC lines (Kaitsuka et al. 2014), indicating that these three factors are sufficient to induce an iβ-cell transcriptional profile in mouse, but not in human. It also highlights that the human system is more complex than a mouse and requires additional factors to induce an iβ-cell transcriptional profile. Likewise, the effect of miR-375 overexpression in the differentiation of iPSCs to iβ-cells was also investigated (Lahmy et al. 2014). Notably, this study induced differentiation of iPSCs into i-β-cells in the absence of any growth factors (Lahmy et al. 2014); miR-375 is the most abundant miR in islets, and its expression is crucial for the regulation of β-cell function, including its differentiation, insulin secretion, and so forth (Eliasson 2017). The increase in miR-375 at the early stages of the differentiation process corresponds to an increase in insulin transcript at the later stages of iβ-cell formation. This strategy facilitated the generation of islet-like cells that secreted insulin in a glucose-dependent manner (Lahmy et al. 2014). The summary of several protocols used for the generation of cells of the pancreatic lineage from human, mouse, and rhesus monkey iPSCs is given in Table 2 (for human) and Table 3 (for mouse and rhesus monkey).

Table 2 Summary of studies investigating differentiation of human iPSCs into insulin-producing cells/clusters

S. No.	iPSCs			Differentiation			Validation		References
	Transcription factors	Starting cell type	Reprogramming approach	Name of stages	Differentiation conditions	Cell type formed	GSIS	Amelioration of hyperglycemia/ in vivo studies	
1	OSKM	Fibroblasts	Retroviral transduction	Definitive endoderm Pancreatic lineages Exocrine/endocrine cells Insulin-producing cells	Serum-free in vitro differentiation	Insulin-producing islet-like clusters	Yes	Not mentioned	Tateishi et al. (2008)
2	OSK	Fibroblasts	Lentiviral transduction	Definitive endoderm Pancreatic specialization Progenitor expansion Maturation	In vitro differentiation	Insulin-producing cells	Yes	Not mentioned	Zhang et al. (2009)
3	OSKM	Fibroblasts	Lentiviral transduction	Definitive endoderm Gut tube endoderm Pancreatic endoderm Hormone-expressing cells	In vitro differentiation	Insulin-producing islet-like clusters	Yes	Not mentioned	Thatava et al. (2010)
4	OSKM	Fibroblasts	Sendai viral transduction	Definitive endoderm Pancreatic and endocrine progenitors Insulin-producing cells	High oxygen condition facilitates insulin-producing cell by inhibiting Notch signaling and activation of Wnt signaling	Insulin-producing cells	No	No	Hakim et al. (2014)
5	OSKM	Fibroblasts	Lentiviral transduction	Not mentioned	Overexpression of miR-375 in human iPSCs	Insulin-producing cells	Yes	No	Lahmy et al. (2014)
6	OSKM	Embryonic lung cells	Retroviral transduction	Definitive endoderm Primitive gut tube Pancreatic progenitors Endocrine progenitors Insulin-expressing cells	Xeno-free culture system and used synthetic scaffold and serum-free media containing humanized and/or recombinant supplements and growth factors	Insulin-producing cells	Yes	No	Shahjalal et al. (2014)
7	OSK	Fibroblasts	Retroviral transduction	Definitive endoderm Posterior foregut Pancreatic endoderm Endocrine cells	In vitro differentiation	Insulin-producing cells	Yes	Yes	Pellegrini et al. (2015)
8	OSKML	Fibroblasts	Retroviral transduction	Embryoid body formation Plating of embryoid body Pancreatic progenitors Insulin-producing cells	In vitro differentiation	Insulin-producing cells	Yes	No	Shaer et al. (2015)

(continued)

Table 2 (continued)

S. No.	iPSCs			Differentiation			Validation		References
	Transcription factors	Starting cell type	Reprogramming approach	Name of stages	Differentiation conditions	Cell type formed	GSIS	Amelioration of hyperglycemia/ in vivo studies	
9	OSKL^{m}L + mp53DD + EBNA1	Fibroblasts or epithelial cells	Episomal vectors	Definitive endoderm Generation of insulin-producing progenitors Maturation of insulin-producing cells	Induction of PDX1 and NKX6.1 expression during in vitro differentiation process	Insulin-producing cells	Yes	No	Walczak et al. (2016)
10	OSKM	Fibroblasts	Sendai viral	Definitive endoderm Posterior foregut Pancreatic endoderm Endocrine cells Pancreatic endocrine cells	In vitro differentiation	Insulin-producing cells	Yes	No	Pellegrini et al. (2018)
11	Not mentioned	Not mentioned	Not mentioned	Definitive endoderm Posterior foregut Pancreatic endoderm Endocrine cells Insulin-producing cells	Microencapsulated pancreatic progenitor cells were transplanted in vivo in an immunocompromised mouse	Insulin-producing cells	Yes	Yes	Haller et al. (2019)
12	Not mentioned	Fibroblasts	Not mentioned	Definitive endoderm Primitive gut tube Pancreatic progenitors Endocrine progenitors Insulin-producing cells	In vitro differentiation	Insulin-producing cells	Yes	No	Wang et al. (2019)
13	OSKM	Fibroblasts	Sendai viral	Definitive endoderm Primitive gut tube Posterior foregut Pancreatic progenitors Endocrine progenitors Insulin-producing cells	In vitro differentiation	Insulin-producing cells	Yes	Yes	Yabe et al. (2019)

iPSCs induced pluripotent stem cells, *O* OCT4, *S* SOX2, *K* KLF4, *M* c-MYC, *L* LIN28, *L^{m}* L-MYC, *mp53DD* dominant negative mutation of the p53 protein, *EBNA1* Epstein–Barr nuclear antigen 1, *3D* three-dimensional, *miR* microRNA, *GSIS* glucose-stimulated insulin secretion

Table 3 Summary of studies investigating differentiation of mouse and rhesus monkey iPSCs into insulin-producing cells

	iPSCs			Differentiation			Validation		
S. No.	Transcription factors	Starting cell type	Reprogramming approach	Name of stages	Differentiation conditions	Cell type formed	GSIS	Amelioration of hyperglycemia/ in vivo studies	References
1	OSKM	Mouse fibroblasts	Retroviral	Embryoid body formation Embryoid body detachment Generation of multilineage progenitors and their differentiation	In vitro differentiation	Insulin-producing cells	Yes	Yes	Alipio et al. (2010)
2	OSKM	Rhesus monkey fibroblasts	Retroviral	Definitive endoderm Pancreatic progenitors Endocrine precursors Insulin-producing cells	In vitro differentiation	Insulin-producing cells	Yes	Yes	Zhu et al. (2011)
3	OSKM	Mouse pancreas-derived epithelial cells or mouse embryonic fibroblasts	Lentiviral	Definitive endoderm Pancreatic progenitors Progenitor expansion Insulin-producing cells	In vitro differentiation	Insulin-producing cells	Yes	Yes	Jeon et al. (2012)
4	OSKM	Mouse embryonic fibroblasts	Lentiviral	Not mentioned	Adenoviral transfection of PDX1, NEUROD1, MafA transcription factors in iPSCs and their subsequent in vitro differentiation	Insulin-producing cells	Yes	Yes	Wang et al. (2014)

iPSCs induced pluripotent stem cells, *O* OCT4, *S* SOX2, *K* KLF4, *M* c-MYC, *GSIS* glucose-stimulated insulin secretion

5 Influence of Microenvironment on the Differentiation of iPSCs to iβ-cells

The in vivo environment of islets presents a complex and dynamic system consisting of a myriad of biophysical, biochemical, and cellular factors that directs the differentiation of β-cells (Galli et al. 2020). In the islet niche, β-cells interact with the extracellular matrix and various cells, namely vascular endothelial cells, endocrine cells, neuronal cells, and immune cells, via signaling molecules and growth factors. These interactions are essential in modulating β-cell proliferation, differentiation, and maturation (Aamodt and Powers 2017; Gheibi et al. 2020b). Furthermore, β-cells also respond to the mechanical cues, including stiffness, shear stress, topography (e.g., roughness), and geometry of each component of its niche (Galli et al. 2020). Moreover, intricate cell-to-cell communications among β-cells and in their microenvironment are essential in facilitating insulin response and viability (Nyitray et al. 2014). The classical two-dimensional (2D) monolayer differentiation cultures are inefficient in providing structural and biophysical characteristics similar to those observed in vivo (Galli et al. 2020). On the other hand, three-dimensional (3D) cultures closely mimic the in vivo microenvironment by preserving and maintaining the physiological structure, cell-to-cell interactions, and cell-to-matrix interactions (Amer et al. 2014). Moreover, 3D culture induction enhances β-cell differentiation, yield, and maturation, in such a way that it can restore euglycemia (Galli et al. 2020; Pagliuca et al. 2014; Wang et al. 2019). Few studies also illustrated the use of 3D culturing using the suspension, polymer scaffold, and microencapsulation-based techniques for differentiation of iPSCs into β-cells (Enderami et al. 2018; Haller et al. 2019; Pagliuca et al. 2014; Wang et al. 2019). The various 3D induction techniques with their benefits and their applications are illustrated in Fig. 4. The suspension-based 3D culture is developed to induce aggregated spherical clusters of endocrine cells, further promoting the formation of mature glucose-responsive β-cells (Amer et al. 2014; Haller et al. 2019; Pagliuca et al. 2014). Pagliuca et al. demonstrated that β-cells generated in large-scale 3D suspension culture systems showed rapid and glucose-responsive insulin production (Pagliuca et al. 2014). Besides, the insulin content and Ca^{2+} fluxes of iPSC-derived β-cells are similar to bona fide adult human β-cells (Pagliuca et al. 2014). Likewise, the effect of 3D culture on the induction of pancreatic progenitor specification and iβ-cell generation was also investigated. The results signified that the expression levels of Pdx1 and other β-cell-specific genes such as Nk6.1, Ngn3, MAFA, Kir6.2, Sur1, and Glut2 and glucose-stimulated insulin secretion (GSIS) were significantly higher in 3D culture compared with 2D culture (Wang et al. 2019). Similar results were obtained when poly-L-lactic acid and polyvinyl alcohol (PLLA/PVA) polymer-based scaffolds were used for differentiating iPSCs to β-cells. These scaffolds mimic the in vivo conditions, facilitating cell-to-matrix and cell-to-cell interactions that promote islet-like aggregation and functional β-cell formation (Enderami et al. 2018). Furthermore, Haller et al. established differentiation of human iPSCs to pancreatic progenitors followed by their microencapsulation and implantation in mice (Haller et al. 2019). These implanted cells were further differentiated into islet-like cells that maintained euglycemia and protected the mice from streptozotocin-induced hyperglycemia (Haller et al. 2019). All the above studies recapitulated in vivo microenvironment via 3D culturing to facilitate the induction of environmental cues for enhanced β-cell development and maturation.

6 Disease Modeling and Drug Discovery

The patient-specific iPSC generation and differentiation into a particular cell type hold a great promise in the recapitulation of disease-specific phenotypes in vitro (Hosokawa et al. 2018; Johannesson et al. 2015). The iPSCs generated

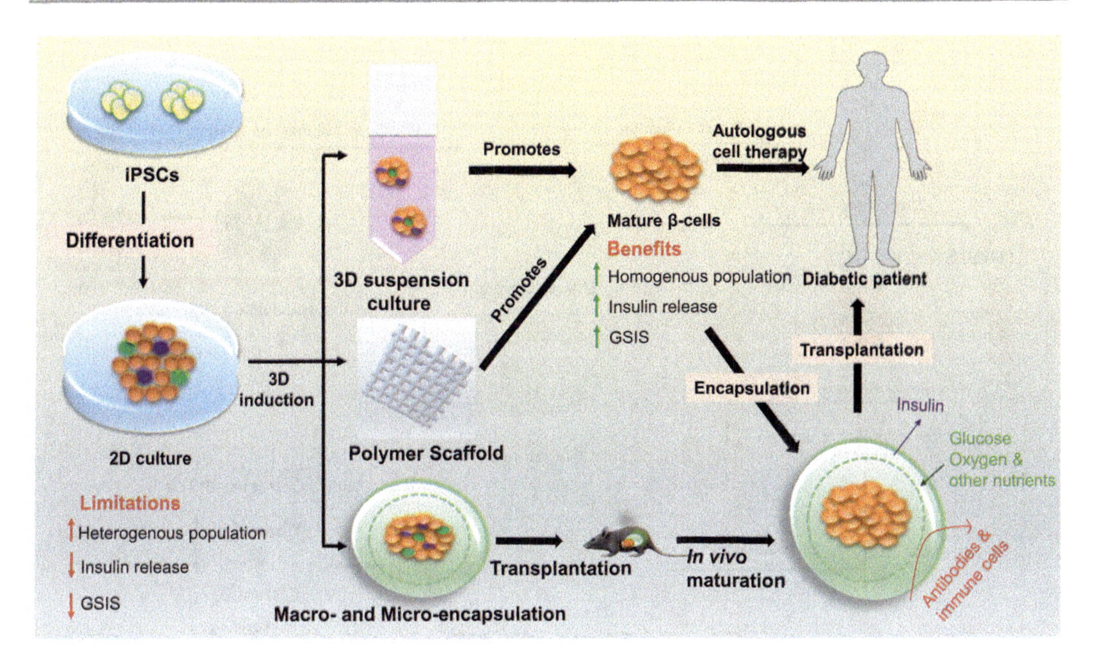

Fig. 4 Influence of culture conditions for iPSCs to β-cell differentiation. The iPSCs to β-cell differentiation protocols utilize both 2D and 3D culture conditions. The 2D culture conditions generate heterogeneous populations of β-like cells (golden), α-cells (green), and δ-cells (purple), as shown in the figure. Moreover, β-like cells induced from these culture conditions show limited insulin release upon glucose stimulation. On the contrary, the 3D culture conditions have proven to be highly efficient in generating pure populations of mature iβ-cells with significantly higher insulin release upon glucose stimulation. These culture conditions mimic in vivo microenvironment by facilitating improved cellular interactions. Various 3D culture induction conditions used for differentiation are mentioned in the figure. More specifically, macroencapsulation and microencapsulation of β-like cells followed by transplantation in mice promote the generation of mature iβ-cells. This strategy allows easy in-and-out of essential nutrients and insulin and protects these cells from an immune response, making them suitable for autologous cell transplantation

from patient cells carry their own genetic architecture, facilitating autologous cell replacement therapy for numerous diseases (Fujikura et al. 2012; Kondo et al. 2018; Millman et al. 2016). Further, the absence of a good animal model that accurately mimics the human conditions has compelled scientists to look for better alternative solutions such as iPSC-based human disease models (Vethe et al. 2017). Substantial efforts are undertaken to understand the disease pathophysiology and development of drugs and cell therapies using iPSC-based human disease models (Kondo et al. 2018). Similarly, iPSCs derived from diabetic patients to generate iβ-cells have opened new avenues to study the pathophysiological conditions of diabetes (Fig. 5). Multiple studies reported that human iPSCs can be derived from patients with T1DM, T2DM, and maturity-onset diabetes of the young (MODY) conditions and be further differentiated to iβ-cells (Hosokawa et al. 2018; Kim et al. 2020; Leite et al. 2020; Maehr et al. 2009; Millman et al. 2016; Teo et al. 2016; Vethe et al. 2017). The summary of various studies describing the generation of iβ-cells from diabetic patients to investigate the pathophysiology of these diseases is given in Table 4.

T1DM is an autoimmune disorder resulting from the destruction of iβ-cells (Kakleas et al. 2015). Various studies demonstrated the derivation of T1DM-specific, insulin-producing, glucose-responsive β-cells from human iPSCs (Kim et al. 2020; Maehr et al. 2009; Manzar et al. 2017; Millman et al. 2016; Thatava et al.

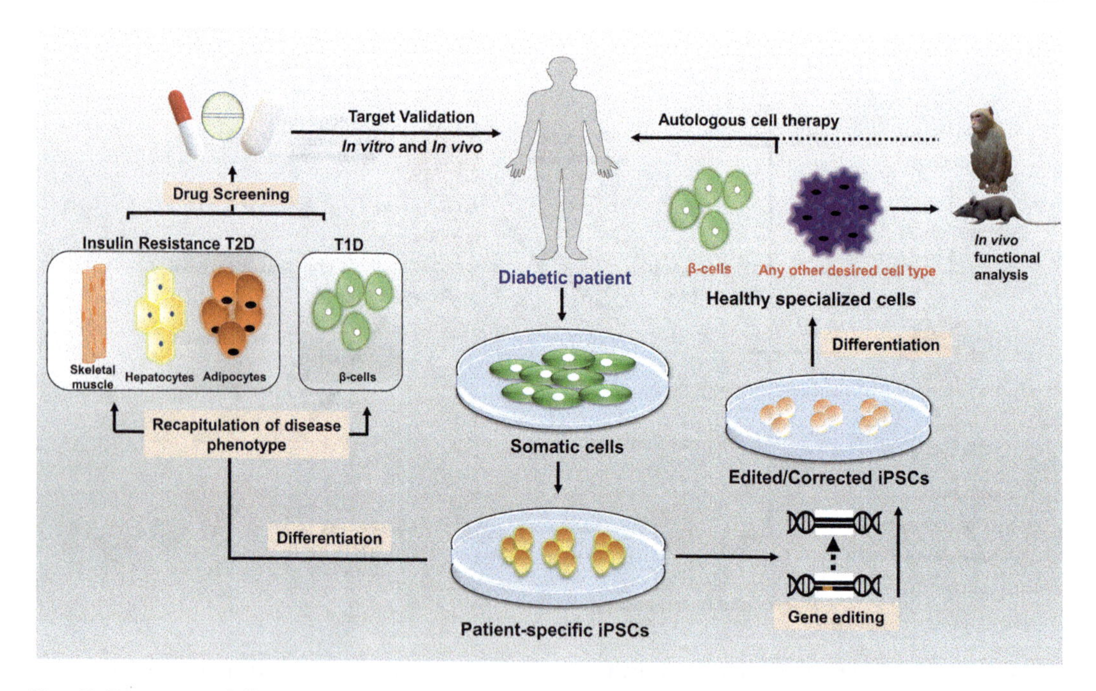

Fig. 5 Disease modeling and genome editing of patient-derived iPSCs. Patient-specific somatic cells can be reprogrammed to iPSCs and can be used to understand the disease pathophysiology or can be corrected via genome editing technologies to obtain transplantable healthy (corrected/edited) iβ-cells. In the case of diabetes, the patient-specific iPSCs can be differentiated to iβ-cells or insulin target cell types to recapitulate the diseased conditions of T1DM and T2DM (insulin resistance), respectively. Understanding the pathophysiology can pave the way for drug screening, further allowing its validation, which will benefit in the mitigation of these diseased conditions. Moreover, genome-edited patient-specific iPSCs can also be differentiated to iβ-cells or any specialized cell type of choice for autologous cell therapy post–successful in vivo functional analysis

2010). Most of these studies have shown that the generated T1DM and nondiabetic β-cells are functionally similar. However, it has also been speculated that T1DM β-cells are different from nondiabetic ones on account of aging and immune system interaction, but these differences are not discussed (Kim et al. 2020; Millman et al. 2016). For a comprehensive understanding of T1DM immunopathogenesis, the reconstruction of autologous immune cell response with T1DM-specific β-cells is essential. This would further benefit in the effective recapitulation of the diseased condition of T1DM and its progression (Leite et al. 2020; Thatava et al. 2013). In support of this, different studies have shown immune cell response with T1DM-specific β-cells via endoplasmic reticulum (ER) stress. ER stress causes islet inflammation by promoting T-cell interaction with β-cells, further triggering its destruction (Leite et al. 2020). In a parallel study, Hosokawa et al. differentiated β-like cells from fulminant T1DM-patient-specific iPSCs (Hosokawa et al. 2018). Further, they demonstrated that these β-like cells were highly susceptible to cytotoxic cytokines due to the upregulation of apoptotic genes. Additionally, the differential expression analysis of immunoregulation-related genes signified abnormal immune function as a potent cause of fulminant T1DM disease progression (Hosokawa et al. 2018).

The most prevalent form of diabetes is T2DM, and this polygenic disease is mediated by insulin resistance and β-cell dysfunction (Elsayed et al. 2021). Insulin resistance is a condition where impaired insulin sensitivity in its target tissues, namely skeletal muscles, adipose tissues, and liver, is observed (Elsayed et al. 2021; Gheibi

Table 4 Summary of patient-specific iPSC-derived pancreatic progenitors and insulin-producing cells for disease modeling and other applications

S. No.	Disease	Mutation	Cell type formed	GSIS	In vivo transplantation	Major finding(s)	References
1	T1DM	N/A	Insulin-producing cells	Yes	No	β-like cells were functional in in vitro studies	Maehr et al. (2009)
2	T2DM	N/A	Insulin-producing cells	No	No	Senescent-related genes were suppressed in keratinocytes of elderly patients and these cells were reprogrammed to iPSCs, which were then differentiated to β-like cells	Ohmine et al. (2012)
3	T1DM	N/A	Insulin-producing cells	Yes	No	iPSCs generated in vitro functional β-like cells upon their guided differentiation	Thatava et al. (2013)
4	Wolfram syndrome	*WFS1*	Insulin-producing cells	Yes	Yes	Consequences of endoplasmic reticulum stress on β-cell function were studied	Shang et al. (2014)
5	T1DM	N/A	Insulin-producing cells	Yes	Yes	β-cells were functionally similar to healthy individuals both in vitro an in vivo	Millman et al. (2016)
6	MODY5	*HNF1B*	Pancreatic progenitors	No	No	Effect of HNF1B mutation on pancreatic hypoplasia was investigated in the MODY5 patient-derived iPSCs	Teo et al. (2016)
7	T1DM	N/A	Insulin-producing cells	Yes	Yes	β-cells derived via 3D induction were functional in vitro and in vivo, and insulin granules in these β-cells were identical to cadaveric β-cells	Manzar et al. (2017)
8	MODY1	*HNF4A*	Insulin-producing cells	Yes	Yes	Investigated the role of HNF4A mutation in the disease progression of MODY1	Vethe et al. (2017)
9	Fulminant T1DM	N/A	Insulin-producing cells	Yes	No	β-like cells formed from iPSCs were used to elucidate the disease mechanisms of fulminant T1DM	Hosokawa et al. (2018)
10	T1DM and T2DM	N/A	Insulin-producing cells	Yes	No	β-cells generated from T1DM and T2DM patients were functionally similar to β-cells of nondiabetic individuals	Kim et al. (2020)
11	T1DM	N/A	Insulin-producing cells	No	No	β-cells formed were used to elucidate the disease mechanisms of T1DM demonstrating immune cells response against β-cells	Leite et al. (2020)

iPSCs induced pluripotent stem cells, *T1DM* type 1 diabetes mellitus, *T2DM* type 2 Diabetes mellitus, *MODY* maturity-onset diabetes of the young, *GSIS*, glucose-stimulated insulin release

et al. 2020a, b). Over time, insulin resistance of target cells reduces insulin secretion ability of β-cells, leading to hyperglycemia and T2DM development (Elsayed et al. 2021; Prentki et al. 2006). Various studies have derived iPSCs from T2DM patients and T2DM-risk patients subjected to insulin resistance to understand the cardinal molecular mechanisms involved (Burkart et al. 2016; Iovino et al. 2014, 2016; Kim et al. 2020; Kudva et al. 2012; Ohmine et al. 2012). Further, these patient-specific iPSCs were used to generate iβ-cells and insulin target cells, including adipocytes, hepatocytes, and skeletal myotubes as well as mesenchymal progenitor cells (Ohmine et al. 2012; Balhara et al. 2015; Iovino et al. 2016; Kim et al. 2020; Noguchi et al. 2013; Carpentier et al. 2016). These patient-derived autologous cells facilitate cell-based therapies and also help in understanding the genetic and cellular defects involved in the disease progression (Iovino et al. 2016; Kim et al. 2020; Ohmine et al. 2012; Gheibi et al. 2020b). For example, Kim et al. generated functional iβ-cells from human iPSCs in a 2-week protocol and demonstrated that these insulin-producing cells produced similar levels of the islet-specific marker, insulin mRNA level, and C-peptide release. This group further suggested that these iβ-cells can facilitate autologous transplantation-based cell therapy in T2DM patients (Kim et al. 2020). Moreover, another group created a novel patient-specific disease model to recapitulate the effect of skeletal muscle insulin resistance in glucose metabolism and T2DM. These genetic insulin-resistant mutant myotubes showed similar defects, namely impaired insulin signaling, insulin-stimulated glucose uptake, etc., as observed in in vivo myotubes (Iovino et al. 2016). Thus, these studies provide a treatment option for T2DM as well as dissect the genetic and phenotypical features of T2DM for a better understanding of the disease (Fig. 5).

A monogenic form of the disease that develops in young adults is termed as maturity-onset diabetes of the young (MODY). It results from an autosomal dominant mutation in genes associated with β-cells development or function (Johannesson et al. 2015; Murphy et al. 2008). Few studies have modeled the disease mechanism of MODY by differentiating iPSCs of patients carrying HNF1B and HNF4A mutations into pancreatic progenitor cells and β-cells (Teo et al. 2016; Vethe et al. 2017). In particular, Teo et al. differentiated human iPSCs of MODY5 patients carrying HNF1B mutation to demonstrate the molecular mechanisms of pancreatic hypoplasia (Teo et al. 2016). This study demonstrated that the perturbations in the transcription factor network, including upregulation of PDX1, PTF1A, GATA4, and GATA6 and downregulation of PAX6 gene expression (Teo et al. 2016), accounting for islet dysfunction and contributed to early-onset diabetes. Similarly, another study explained the molecular mechanisms of the MODY1 disease phenotype via human iPSCs to β-cell differentiation model (Vethe et al. 2017). The results demonstrated that the β-cells generated from these human iPSCs showed similar physiological levels of GSIS and expression of β-cell markers. Thus, it was suggested that the MODY1 disease phenotype is independent of β-cell differentiation and resulted from mechanisms such as stress-induced cell death, which occurred post-β-cell formation (Vethe et al. 2017).

In addition, insulin-producing cells were also differentiated from patients with Wolfram syndrome (WFS). Wolfram syndrome is caused by a recessive mutation in the *WFS1* gene and is characterized by an abnormal increase in endoplasmic reticulum stress molecules and a decrease in insulin secretion (Inoue et al. 1998; Shang et al. 2014). The human iPSCs generated from patients suffering from this syndrome were differentiated into iβ-cells (Shang et al. 2014). Interestingly, these iβ-cells displayed a reduced insulin secretion in the presence of abnormally high ER stress molecule (Shang et al. 2014), indicating that these iPSCs can be used as a clinical model to study the disease pathophysiology and in the identification of its therapeutic drug. In line with this, the authors observed that high ER stress is related to the amplification of unfolded protein responses such as GRP78 and XBP-1 and reduction in ubiquitination of ATF6α. The authors used this iPSC disease model to identify 4-phenyl butyric acid (4PBA), a chemical chaperone, that reduced the activity of unfolded protein response pathway and thus restored the insulin secretion levels (Shang et al. 2014). To conclude, these patient-specific iPSC-derived pancreatic progenitors and iβ-cells offer an incredible platform for disease modeling and drug discovery to treat DM.

7 Applications of Genome Editing in the Promotion of iPSCs to iβ-cells

The use of iPSC-derived iβ-cells to tackle the growing prevalence of DM is undoubtedly a promising platform and will prove beneficial to the whole diabetic community. However, the underlying etiology of DM shows a mutation in multiple genes and a complicated network of signaling pathways that need to be corrected before the transformed cells can be transplanted back to the patient. The rapid advances in genome editing techniques (e.g., CRISPR-Cas system) have facilitated a whole new domain in research where human genome sequence can be precisely manipulated to achieve a therapeutic effect. This includes correction of a specific gene that is responsible for causing the disease and addition or deletion of certain gene(s) to reverse the effect of disease (Fig. 5).

Various mutations at the genetic level are held responsible for impaired glucose secretion and/or sensing. For instance, mutations in GCK are associated with MODY (Velho et al. 1992, 1996). Similarly, mutations in the *WFS1* gene are known to cause Wolfram syndrome (Inoue et al. 1998; Shang et al. 2014). In a recent study, the *CRISPR-Cas9* gene editing strategy was employed to correct the defect in the *WFS1* gene in three individual patients' iPSCs to derive autologous β-cells (Maxwell et al. 2020). In this study, iPSCs were generated from the skin fibroblasts and differentiated to glucose-responsive cells using growth factors and small molecules (Maxwell et al. 2020). They compared these unedited iPSC-derived cells to their gene-edited counterparts and showed that the edited ones had better glucose responsiveness and higher insulin secretion (Maxwell et al. 2020). Furthermore, the glucose-responsive test using iPSC-derived unedited β-cells, their *CRISPR-Cas9*-edited β-cells, and healthy islets from a cadaveric donor in streptozotocin-treated mice was performed. The corrected or edited β-cells were at par with the healthy islets and reversed normoglycemic condition within 2 weeks (Maxwell et al. 2020). In another study, permanent neonatal DM patients were observed to have a mutation in the start codon of the *INS* gene, which severely affected their insulin sensitivity (Ma et al. 2018). This mutation was corrected using the *CRISPR-Cas9* gene editing tool in permanent neonatal DM patient–specific iPSCs, and these corrected iPSCs were subsequently differentiated into pancreatic endocrine cells. These differentiated cells showed similar insulin secretion levels compared with healthy islets (Ma et al. 2018). Moreover, genome editing has also been used to understand the underlying mechanisms in pancreatic development and DM. For instance, gene editing strategies demonstrated that *RFX6* gene mutation highly affects pancreatic progenitor development and its maturation to endocrinal cells in permanent neonatal DM patients (Zhu et al. 2016). Thus, these studies demonstrated the potential of iPSCs in combination with genome editing technologies (for the correction of gene defects) in producing highly functional β-cells that will aid in patient-specific diabetes cell therapy (Fig. 5).

8 Current Challenges and Future Prospects

Although recent advances in the field of β-cell restoration have opened up a state-of-the-art arena in the field of DM, meticulous attention must be paid to the quality of the generated cells before they can be used for treatment purposes. The shortcomings associated with respect to clinical aspects are summarized below.

Generating safer iPSCs that are devoid of integrated foreign transgenes is considered a prime hurdle that needs to be addressed. In most cases, iPSCs are generated using viral methods such as γ-retro- and lentiviral vectors due to their higher efficiency compared with nonviral counterparts (Hu 2014; Dey et al. 2021). iPSCs generated through viral-based approaches might cause insertional mutagenesis resulting in tumor formation. Integration-free reprogramming approaches like episomal vectors, Sendai virus, miRs, synthetic messenger RNAs, small molecules, and recombinant proteins can generate iPSCs that are safe and have the potential of clinical applicability (Borgohain et al. 2019; Dey et al. 2021). However, lower efficiency, slow kinetics, and cumbersome protocols often mar its usage. Moreover, the source of starting cell type and genetic makeup of the donor also play a crucial role. In 2012, Kajiwara et al. analyzed 28 iPSC colonies that originated from different somatic cell sources. They showed that iPSCs derived from peripheral blood cells had better differentiation capability than adult fibroblasts when derived from different donors. However, when iPSCs generated from peripheral blood cells and adult fibroblasts of the same donor were compared, it showed no significant difference (Kajiwara et al. 2012). Moreover, the differentiation techniques taken to specify these iPSCs to iβ-cells should be well-defined, simple, reproducible, and effective and should not leave a single cell in an undifferentiated state as this might cause teratoma formation (Soejitno and Prayudi 2011).

To date, different groups have tried to produce glucose-responsive mature iβ-cells or pancreatic progenitor cells by the differentiation of reprogrammed iPSCs (Enderami et al. 2018; Millman et al. 2016; Pellegrini et al. 2015; Shahjalal et al. 2014; Yabe et al. 2019; Zhang et al. 2009; Zhu et al. 2011; Tateishi et al. 2008). However, often these generated iβ-cells are immature in nature and have a limited glucose response (Tateishi et al. 2008; Thatava et al. 2010; Wang et al. 2019). Moreover, it is necessary that these cells express mature β-cell markers in vivo like MafA, Nkx6.1, Isl1, and C-peptide and should not coexpress α-, δ-, and PP-cell-specific markers, delineating their immature and heterogeneous characteristics (Soejitno and Prayudi 2011). Additionally, these differentiated cells are heterologous in nature, containing a mixture of bihormonal (insulin$^+$/glucagon$^+$) cells along with progenitor cells (Vethe et al. 2017). It has also been noted that the reprogrammed cells are unable to secrete insulin upon varying the concentrations of glucose (Tateishi et al. 2008). For example, Yabe et al. observed that spheroidal cells, when grown in culture dishes, successfully secreted insulin at low glucose concentrations but failed to secrete insulin at higher concentrations. Furthermore, the glucose responsiveness of the differentiated cells was inconsistent in vivo (Yabe et al. 2019). The same group tried 3D culturing of the cells and observed certain C-peptide secretion at low glucose concentrations, but these cells also failed at higher concentrations (Yabe et al. 2019). Additionally, another group observed that the clonal variation exists, and cells differentiated from different clones behaved differently in terms of glucose response (Tateishi et al. 2008). Further, these cells fail to mimic adult β-cells completely. Moreover, in the case of clinical applicability, a large number of pure populations of functional iβ-cells are required for transplantation, which can survive in in vivo conditions to reverse hyperglycemia. Hence, further research must be performed to understand molecular mechanisms involved that can lead to effective differentiation protocols, resulting in a higher number of a functional and homogeneous population of iβ-cells that can be used in cell therapy.

The immune reaction of transplanted β-cells is another concern that limits the usage of iβ-cells for therapeutic purposes. These cells often succumb to autoimmune destruction, and patients rely on immunosuppressive drugs to suppress immune rejection, which can be beneficial to serious opportunistic infections (Aguayo-Mazzucato and Bonner-Weir 2010). Encapsulation of iβ-cells in semipermeable membrane devices has been reported to facilitate maturation and function in vivo, which in turn reduced the use of immunosuppressive drugs (Wan et al. 2017). Addressing these key points will prove iPSC-based iβ-cells derivation as a superior alternative strategy to treat DM.

Acknowledgments We thank all the members of the Laboratory for Stem Cell Engineering and Regenerative Medicine (SCERM) for their critical reading and excellent support. This work was financially supported by the Ministry of Science and Technology, Govt. of India (Ref. No.: BT/COE/34/SP28408/2018) under the North East Center for Biological Sciences and Healthcare Engineering (NECBH) outreach program hosted by Indian Institute of

Technology Guwahati (IITG), Guwahati, Assam, sponsored by the Department of Biotechnology (DBT), Govt. of India (NECBH/2019-20/136). This work was partially funded by IIT Guwahati Institutional Top-Up on Start-Up Grant.

Declarations

Conflict of Interest The authors declare that they have no potential conflict of interest.

Ethics Approval Not applicable

Informed Consent Not applicable

Research Involving Human Participants and/or Animals None

Availability of Data and Material Not applicable

Author Contribution Akriti Agrawal and Gloria Narayan were responsible for conception and design, collection and assembly of data, data analysis and interpretation, manuscript writing, and final approval of the manuscript. Ranadeep Gogoi was responsible for data analysis and interpretation, manuscript writing, and final approval of the manuscript, and Rajkumar P Thummer was responsible for conception and design, collection and assembly of data, data analysis and interpretation, manuscript writing, final approval of the manuscript, and financial support. All the authors gave consent for publication.

References

Aamodt KI, Powers AC (2017) Signals in the pancreatic islet microenvironment influence β-cell proliferation. Diabetes Obes Metab 19:124–136

Aguayo-Mazzucato C, Bonner-Weir S (2010) Stem cell therapy for type 1 diabetes mellitus. Nat Rev Endocrinol 6(3):139–148

Alipio Z, Liao W, Roemer EJ, Waner M, Fink LM, Ward DC, Ma Y (2010) Reversal of hyperglycemia in diabetic mouse models using induced-pluripotent stem (iPS)-derived pancreatic β-like cells. Proc Natl Acad Sci U S A 107(30):2–7. https://doi.org/10.1073/pnas.1007884107

Amer LD, Mahoney MJ, Bryant SJ (2014) Tissue engineering approaches to cell-based type 1 diabetes therapy. Tissue Eng Part B Rev 20(5):455–467. https://doi.org/10.1089/ten.teb.2013.0462

Ameri J, Ståhlberg A, Pedersen J, Johansson JK, Johannesson MM, Artner I, Semb H (2010) FGF2 specifies hESC-derived definitive endoderm into foregut/midgut cell lineages in a concentration-dependent manner. Stem Cells 28(1):45–56

Amour KAD, Bang AG, Eliazer S, Kelly OG, Agulnick AD, Smart NG, Moorman MA, Kroon E, Carpenter MK, Baetge EE (2006) Production of pancreatic hormone – expressing endocrine cells from human embryonic stem cells. Nat Biotechnol 24 (11):1392–1401. https://doi.org/10.1038/nbt1259

Balhara B, Burkart A, Topcu V, Lee YK, Cowan C, Kahn CR, Patti ME (2015) Severe insulin resistance alters metabolism in mesenchymal progenitor cells. Endocrinology 156(6):2039–2048. https://doi.org/10.1210/en.2014-1403

Borgohain MP, Haridhasapavalan KK, Dey C, Adhikari P, Thummer RP (2019) An insight into DNA-free reprogramming approaches to generate integration-free induced pluripotent stem cells for prospective biomedical applications. Stem Cell Rev Rep 15(2):286–313. https://doi.org/10.1007/s12015-018-9861-6

Burkart AM, Tan K, Warren L, Iovino S, Hughes KJ, Kahn CR, Patti ME (2016) Insulin resistance in human iPS cells reduces mitochondrial size and function. Sci Rep 6:1–12. https://doi.org/10.1038/srep22788

Carmines PK (2010) The renal vascular response to diabetes. Curr Opin Nephrol Hypertens 19(1):85

Carpentier A, Nimgaonkar I, Chu V, Xia Y, Hu Z, Liang TJ (2016) Hepatic differentiation of human pluripotent stem cells in miniaturized format suitable for high-throughput screen. Stem Cell Res 16(3):640–650

Cichosz SL, Frystyk J, Tarnow L, Fleischer J (2017) Are changes in heart rate variability during hypoglycemia confounded by the presence of cardiovascular autonomic neuropathy in patients with diabetes? Diabetes Technol Ther 19(2):91–95

Dey C, Narayan G, Krishna Kumar H, Borgohain MP, Lenka N, Thummer RP (2016) Cell-penetrating peptides as a tool to deliver biologically active recombinant proteins to generate transgene-Free induced pluripotent stem cells. Stud Stem Cells Res Ther 3(1):6–15

Dey C, Raina K, Haridhasapavalan KK, Thool M, Sundaravadivelu PK, Adhikari P, Gogoi R, Thummer RP (2021) An overview of reprogramming approaches to derive integration-free induced pluripotent stem cells for prospective biomedical applications. In: Recent advances in IPSC technology. Academic, Amsterdam, pp 231–287

Ebrahimi B (2015) Reprogramming barriers and enhancers: strategies to enhance the efficiency and kinetics of induced pluripotency. Cell Regeneration 4 (1):1–12

Eliasson L (2017) The small RNA miR-375 – a pancreatic islet abundant miRNA with multiple roles in endocrine beta cell function. Mol Cell Endocrinol 456:95–101

Elsayed AK, Vimalraj S, Nandakumar M, Abdelalim EM (2021) Insulin resistance in diabetes: the promise of using induced pluripotent stem cell technology. World J Stem Cells 13(3):221

Enderami SE, Kehtari M, Abazari MF, Ghoraeian P, Nouri Aleagha M, Soleimanifar F, Soleimani M, Mortazavi Y, Nadri S, Mostafavi H, Askari H (2018)

Generation of insulin-producing cells from human induced pluripotent stem cells on PLLA/PVA nanofiber scaffold. Artifi Cells Nanomed Biotechnol 46(sup1):1062–1069. https://doi.org/10.1080/21691401.2018.1443466
Ferreira LMR, Mostajo-Radji MA (2013) How induced pluripotent stem cells are redefining personalized medicine. Gene 520(1):1–6
Fong DS, Aiello L, Gardner TW, King GL, Blankenship G, Cavallerano JD, Ferris FL, Klein R (2004) Retinopathy in diabetes. Diabetes Care 27(suppl 1):s84–s87
Fujikura J, Nakao K, Sone M, Noguchi M, Mori E, Naito M, Taura D, Harada-Shiba M, Kishimoto I, Watanabe A, Asaka I, Hosoda K, Nakao K (2012) Induced pluripotent stem cells generated from diabetic patients with mitochondrial DNA A3243G mutation. Diabetologia 55(6):1689–1698. https://doi.org/10.1007/s00125-012-2508-2
Galli A, Algerta M, Marciani P, Schulte C, Lenardi C, Milani P, Maffioli E, Tedeschi G, Perego C (2020) Shaping pancreatic β-cell differentiation and functioning: the influence of mechanotransduction. Cell 9(2). https://doi.org/10.3390/cells9020413
Gheibi S, Samsonov AP, Gheibi S, Vazquez AB, Kashfi K (2020a) Regulation of carbohydrate metabolism by nitric oxide and hydrogen sulfide: implications in diabetes. Biochem Pharmacol 176:113819
Gheibi S, Singh T, da Cunha JPMCM, Fex M, Mulder H (2020b) Insulin/glucose-responsive cells derived from induced pluripotent stem cells: disease modeling and treatment of diabetes. Cell 9(11):2465
Hakim F, Kaitsuka T, Raeed JM, Wei FY, Shiraki N, Akagi T, Yokota T, Kume S, Tomizawa K (2014) High oxygen condition facilitates the differentiation of mouse and human pluripotent stem cells into pancreatic progenitors and insulin-producing cells. J Biol Chem 289(14):9623–9638. https://doi.org/10.1074/jbc.M113.524363
Haller C, Piccand J, De Franceschi F, Ohi Y, Bhoumik A, Boss C, De Marchi U, Jacot G, Metairon S, Descombes P, Wiederkehr A, Palini A, Bouche N, Steiner P, Kelly OG, Kraus R-C, M. (2019) Macroencapsulated human iPSC-derived pancreatic progenitors protect against STZ-induced hyperglycemia in mice. Stem Cell Rep 12(4):787–800. https://doi.org/10.1016/j.stemcr.2019.02.002
Haridhasapavalan KK, Borgohain MP, Dey C, Saha B, Narayan G, Kumar S, Thummer RP (2019) An insight into non-integrative gene delivery approaches to generate transgene-free induced pluripotent stem cells. Gene 686:146–159. https://doi.org/10.1016/j.gene.2018.11.069
Haridhasapavalan KK, Raina K, Dey C, Adhikari P, Thummer RP (2020) An insight into reprogramming barriers to iPSC generation. Stem Cell Rev Rep 16 (1):56–81
Herrera B, Inman GJ (2009) A rapid and sensitive bioassay for the simultaneous measurement of multiple bone morphogenetic proteins. Identification and quantification of BMP4, BMP6 and BMP9 in bovine and human serum. BMC Cell Biol 10(1):1–11
Holman RR, Sourij H, Califf RM (2014) Cardiovascular outcome trials of glucose-lowering drugs or strategies in type 2 diabetes. Lancet 383(9933):2008–2017
Hosokawa Y, Toyoda T, Fukui K, Baden MY, Funato M, Kondo Y, Sudo T, Iwahashi H, Kishida M, Okada C, Watanabe A, Asaka I, Osafune K, Imagawa A, Shimomura I (2018) Insulin-producing cells derived from 'induced pluripotent stem cells' of patients with fulminant type 1 diabetes: vulnerability to cytokine insults and increased expression of apoptosis-related genes. J Diabetes Invest 9(3):481–493. https://doi.org/10.1111/jdi.12727
Hossain MK, Dayem AA, Han J, Saha SK, Yang GM, Choi HY, Cho SG (2016) Recent advances in disease modeling and drug discovery for diabetes mellitus using induced pluripotent stem cells. Int J Mol Sci 17 (2):1–17. https://doi.org/10.3390/ijms17020256
Hu K (2014) All roads lead to induced pluripotent stem cells: the technologies of iPSC generation. Stem Cells Dev 23(12):1285–1300. https://doi.org/10.1089/scd.2013.0620
Inoue H, Tanizawa Y, Wasson J, Behn P, Kalidas K, Bernal-Mizrachi E, Mueckler M, Marshall H, Donis-Keller H, Crock P, others (1998) A gene encoding a transmembrane protein is mutated in patients with diabetes mellitus and optic atrophy (Wolfram syndrome). Nat Genet 20(2):143–148
Iovino S, Burkart AM, Kriauciunas K, Warren L, Hughes KJ, Molla M, Lee YK, Patti ME, Kahn CR (2014) Genetic insulin resistance is a potent regulator of gene expression and proliferation in human iPS cells. Diabetes 63(12):4130–4142. https://doi.org/10.2337/db14-0109
Iovino S, Burkart AM, Warren L, Patti ME, Kahn CR (2016) Myotubes derived from human-induced pluripotent stem cells mirror in vivo insulin resistance. Proc Natl Acad Sci U S A 113(7):1889–1894. https://doi.org/10.1073/pnas.1525665113
Jennings RE, Berry AA, Kirkwood-Wilson R, Roberts NA, Hearn T, Salisbury RJ, Blaylock J, Hanley KP, Hanley NA (2013) Development of the human pancreas from foregut to endocrine commitment. Diabetes 62(10):3514–3522
Jeon K, Lim H, Kim J, Van Thuan N, Park SH, Lim Y, Choi H, Lee E, Kim J, Lee M, Cho S (2012) Differentiation and transplantation of functional pancreatic beta cells generated from induced pluripotent stem cells derived from a type 1 diabetes mouse model. Stem Cells Dev 21(14):2642–2655. https://doi.org/10.1089/scd.2011.0665
Jiang L (2011) γ-secretase inhibitor, DAPT inhibits self-renewal and stemness maintenance of ovarian cancer stem-like cells in vitro. Chin J Cancer Res 23 (2):140–146. https://doi.org/10.1007/s11670-011-0140-1
Johannesson M, Ståhlberg A, Ameri J, Sand FW, Norrman K, Semb H (2009) FGF4 and retinoic acid

direct differentiation of hESCs into PDX1-expressing foregut endoderm in a time- and concentration-dependent manner. PLoS One 4(3):e4794
Johannesson B, Sui L, Freytes DO, Creusot RJ, Egli D (2015) Toward beta cell replacement for diabetes. EMBO J 34(7):841–855
Kahn R (2003) Follow-up report on the diagnosis of diabetes mellitus: the expert committee on the diagnosis and classifications of diabetes mellitus. Diabetes Care 26(11):3160
Kaitsuka T, Noguchi H, Shiraki N, Kubo T, Wei F-Y, Hakim F, Kume S, Tomizawa K (2014) Generation of functional insulin-producing cells from mouse embryonic stem cells through 804G cell-derived extracellular matrix and protein transduction of transcription factors. Stem Cells Transl Med 3(1):114–127
Kajiwara M, Aoi T, Okita K, Takahashi R, Inoue H, Takayama N, Endo H, Eto K, Toguchida J, Uemoto S, others (2012) Donor-dependent variations in hepatic differentiation from human-induced pluripotent stem cells. Proc Natl Acad Sci 109 (31):12538–12543
Kakleas K, Soldatou A, Karachaliou F, Karavanaki K (2015) Associated autoimmune diseases in children and adolescents with type 1 diabetes mellitus (T1DM). Autoimmun Rev 14(9):781–797. https://doi.org/10.1016/j.autrev.2015.05.002
Katsarou A, Gudbjörnsdottir S, Rawshani A, Dabelea D, Bonifacio E, Anderson BJ, Jacobsen LM, Schatz DA, Lernmark Å (2017) Type 1 diabetes mellitus. Nat Rev Dis Primers 3(1):1–17
Kim MJ, Lee EY, You YH, Yang HK, Yoon KH, Kim JW (2020) Generation of iPSC-derived insulin-producing cells from patients with type 1 and type 2 diabetes compared with healthy control. Stem Cell Res 48 (July):101958. https://doi.org/10.1016/j.scr.2020.101958
Kondo Y, Toyoda T, Inagaki N, Osafune K (2018) iPSC technology-based regenerative therapy for diabetes. J Diabetes Invest 9(2):234–243. https://doi.org/10.1111/jdi.12702
Kudva YC, Ohmine S, Greder LV, Dutton JR, Armstrong A, De Lamo JG, Khan YK, Thatava T, Hasegawa M, Fusaki N, others (2012) Transgene-free disease-specific induced pluripotent stem cells from patients with type 1 and type 2 diabetes. Stem Cells Transl Med 1(6):451–461
Kunisada Y, Tsubooka-Yamazoe N, Shoji M, Hosoya M (2011) Small molecules induce efficient differentiation into insulin-producing cells from human induced pluripotent stem cells. Stem Cell Res. https://doi.org/10.1016/j.scr.2011.10.002
Lahmy R, Soleimani M, Sanati MH, Behmanesh M, Kouhkan F, Mobarra N (2014) MiRNA-375 promotes beta pancreatic differentiation in human induced pluripotent stem (hiPS) cells. Mol Biol Rep 41 (4):2055–2066. https://doi.org/10.1007/s11033-014-3054-4
Leite NC, Sintov E, Meissner TB, Brehm MA, Greiner DL, Harlan DM, Melton DA (2020) Modeling type 1 diabetes in vitro using human pluripotent stem cells. Cell Rep 32(2):107894. https://doi.org/10.1016/j.celrep.2020.107894
Leon BM, Maddox TM (2015) Diabetes and cardiovascular disease: epidemiology, biological mechanisms, treatment recommendations and future research. World J Diabetes 6(13):1246
Liu H, Wang J, He T, Becker S, Zhang G, Li D, Ma X (2018) Butyrate: a double-edged sword for health? Adv Nutr 16:21–29. https://doi.org/10.1093/advances/nmx009
Lorenzo PI, Juárez-Vicente F, Cobo-Vuilleumier N, García-Domínguez M, Gauthier BR (2017) The diabetes-linked transcription factor PAX4: from gene to functional consequences. Genes 8(3):101
Loskill P, Huebsch N (2019) Engineering tissues from induced pluripotent stem cells. Tissue Eng A 25 (9–10):707–710
Ma S, Viola R, Sui L, Cherubini V, Barbetti F, Egli D (2018) β cell replacement after gene editing of a neonatal diabetes-causing mutation at the insulin locus. Stem Cell Rep 11(6):1407–1415
Maehr R, Chen S, Snitow M, Ludwig T, Yagasaki L, Goland R, Leibel RL, Melton DA (2009) Generation of pluripotent stem cells from patients with type 1 diabetes. Proc Natl Acad Sci 106(37):15768–15773
Manzar GS, Kim EM, Zavazava N (2017) Demethylation of induced pluripotent stem cells from type 1 diabetic patients enhances differentiation into functional pancreatic β cells. J Biol Chem 292(34):14066–14079. https://doi.org/10.1074/jbc.M117.784280
Maxwell KG, Augsornworawat P, Velazco-Cruz L, Kim MH, Asada R, Hogrebe NJ, Morikawa S, Urano F, Millman JR (2020) Gene-edited human stem cell–derived β cells from a patient with monogenic diabetes reverse preexisting diabetes in mice. Sci Transl Med 12 (540). https://doi.org/10.1126/scitranslmed.aax9106
Merani S, Shapiro AMJ (2006) Current status of pancreatic islet transplantation. Clin Sci 110(6):611–625
Millman JR, Xie C, Van Dervort A, Gürtler M, Pagliuca FW, Melton DA (2016) Generation of stem cell-derived β-cells from patients with type 1 diabetes. Nat Commun 7. https://doi.org/10.1038/ncomms11463
Modi H, Jacovetti C, Tarussio D, Metref S, Madsen OD, Zhang F, Rantakari P, Poutanen M, Nef S, Gorman T (2015) Autocrine action of IGF2 regulates adult β-cell mass and function. Diabetes 64(August):4148–4157. https://doi.org/10.2337/db14-1735
Molitch ME, DeFronzo RA, Franz MJ, Keane WF, others (2004) Nephropathy in diabetes. Diabetes Care 27:S79
Murphy R, Ellard S, Hattersley AT (2008) Clinical implications of a molecular genetic classification of monogenic β-cell diabetes. Nat Clin Pract Endocrinol Metab 4(4):200–213
Noguchi M, Hosoda K, Nakane M, Mori E, Nakao K, Taura D, Yamamoto Y, Kusakabe T, Sone M,

Sakurai H, others (2013) In vitro characterization and engraftment of adipocytes derived from human induced pluripotent stem cells and embryonic stem cells. Stem Cells Dev 22(21):2895–2905

Nyitray CE, Chavez MG, Desai TA (2014) Mechanosensing and β-catenin signaling. Tissue Eng Part A 20:1888–1895. https://doi.org/10.1089/ten.tea.2013.0692

Ohmine S, Squillace KA, Hartjes KA, Deeds MC, Armstrong AS, Thatava T, Sakuma T, Terzic A, Kudva Y, Ikeda Y (2012) Reprogrammed keratinocytes from elderly type 2 diabetes patients suppress senescence genes to acquire induced pluripotency. Aging 4 (1):60–73. https://doi.org/10.18632/aging.100428

Okita K, Ichisaka T, Yamanaka S (2007) Generation of germline-competent induced pluripotent stem cells. Nature 448(7151):313–317

Pagliuca FW, Millman JR, Gürtler M, Segel M, Van Dervort A, Ryu JH, Peterson QP, Greiner D, Melton DA (2014) Generation of functional human pancreatic β cells in vitro. Cell 159(2):428–439. https://doi.org/10.1016/j.cell.2014.09.040

Pan FC, Wright C (2011) Pancreas organogenesis: from bud to plexus to gland. Dev Dyn 240(3):530–565

Park IH, Arora N, Huo H, Maherali N, Ahfeldt T, Shimamura A, Lensch MW, Cowan C, Hochedlinger K, Daley GQ (2008) Disease-specific induced pluripotent stem cells. Cell 134(5):877–886. https://doi.org/10.1016/j.cell.2008.07.041

Pellegrini S, Ungaro F, Mercalli A, Melzi R, Sebastiani G, Dotta F, Broccoli V, Piemonti L, Sordi V (2015) Human induced pluripotent stem cells differentiate into insulin-producing cells able to engraft in vivo. Acta Diabetol 52(6):1025–1035. https://doi.org/10.1007/s00592-015-0726-z

Pellegrini S, Cantarelli E, Sordi V, Nano R, Piemonti L (2016) The state of the art of islet transplantation and cell therapy in type 1 diabetes. Acta Diabetol 53 (5):683–691

Pellegrini S, Manenti F, Chimienti R, Nano R, Ottoboni L, Ruffini F, Martino G, Ravassard P, Piemonti L, Sordi V (2018) Differentiation of Sendai virus-reprogrammed iPSC into β cells, compared with human pancreatic islets and immortalized β cell line. Cell Transplant 27(10):1548–1560. https://doi.org/10.1177/0963689718798564

Pop-Busui R, Boulton AJM, Feldman EL, Bril V, Freeman R, Malik RA, Sosenko JM, Ziegler D (2017) Diabetic neuropathy: a position statement by the American Diabetes Association. Diabetes Care 40(1):136–154

Powis G, Berggren MM, Gallegos A, Abraham R, Ashendel C, Zalkow L, Matter WF, Dodge J, Grindey G, Vlahos CJ (1994) Wortmannin, a potent and selective inhibitor of Phosphatidylinositol-3-kinase. Cancer Res 54:2419–2424

Prentki M, Nolan CJ (2006) Islet β cell failure in type 2 diabetes. J Clin Invest 116(7):1802–1812

Quinn TJ, Dawson J, Walters MR (2011) Sugar and stroke: cerebrovascular disease and blood glucose control. Cardiovasc Ther 29(6):e31–e42

Russ HA, Parent AV, Ringler JJ, Hennings TG, Nair GG, Shveygert M, Guo T, Puri S, Haataja L, Cirulli V, Blelloch R, Szot GL, Arvan P, Hebrok M (2015) Controlled induction of human pancreatic progenitors produces functional beta-like cells in vitro. EMBO J 34 (13):1759–1772

Saeedi P, Petersohn I, Salpea P, Malanda B, Karuranga S, Unwin N, Colagiuri S, Guariguata L, Motala AA, Ogurtsova K, Shaw JE, Bright D, Williams R (2019) Global and regional diabetes prevalence estimates for 2019 and projections for 2030 and 2045: results from the international diabetes federation diabetes atlas, 9th edition. Diabetes Res Clin Pract 157:107843. https://doi.org/10.1016/j.diabres.2019.107843

Saeedi M, Cao Y, Fadl H, Gustafson H, Simmons D (2021) Increasing prevalence of gestational diabetes mellitus when implementing the IADPSG criteria: a systematic review and meta-analysis. Diabetes Res Clin Pract 172:108642

Saha B, Borgohain M, Dey C, Thummer RP (2018) iPS cell generation: current and future challenges. Ann Stem Cell Res Ther 1(2):1–4

Saxena P, Heng BC, Bai P, Folcher M, Zulewski H, Fussenegger M (2016) A programmable synthetic lineage-control network that differentiates human IPSCs into glucose-sensitive insulin-secreting beta-like cells. Nat Commun 7:1–14. https://doi.org/10.1038/ncomms11247

Shaer A, Azarpira N, Vahdati A, Karimi MH, Shariati M (2015) Differentiation of human-induced pluripotent stem cells into insulin-producing clusters. Exp Clin Transplant 13(1):68–75. https://doi.org/10.6002/ect.2013.0131

Shahjalal HM, Shiraki N, Sakano D, Kikawa K, Ogaki S, Baba H, Kume K, Kume S (2014) Generation of insulin-producing β-like cells from human iPS cells in a defined and completely xeno-free culture system. J Mol Cell Biol 6(5):394–408. https://doi.org/10.1093/jmcb/mju029

Shang L, Hua H, Foo K, Martinez H, Watanabe K, Zimmer M, Kahler DJ, Freeby M, Chung W, LeDuc C, Goland R, Leibel RL, Egli D (2014) β-cell dysfunction due to increased ER stress in a stem cell model of wolfram syndrome. Diabetes 63(3):923–933. https://doi.org/10.2337/db13-0717

Singh VK, Kalsan M, Kumar N, Saini A, Chandra R (2015) Induced pluripotent stem cells: applications in regenerative medicine, disease modeling, and drug discovery. Front Cell Dev Biol 3:2

Soejitno A, Prayudi PKA (2011) The prospect of induced pluripotent stem cells for diabetes mellitus treatment. Ther Adv Endocrinol Metab 2(5):197–210

Takahashi K, Yamanaka S (2006) Induction of pluripotent stem cells from mouse embryonic and adult fibroblast cultures by defined factors. Cell 126(4):663–676

Takahashi K, Tanabe K, Ohnuki M, Narita M, Ichisaka T, Tomoda K, Yamanaka S (2007) Induction of pluripotent stem cells from adult human fibroblasts by defined factors. Cell 131(5):861–872

Tateishi K, He J, Taranova O, Liang G, D'Alessio AC, Zhang Y (2008) Generation of insulin-secreting islet-

like clusters from human skin fibroblasts. J Biol Chem 283(46):31601–31607
Teo AKK, Lau HH, Valdez IA, Dirice E, Tjora E, Raeder H, Kulkarni RN (2016) Early developmental perturbations in a human stem cell model of MODY5/HNF1B pancreatic hypoplasia. Stem Cell Rep 6(3):357–367. https://doi.org/10.1016/j.stemcr.2016.01.007
Thatava T, Nelson TJ, Edukulla R, Sakuma T, Ohmine S, Tonne JM, Yamada S, Kudva Y, Terzic A, Ikeda Y (2010) Indolactam V/GLP-1-mediated differentiation of human iPS cells into glucose-responsive insulin-secreting progeny. Gene Ther 18(3):283–293. https://doi.org/10.1038/gt.2010.145
Thatava T, Kudva YC, Edukulla R, Squillace K, De Lamo JG, Khan YK, Sakuma T, Ohmine S, Terzic A, Ikeda Y (2013) Intrapatient variations in type 1 diabetes-specific iPS cell differentiation into insulin-producing cells. Mol Ther 21(1):228–239. https://doi.org/10.1038/mt.2012.245
Tzoulaki I, Molokhia M, Curcin V, Little MP, Millett CJ, Ng A, Hughes RI, Khunti K, Wilkins MR, Majeed A, others (2009) Risk of cardiovascular disease and all cause mortality among patients with type 2 diabetes prescribed oral antidiabetes drugs: retrospective cohort study using UK general practice research database. BMJ 339:b4731
Velho G, Froguel P, Clement K, Pueyo ME, Zouali H, Cohen D, Passa P, Rakotoambinina B, Robert J-J (1992) Primary pancreatic beta-cell secretory defect caused by mutations in glucokinase gene in kindreds of maturity onset diabetes of the young. Lancet 340 (8817):444–448
Velho G, Petersen KF, Perseghin G, Hwang J-H, Rothman DL, Pueyo ME, Cline GW, Froguel P, Shulman GI, others (1996) Impaired hepatic glycogen synthesis in glucokinase-deficient (MODY-2) subjects. J Clin Invest 98(8):1755–1761
Vethe H, Bjørlykke Y, Ghila LM, Paulo JA, Scholz H, Gygi SP, Chera S, Ræder H (2017) Probing the missing mature β-cell proteomic landscape in differentiating patient iPSC-derived cells. Sci Rep 7(1):1–14. https://doi.org/10.1038/s41598-017-04979-w
Walczak MP, Drozd AM, Stoczynska-Fidelus E, Rieske P, Grzela DP (2016) Directed differentiation of human iPSC into insulin producing cells is improved by induced expression of PDX1 and NKX6.1 factors in IPC progenitors. J Transl Med 14(1):1–16. https://doi.org/10.1186/s12967-016-1097-0
Wan J, Huang Y, Zhou P, Guo Y, Wu C, Zhu S, Wang Y, Wang L, Lu Y, Wang Z (2017) Culture of iPSCs derived pancreatic β-like cells in vitro using decellularized pancreatic scaffolds: a preliminary trial. Biomed Res Int 2017. https://doi.org/10.1155/2017/4276928
Wang L, Huang Y, Guo Q, Fan X, Lu Y, Zhu S, Wang Y, Bo X, Chang X, Zhu M, Wang Z (2014) Differentiation of iPSCs into insulin-producing cells via adenoviral transfection of PDX-1, NeuroD1 and MafA. Diabetes Res Clin Pract 104(3):383–392. https://doi.org/10.1016/j.diabres.2014.03.017
Wang Q, Donelan W, Ye H, Jin Y, Lin Y, Wu X, Wang Y, Xi Y (2019) Real-time observation of pancreatic beta cell differentiation from human induced pluripotent stem cells. Am J Transl Res 11(6):3490–3504
Yabe SG, Fukuda S, Nishida J, Takeda F, Nashiro K, Okochi H (2019) Induction of functional islet-like cells from human iPS cells by suspension culture. Regen Ther 10:69–76. https://doi.org/10.1016/j.reth.2018.11.003
Young W (2012) Patient-specific induced pluripotent stem cells as a platform for disease modeling, drug discovery and precision personalized medicine. J Stem Cell Res Ther 01(S10). https://doi.org/10.4172/2157-7633.s10-010
Yu J, Vodyanik MA, Smuga-Otto K, Antosiewicz-Bourget J, Frane JL, Tian S, Nie J, Jonsdottir GA, Ruotti V, Stewart R, others (2007) Induced pluripotent stem cell lines derived from human somatic cells. Science 318(5858):1917–1920
Zhang D, Jiang W, Liu M, Sui X, Yin X, Chen S, Shi Y, Deng H (2009) Highly efficient differentiation of human ES cells and iPS cells into mature pancreatic insulin-producing cells. Cell Res 19:429–438. https://doi.org/10.1038/cr.2009.28
Zhu FF, Zhang PB, Zhang DH, Sui X, Yin M (2011) Generation of pancreatic insulin-producing cells from rhesus monkey induced pluripotent stem cells. Diabetologia 54:2325–2336. https://doi.org/10.1007/s00125-011-2246-x
Zhu Z, Li QV, Lee K, Rosen BP, González F, Soh CL, Huangfu D (2016) Genome editing of lineage determinants in human pluripotent stem cells reveals mechanisms of pancreatic development and diabetes. Cell Stem Cell 18(6):755–768. https://doi.org/10.1016/j.stem.2016.03.015

Adv Exp Med Biol - Cell Biology and Translational Medicine (2021) 14: 29–43
https://doi.org/10.1007/5584_2021_644

Published online: 12 June 2021

Differentiated Cells Derived from Hematopoietic Stem Cells and Their Applications in Translational Medicine

Sophia S. Fernandes, Lalita S. Limaye, and Vaijayanti P. Kale

Abstract

Hematopoietic stem cells (HSCs) and their development are one of the most widely studied model systems in mammals. In adults, HSCs are predominantly found in the bone marrow, from where they maintain homeostasis. Besides bone marrow and mobilized peripheral blood, cord blood is also being used as an alternate allogenic source of transplantable HSCs. HSCs from both autologous and allogenic sources are being applied for the treatment of various conditions like blood cancers, anemia, etc. HSCs can further differentiate to mature blood cells. Differentiation process of HSCs is being extensively studied so as to obtain a large number of pure populations of various differentiated cells in vitro so that they can be taken up for clinical trials. The ability to generate sufficient quantity of clinical-grade specialized blood cells in vitro would take the field of hematology a step ahead in translational medicine.

S. S. Fernandes and L. S. Limaye
Stem Cell Lab, National Centre for Cell Science, Pune, India

V. P. Kale (✉)
Symbiosis Centre for Stem Cell Research, Symbiosis School of Biological Sciences, Symbiosis International (Deemed University), Pune, India
e-mail: vaijayanti.kale@ssbs.edu.in
; vaijayanti.kale@gmail.com

Keywords

Clinical utility · Hematopoietic stem cells (HSCs) · HSCs in therapeutics · Mature blood cells

Abbreviations

BM	bone marrow
CLP	common lymphoid progenitor
CMP	common myeloid progenitor
DCs	dendritic cells
DMSO	dimethyl sulfoxide
ESCs	embryonic stem cells
FCS	fetal calf serum
GMP	good manufacturing practice
GvHD	graft-vs-host disease
HLA	human leukocyte antigen
HPCs	hematopoietic progenitor cells
HSCs	hematopoietic stem cells
iPSCs	induced pluripotent stem cells
LT-HSCs	long-term-hematopoietic stem cells
MKs	megakaryocytes
MNCs	mononuclear cells
MSCs	mesenchymal stem cells
PB	peripheral blood
PSCs	pluripotent stem cells
RBCs	red blood cells
Tregs	T regulatory cells
UCB	umbilical cord blood

1 Introduction

The concept of "stem cells" was first introduced by Till and McCulloch in 1961 while studying the hematopoietic system in mice. They defined these cells as multipotent cells, which can self-renew (make copies of their own) as well as differentiate into blood cells. The origin of hematopoietic lineage and its homeostasis and differentiation have been some of the most widely studied systems in mammals. This has expedited the use of hematopoietic cells in clinical translation as well (Kumar and Verfaillie 2012). The present chapter focuses on the differentiation of hematopoietic stem cells (HSCs) into functional blood cells such as dendritic cells (DCs), megakaryocytes (MKs), platelets, red blood cells (RBCs), etc. that could potentially be used in translational medicine. This chapter also focuses on the mechanisms involved in the differentiation of HSCs into various functional blood cells.

2 Hematopoietic Stem Cells

2.1 Origin of HSCs and Their Characterization

HSCs are responsible for maintaining homeostasis and replenishment of the blood cells over the individual's life span. During embryogenesis, hematopoietic progenitors first emerge in the yolk sac. Gradually, the development site progressively shifts to the aorta-gonad-mesonephros (AGM) region, the placenta, and the fetal liver, where the fetal HSCs mature into adult ones. At birth, the HSCs populate the bone marrow, making it the primary site of adult hematopoiesis (Mikkola and Orkin 2006; Mahony and Bertrand 2019). The bone marrow microenvironment provides a suitable niche, where the HSCs can either self-renew to generate more HSCs, or differentiate and mature into various types of blood cells depending on the physiological demand. The niche comprises of different stromal cells like mesenchymal stem cells (MSCs), osteoblasts, endothelial cells, reticular cells, adipocytes, etc. that play a very important role in providing the developing HSCs with cues and stimuli to facilitate the production of desired types of functional cells (Pinho and Frenette 2019). HSCs are multipotent in nature and are capable of forming the specialized blood lineage cells through asymmetric cell divisions (Till and McCulloch 1961; Fig. 1). HSCs from murine and human sources show different phenotypic characters (Table 1). Murine HSCs are defined as the cells having a long-term reconstituting ability (LT-HSCs), which can give rise to both myeloid and lymphoid lineages and can be isolated using a flow cytometer as the cells that are lineage negative (Lin^-), Sca-1 positive ($Sca\text{-}1^+$), c-kit positive ($c\text{-}kit^+$), and CD34 low/negative (LSK $CD34^{low/neg}$)(Osawa et al. 1996). On the other hand, very primitive human HSCs are found to be Lin^-, $CD133^+$, $GPI\text{-}80^+$, and $CD34^-$ by marker selection. Interestingly, human HSCs contained in the cord blood source display CD34 positivity as a distinct marker. The ability to home and engraft into the bone marrow of the recipients when injected via intravenous route is the most important characteristic of HSCs and forms the very basis of the experimental and clinical transplantations (Sumide et al. 2018).

2.2 Different Sources of HSCs

In adult individuals, bone marrow (BM) contains the major pool of HSCs. The frequency of CD34-expressing cells in the adult BM typically ranges between 0.5% and 5% (Hordyjewska et al. 2015). There are two methods to harvest the HSCs from this source – the first method consists of isolation of HSCs from BM aspirates, and the second method consists of mobilization of HSCs through apheresis procedure into peripheral blood. BM aspiration is a very invasive and painful procedure for the donor (Chen et al. 2013). On the other hand, apheresis is done by administration of cytokines like G-CSF or pharmacological agents like AMD3100 so that the HSCs mobilize into the bloodstream along with the hematopoietic progenitors, and then these can be separated out based on their CD34 positivity (Domen et al.

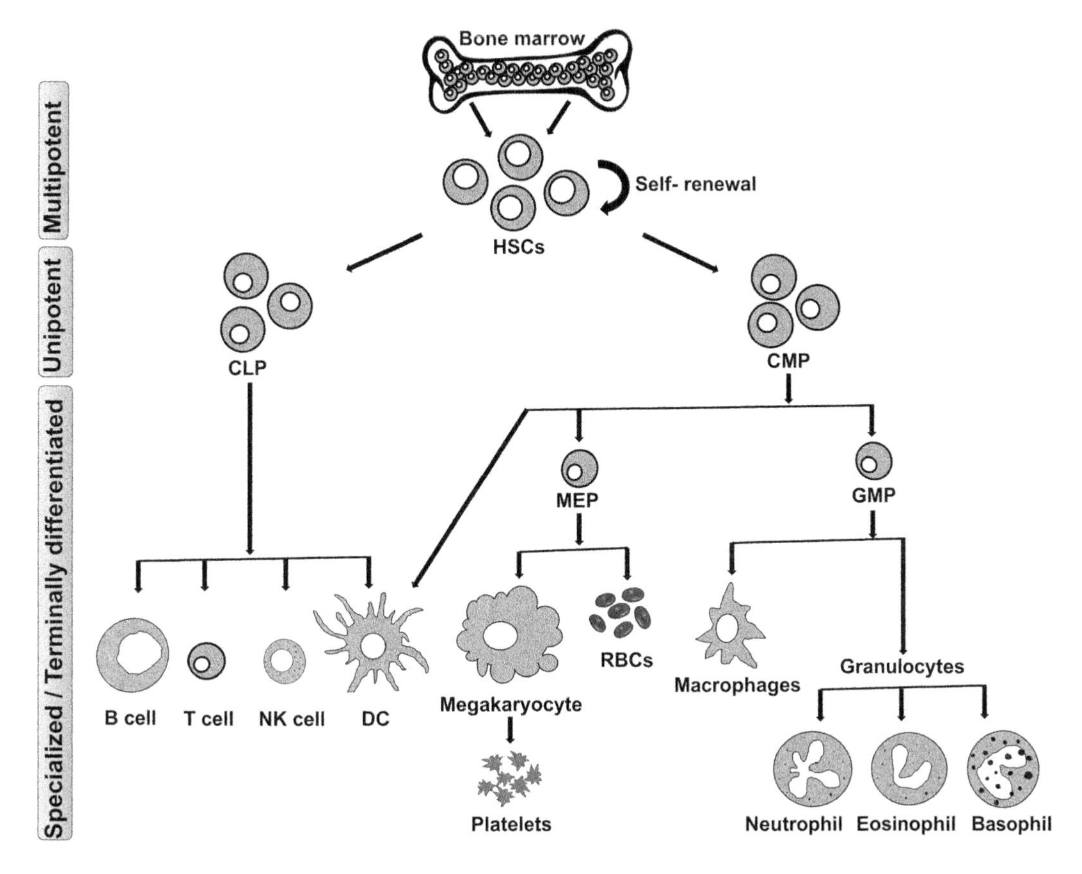

Fig. 1 Hierarchy of maturation occurring in the blood lineage: the step-wise generation of specialized blood cells from hematopoietic stem cells (HSCs) depicting its potency state at each level

2006). Umbilical cord blood (UCB) is yet another rich source of adult HSCs, which has been widely studied and explored (Broxmeyer et al. 2020; Mayani et al. 2020). UCB contains almost 0.02–1.43% of $CD34^+$ cells. The peripheral blood too contains $CD34^+$ cells, but in a very low abundance (<0.01%) (Hordyjewska et al. 2015).

Pluripotent stem cells (PSCs) that comprise of embryonic stem cells (ESCs) and induced pluripotent stem cells (iPSCs) are also being explored for their differentiation potential to hematopoietic lineage due to their unique ability to differentiate into cells of various lineages. One approach consisting of overexpressing a set of seven transcription factors – ERG, HOXA5, HOXA9, HOXA10, LCOR, RUNX1, and SPI1 – via a lentiviral delivery system into human PSCs converts these cells to HSCs and hematopoietic progenitor cells (HPCs). When these cells were infused in mouse models, these PSC-derived HSCs were found to give rise to both the myeloid and the lymphoid cells and also survive through primary and secondary transplants. However, the molecular signature of these cells is still distinct from UCB-HSCs, suggesting that this approach needs further fine-tuning (Sugimura et al. 2017). The beneficial role of Runx1a overexpression has been identified in the differentiation of PSCs into HPCs (Ran et al. 2013). Use of mouse stromal cell line-OP9 has also been reported to enhance HPC differentiation from PSCs, as this cell line mimics the BM niche and secretes factors that support the development into HPSCs (Demirci et al. 2020). For better clinical utility, attempts have also been initiated to formulate serum-free

Table 1 Phenotypic characteristics depicting distinctive states from primitive HSCs, HSCs to MPPs, CLP, and CMP occurring during hematopoiesis in both murine and human models

	Murine	Human
Primitive HSCs	Lin^-, $c\text{-}kit^{+/high}$, $Sca\text{-}1^+$, $CD34^{low/neg}$, $CD150^+$ and $CD48^-$	Lin^-, $CD133^+$, $GPI\text{-}80^+$ and $CD34^-$
HSCs	Lin^-, $c\text{-}kit^+$, $Sca\text{-}1^+$, $CD34^-$, $CD150^+$, $CD48^-$ and $Flt3^-$	Lin^-, $CD34^+$, $CD38^-$, $CD90^+$, $CD49f^+$, $Flt3^+$ and $CD45RA^-$
Multipotent progenitors	Lin^-, $c\text{-}kit^+$, $Sca\text{-}1^+$, $CD34^+$, $CD150^-$ and $Flt3^+$	Lin^-, $CD34^+$, $CD38^-$, $Flt3^+$, $CD90^-$ and $CD45RA^-$
CLP	Lin^-, $c\text{-}kit^{low}$, $Sca\text{-}1^+$, $Flt3^{Hi}$, $IL7RA^+$	Lin^-, $CD34^+$, $CD38^+$, $CD19^-$ and $CD10^+$
CMP	Lin^-, $Sca\text{-}1^{low}$, $cKit^{high}$, $CD34^{high}$, $CD16/32^{low}$	Lin^-, $CD34^+$, $CD38^+$, $IL3Ra^{low}$ and $CD45RA^-$

Reference: Seita and Weissman 2010; Sumide et al. 2018; Tajer et al. 2019, Osawa et al. 1996, Ichii et al. 2014, Cheng et al. 2020

defined culture media for such differentiation (Chicha et al. 2011, Niwa et al. 2011). However, the HSC-like cells derived from iPSCs in vitro lack important functions like homing and engraftment, and these mostly remain adherent under culture conditions. On further investigation it was found that the overexpression of the epithelial-to-mesenchymal transition inhibiting miRNA in these differentiated progeny was the causative factor of these defects (Risueño et al. 2012; Meader et al. 2018). If the reprogramming genes fail to silence or inactivate after differentiation is induced, they lead to inefficient generation of the differentiated progeny (Ramos-Mejía et al. 2012). Although various groups show differentiation of iPSCs to hematopoietic lineages, there are still several aspects that need improvement before the data can actually be extrapolated to regenerative medicine and the generated cells can be tested in clinical trials.

2.3 Phenotypic Analysis of Different Stages of HSCs' Differentiation and Maturation

HSCs were identified by Harrison and group as the pool of cells having an ability to repopulate the lymphoid and the myeloid compartments following a transplant (Harrison and Zhong 1992). In humans, HSCs are identified as cells having Lin^- $CD34^+$ $CD38^-$ $CD90^+$ $CD45RA^-$ phenotype. As these cells progress to a state where they form multipotent progenitors, their phenotype changes to Lin^- $CD34^+$ $CD38^-$ $CD90^-$ $CD45RA^-$. These can further form common lymphoid progenitors (CLPs) or common myeloid progenitors (CMPs) depending on the physiological demand for differentiated cells. The CLPs are characterized as Lin^- $CD34^+$ $CD38^+$ $CD10^+$, and these progenitors can further form T cells, B cells, natural killer cells (NKs), or DCs based on the cytokine stimulus they are exposed to during maturation. The CMPs display a Lin^- $CD34^+$ $CD38^+$ $IL3Ra^{low}$ $CD45RA^-$ phenotype. These CMPs can then take up either the megakaryocyte/erythrocyte progenitor (MEP), granulocyte/macrophage progenitor (GMP), or pro-DC fate. The MEPs are Lin^- $CD34^+$ $CD38^+$ $IL3Ra^-$ $CD45RA^-$ which can further specialize into mature platelets or erythroid cells. The GMPs are characterized as Lin^- $CD34^+$ $CD38^+$ $IL3Ra^+$ $CD45RA^-$, which can differentiate into either granulocytes or macrophages. Granulocytes comprise of the protein granule-containing cells, viz., the neutrophils, basophils, and eosinophils (Table 1; Seita and Weissman 2010; Sumide et al. 2018; Tajer et al. 2019). The hematopoietic progenitor and HSC states are very tightly regulated by an array of cytokines that help these cells in performing their optimal functions (Ikonomi et al. 2020). Even a small perturbation in this development can cause serious disease conditions like anemia, thrombocytopenia, neutropenia, leukemia, etc. (Tunstall-Pedoe et al. 2008; Roy et al. 2012). The quiescence and the activation of HSCs during homeostasis need to be appropriately controlled; a non-responsive HSC pool could lead to severe deficiencies of mature blood cells, whereas a hyperactive HSC pool could lead to a premature exhaustion of the HSC population (Wilson et al. 2004). Apart from the primary role of BM progenitors in regulating hematopoietic homeostasis, very recently, they have also been identified in the formation of non-hematopoietic decidual cells, aiding in the formation and functioning of decidual stroma during pregnancy (Tal et al. 2019).

3 Challenges Faced While Using In Vitro Generated Differentiated Cells as an Alternative to Those Directly Obtained from UCB or Peripheral Blood (PB): Attempts Made to Improve the Process

For the treatment of certain blood-related disorders, apheresis procedures are conducted to suffice the need for healthy blood cells (Adamski et al. 2018). For such transplantation purposes, cells need to be isolated from suitable donors. This procedure is limited by several factors such as unavailability of HLA-matched donors, risk of transmissible diseases, use of high doses of cytokines, variable efficacy of the yield, etc. (Mcmanus and Mitchell 2014). As an alternative to this, in vitro differentiation of blood cells from the HSCs was proposed to meet the increasing demand for such cells (Devine et al. 2010; Kumar and Geiger 2017). HSCs isolated from BM or UCB can serve as a starting material for obtaining these differentiated cells. However, to obtain them in sufficient quantities required for transplantation, there is a need to first expand the HSCs. Several attempts have been made in this direction. Use of cytokines and small molecules such as insulin-like growth factor-binding protein 2 (IGFBP2), angiopoietin-like proteins, stemregenin-1 (aryl hydrocarbon receptor inhibitor), nicotinamide (SIRT1 inhibitor), UM171 (pyrimidoindole derivative), stem cell factor, thrombopoietin, Fms-related tyrosine kinase 3 ligand, interleukin-6, interleukin-3, resveratrol, etc. has been successful in the expansion of HSCs in vitro (Zhang et al. 2008; Peled et al. 2012; Flores-Guzmán et al. 2013; Fan et al. 2014; Farahbakhshian et al. 2014; Fares et al. 2014; Heinz et al. 2015; Huang et al. 2019). Use of apoptosis pathway modulators like caspase and calpain inhibitors during in vitro HSC expansion has also shown promising results in engraftment of UCB-HSCs when transplanted in mouse models (Sangeetha et al. 2010; Sangeetha et al. 2012). Co-culture of HSCs with human mesenchymal stem/stromal cells (hMSCs) has been shown to result in an efficient expansion of HSCs (Kadekar et al. 2015; Perucca et al. 2017; Papa et al. 2020). But the conditions provided to the HSCs in vitro for proliferation and expansion usually result in pushing the HSCs into a cycling phase. However, for the HSCs to sustain their multipotency during longer duration, maintenance of a quiescent state is very important. Hence, expansion of HSCs without compromising their "stemness" has still remained a huge challenge in the field. In a recent study, HSCs could be held in a steady state in vitro by modulating the culture conditions to hypoxia, low cytokines, and high fatty acid composition in the culture media (Kobayashi et al. 2019).

Cells developed in vitro mostly rely upon 2D culturing methods which acclimatize the cells to specific substrates and culture conditions that result in low transplant efficiencies with limited functionality of the cells. To overcome this, 3D culture methods have been proposed for maintaining and differentiating cells from various stem cells. In one approach, hMSCs entrapped in hydrogels were used as a 3D substrate so as to recapitulate the in vivo niche for the human BM-derived HSCs. The resultant HSCs depicted superior functionality and stem cell attributes, as compared to those cultured under 2D conditions (Sharma et al., 2012). Another study depicts the beneficial effects of 3D fibrin scaffolds in combination with MSCs to expand UCB-HSCs (Ferreira et al. 2012). In a recent report, use of 3D zwitterionic hydrogel cultures was found to be beneficial for the expansion of both UCB- and BM-HSCs (Bai et al. 2019). Although these in vitro expanded cells are closer to the native state in terms of growth conditions and functionality, it is difficult to reproducibly obtain them in large numbers that would be sufficient for transplant purposes (McKee and Chaudhry 2017). When a comparison was performed between HSCs and progenitor cells cultured in vitro from UCB-HSCs and their freshly isolated counterparts, it was observed that although the in vitro cultured cells were immunophenotypically similar to those present in vivo, they lacked the functionality and genetic signature of their in vivo counterparts, i.e., freshly isolated UCB-HSCs. The in vitro cultured cells displayed a distinct myeloid bias, and additionally, the long-term-culture-initiating cells (LTC-IC), which are considered to be the most primitive HSCs during transplantation setting, were also found to be in lower numbers in the in vitro generated cells, as opposed to those present in the fresh UCB. To improve on this, a strategy of co-infusing in vitro expanded HSCs with the native freshly isolated HSCs was used; however, even in this situation, the cells that survived longer during the transplant in the host were the freshly isolated ones, thereby indicating that the in vitro manipulated HSCs lack the longevity and functionality as the fresh ones (Dircio-Maldonado et al. 2018). Despite these challenges, efforts to efficiently expand the UCB units are ongoing so that they can be utilized for clinical applications (Kiernan et al. 2017; Mayani 2019). A few phase I/II clinical trials make use of HSCs that have been expanded in vitro using different compounds like UM171 (Cohen et al. 2020), nicotinamide (Horwitz et al. 2019), StemRegenin-1 (Wagner Jr et al. 2016), and copper chelator-tetraethylenepentamine (De Lima et al. 2008). In these trials, the outcome has been promising as the expanded HSCs did engraft efficiently in the recipient after the transplant.

4 Cord Blood Is a Superior Source of In Vitro Generated HSCs and HPCs

During ex vivo expansion, UCB-HSCs exhibited better expansion ability and yielded a much higher number of progenitor colonies, as compared to BM-HSCs under serum-free conditions with cytokine stimulus (Kim et al. 2005). In a study employing in vitro priming of HSCs on OP9 stroma, the UCB-HSCs displayed better reconstitution of the T-lymphoid fraction, as compared to BM-HSCs, whereas the myeloid fraction remained unchanged (De Smedt et al. 2011). To address the issue of graft-vs-host disease (GvHD), expansion of T regulatory (Tregs) cells in vitro from either UCB or adult PB is an approach before using them in cell therapy. In such cases, Tregs having better immunomodulatory activity could be obtained in higher numbers from UCB samples, as compared to PB sample (Fan et al. 2012). UCB samples have certain advantages like the naïve state of their HSCs, tolerability in terms of HLA mismatches (a 4–6/6 is accepted in the case of UCB donors), ready availability in UCB banks due to easy collection and harvesting procedures, etc. (Schönberger et al. 2004; Beksac 2016). In a pediatric clinical trial assessment, UCB stem cell transplants yielded better engraftment in the recipients with early and committed progenitors, as compared to BM transplant (Frassoni et al. 2003). Recent strategies employ depletion of TCR$\alpha\beta^+$/CD19$^+$ in donor cells, which helps reduce GvHD, thus enhancing survivability on UCB transplant in a HLA mismatch recipient (Elfeky et al. 2019). Even in the aspect of obtaining differentiated progeny, specifically the megakaryocyte lineage, the UCB source proves to be superior, as compared to BM (Tao et al. 1999) or even mobilized PB (Bruyn et al. 2005). Domogala and colleagues observed that NK cells differentiated from UCB-CD34 cells in vitro could be obtained in higher numbers and were comparable to those isolated directly from UCB and PB samples for functional and phenotypic characteristics to be used in immunotherapy (Domogala et al. 2017). Furthermore, both sources, PB-MNCs and UCB-MNCs, displayed comparable phenotypic and functional attributes when primed toward the DC lineage (Kumar et al. 2015).

5 In Vitro Differentiated Hematopoietic Cells for Clinical Applications

The ultimate goal of in vitro differentiation studies of stem cells is to utilize the protocols for applications in cell therapy. Here we describe some such studies which have potential clinical applications in the near future.

5.1 T Cells

An approach of modulation of the hematopoietic niche to stimulate Notch signaling, which in turn leads to an enhanced T cell progenitor generation, can be extrapolated for clinical studies (Shukla et al. 2017). Olbrich et al. formulated a retroviral construct in PB- and CB-CD34$^+$ cells that overexpress chimeric antigen receptor (CAR) targeted against the human cytomegalovirus (HCMV) glycoprotein B that is known to be dominant during viral reactivation. On differentiating such transduced CD34$^+$ to T cells, the generated T cells exhibit remarkable on-target specificity and cytotoxicity to HCMV infections under in vitro conditions and in mouse models, thus proving to be a good target for further clinical studies (Olbrich et al. 2020). Use of StemRegenin-1 (SR-1), a purine derivative that antagonizes aryl hydrocarbon receptor, in expansion medium results in >250-fold expansion of HPCs in vitro. Due to this property it is now being tested in clinical trials. Expanded HPCs obtained by this method have further been evaluated for differentiation into pro-T cells in vitro which also display effective T cell functions on being infused in vivo (Singh et al. 2019).

5.2 Dendritic Cells

Since DCs are present in very low numbers in peripheral blood, obtaining them in sufficient numbers for immune therapies poses difficulty. The SR-1 expanded HPCs were differentiated to DCs that resulted in large-scale generation with functional properties comparable to the circulating DCs (Thordardottir et al. 2014). Plantinga et al. proposed a method of generation of conventional DCs from CB-CD34^{+} cells by selection based on CD115 positivity. These cells generated DCs with 75–95% purity and were functional as a DC vaccine even on maturation and stimulation (Plantinga et al. 2019). The findings were taken a step ahead by generating GMP-grade DCs from UCB-CD34 cells that specifically expressed Wilms' tumor 1 protein as a potent DC vaccine candidate for the use in pediatric acute myeloid leukemia (AML) patients (Plantinga et al. 2020). In the case of multiple myeloma cancer patient samples, DCs derived from HSCs were found to be a potent source for generation of DC vaccine. The phenotypic characters, antigen uptake, migration, and T cell stimulation were comparable to those derived from healthy donors. These DCs could also raise a tumor-specific cytotoxic T lymphocyte (CTL) reaction, but it was less robust when compared to the healthy donor source due to T cell exhaustion seen in multiple myeloma samples. Therefore, if CTL-4 blocking was used in combination with such DCs, it could serve as a promising candidate for immunotherapy in multiple myeloma (Shinde et al. 2019).

5.3 Megakaryocytes and Platelets

Large-scale platelets could be obtained in vitro from UCB-HSCs using a three-step method, comprising of adherent and liquid cultures. These platelets were similar to PB platelets and the yield obtained was approximately 3.4 units of platelets from 1 unit of CB used (Matsunaga et al. 2006). With an aim to reduce incidences of graft rejection, platelets that lacked β2-microglobulin gene have been differentiated from hiPSCs. The resultant platelets produced in high numbers were also HLA-ABC negative, and hence, such approaches could be explored from the therapeutic aspect (Feng et al. 2014; Norbnop et al. 2020). Improvements have also been tried on the culture method aspects, wherein a gas-permeable silicone membrane was used instead of the regular in vitro 2D culture dishes. The enhanced gas and media exchange resulted in the generation of MKs and platelets from mobilized PB-CD34 cells in higher numbers with functional attributes that made the process more clinically relevant (Martinez and Miller 2019). In another attempt to produce GMP-grade MKs and platelets, a roller bottle culture system was used that generated a large number of MKs from UCB-CD34^{+} cells that retained all standard characteristics coupled with high purity. The expansion was in the order of 2.5×10^{4} MKs from one CD34^{+} cell, and the optimal temperature for short-term storage to be applied in clinics was observed to be 22 °C when stored in saline supplemented with 10% human serum albumin (Guan et al. 2020).

5.4 Natural Killer Cells

Cryopreserved UCB units were used to purify HSCs, which were then differentiated into NKs in a closed bioreactor system that allowed a clinically relevant scaling up for their use in adoptive immunotherapy. This method yielded almost 2000-fold expansion of the cells with >90% purity while retaining their functional attributes

(Spanholtz et al. 2011). To eliminate the cytotoxicity induced due to chemotherapy given for various blood cancer treatment regimes, a strategy was devised for acute lymphoblastic leukemia (ALL) to increase the efficacy of cell-based therapies. UCB-CD34$^+$-derived CD16$^+$ NK cells when used in combination with anti-CD47 antibody were found to be potent in building up an anti-tumor immune response and reducing the load of ALL (Valipour et al. 2020). In another phase-1 clinical trial, NK cells were differentiated from PB mononuclear cells (MNCs) and sufficient expansion was achieved ex vivo on stromal cells expressing IL-21. These cells were then infused to leukemic patients undergoing haploidentical hematopoietic stem cell transplant (HSCT) at different time points from the same MNC donor. The leukemic patients that received NK cell transplant in addition to HSCT showed lower rate of relapse and mortality, lower viral infections, and reduced GvHD (Ciurea et al. 2017).

5.5 Red Blood Cells

To eliminate the batch-to-batch variation in experimental methods, RBCs were differentiated from UCB-MNCs in a xeno-free controlled environment, where not only the yield was high, but the method could also be scaled up for clinical applications (Rallapalli et al. 2019). On similar lines, PB-MNCs (without the need for CD34 isolation step) could be differentiated to erythroid cultures under GMP conditions in vitro, which also could be scaled up in bioreactors for a large-scale RBC production for use in clinics. More than 90% of them achieved enucleation and the functional attributes like oxygen-binding ability were comparable to their in vivo counterparts (Heshusius et al. 2019). Production of clinical-grade RBCs along with their expansion was achieved from CB-CD34$^+$ cells by in vitro stimulus using cytokines followed by expansion on CB-derived MSCs (Baek et al. 2008). The in vitro culture duration of obtaining RBCs from HSCs was accelerated by 3 days on supplementing TGF-β1 during the terminal erythroid maturation step (Kuhikar et al. 2020). Functional erythrocytes on a large scale were obtained from CB-CD34$^+$ cells ex vivo using a bottle-turning device culture method. About 200 million erythrocytes with more than 90% positivity for RBC marker CD235a could be obtained from one CB-CD34$^+$ cell. Safety and efficacy studies were validated on murine and non-human primate model, and therefore, this approach can provide an alternative to the traditional transplants in clinics (Zhang et al. 2017).

6 Efficient Cryopreservation of In Vitro Generated Differentiated Blood Cells Is Crucial for Their Therapeutic Use

There are some reports in the literature that attempt to achieve optimal cryopreservation of specialized blood cells from HSCs that can be taken to clinical trials. Different freezing agents and methods have been employed for the purpose; however, DMSO still remains the most widely used cryoprotectant (Li et al. 2019). A DC-based cellular vaccine derived from CB-CD34$^+$ cells for use as immunotherapy for esophageal cancer displayed effective cryopreservation in 2.5% DMSO, 2.5% glucose, and 10% FCS. The thawed cellular vaccine maintained viability, immunophenotype, T cell activation, and specific cytotoxic T lymphocyte activity (Yu et al. 2016). Balan et al. demonstrated that both fresh and frozen units of UCB could generate DCs with equivalent morphological and functional characteristics (Balan et al. 2010). Similar findings were reported in the context of freezing NK cells derived from CB-CD34$^+$ cells, where the cells did not show diminished morphological or functional attributes upon thawing (Domogala et al. 2016). Surprisingly, Luevano et al. observed that cryopreserved units of CB-CD34$^+$ cells gave rise to a higher number of NK cells without compromising their functionality after differentiation in vitro, when compared to their freshly isolated counterparts from mobilized PB or CB (Luevano et al. 2014). In another preclinical study, a commercial solution, CryoStor, containing 10% DMSO was used to cryopreserve CB-CD34$^+$ cell-derived MKs. The results indicated an efficient MK recovery and platelet production post-thaw (Patel et al. 2019).

Mobilized peripheral blood HSCs demonstrated comparable viability, i.e., > 95%, and functionality when cryopreserved either for 5 weeks or for 10 years in 5% DMSO as the freezing agent (Abbruzzese et al. 2013). For autologous or allogenic transplants, units of BM or PB are usually collected just before a planned infusion; hence, they seldom undergo a long-term storage before use. On the other hand, UCB units need a long-term storage as their use depends on the demand for an allogenic HLA-matched transplant (Watt et al. 2007). Thus, its efficient cryopreservation is of prime importance if these cells need to be explored for clinical applications. In a fresh unit of UCB, cells are viable for a maximum of few days when stored at 4 °C. Cryopreservation increases its availability by providing easy transport and enhancing its shelf life. Dimethyl sulfoxide (DMSO) has been the most commonly used freezing agent for cryopreservation of HSCs (Hornberger et al. 2019). DMSO in the range of 7.5–10% was found to be optimal for cryopreservation of UCB cells when added not more than 1 h before freezing and washed out before 30 min post-thaw (Fry et al. 2015). In addition, 5% DMSO with 6% pentastarch and 25% human albumin were found to cryopreserve UCB-MNCs in a better way, as compared to 10% DMSO alone (Hayakawa et al. 2010). Inclusion of trehalose – a membrane stabilizer and catalase – an antioxidant to the 10% DMSO freezing mixture significantly enhanced recovery of UCB-HSCs post-thaw (Limaye and Kale 2001; Sasnoor et al. 2005). Mitchell et al. monitored the recovery of total nucleated cells post-thaw in UCB units stored for a long time and did not find a difference in cell recovery after a storage period of 10 years (Mitchell et al. 2015). Hence, UCB cells can be effectively stored in DMSO-containing freezing media with a minimal loss of viability for longer duration as well (Jaing et al. 2018). Ice recrystallization during freezing has been one of the major causes of damage to the recovery of UCB progenitors. Use of ice recrystallization inhibitor (IRI)-N-(2-fluorophenyl)-D-gluconamide during freezing improves the percentage of engraftable cells post-thaw (Jahan et al. 2020). In an attempt to find alternatives to DMSO, so as to avoid the toxicity associated with it, compounds like pentaisomaltose – a carbohydrate – were used and found to effectively cryopreserve UCB stem cells, when compared to DMSO (Svalgaard et al. 2016).

7 Future Prospects of Stem Cell-Derived Differentiated Cells in Therapeutics

Blood cells obtained by apheresis (enriched in HSCs) are already applied therapeutically in autologous and allogenic settings. However, use of specialized cells generated in vitro from the HSCs is still at an early stage of research and preclinical studies. The field of regenerative medicine and transplantation is in dire need of alternative and novel sources and methods that can yield a large number of functional blood cells needed for therapeutic purposes. Recent attempts aim at producing a large-scale generation of differentiated cells from HSCs as well as PSCs for their application in regenerative medicine, and immunotherapies have shown promising results and hence can soon lead to their clinical applications (Xue and Milano 2020). However, before these methodologies can be applied clinically, strict ethical regulations and guidelines need to be formulated so as to channelize their use for transplant purposes (Kline and Bertolone 1998).

Conflict of Interest The authors declare no conflict of interests.

Ethical Approval The authors declare that this article does not contain any direct studies with animals or human participants.

References

Abbruzzese L et al (2013) Long term cryopreservation in 5% DMSO maintains unchanged CD 34+ cells viability and allows satisfactory hematological engraftment after peripheral blood stem cell transplantation. Vox Sang 105:77–80. https://doi.org/10.1111/vox.12012

Adamski J, Ipe TS, Kinard T (2018) Therapeutic and donor apheresis. In: Transfusion medicine, apheresis, and hemostasis. Academic, pp 327–351. https://doi.org/10.1016/B978-0-12-803999-1.00014-6

Baek EJ, Kim HS, Kim S, Jin H, Choi TY, Kim HO (2008) In vitro clinical-grade generation of red blood cells from human umbilical cord blood CD34+ cells. Transfusion 48:2235–2245. https://doi.org/10.1111/j.1537-2995.2008.01828.x

Bai T et al (2019) Expansion of primitive human hematopoietic stem cells by culture in a zwitterionic hydrogel. Nat Med 25:1566–1575. https://doi.org/10.1038/s41591-019-0601-5

Balan S, Kale VP, Limaye LS (2010) A large number of mature and functional dendritic cells can be efficiently generated from umbilical cord blood–derived mononuclear cells by a simple two-step culture method. Transfusion 50:2413–2423. https://doi.org/10.1111/j.1537-2995.2010.02706.x

Beksac M (2016) Is there any reason to prefer cord blood instead of adult donors for hematopoietic stem cell transplants? Front Med 2:95. https://doi.org/10.3389/fmed.2015.00095

Broxmeyer HE, Cooper S, Capitano ML (2020) Enhanced collection of phenotypic and engrafting human cord blood hematopoietic stem cells at 4° C. Stem Cells 38:1326–1331. https://doi.org/10.1002/stem.3243

Bruyn CD, Delforge A, Martiat P, Bron D (2005) Ex vivo expansion of megakaryocyte progenitor cells: cord blood versus mobilized peripheral blood. Stem Cells Dev 14:415–424. https://doi.org/10.1089/scd.2005.14.415

Chen SH, Wang TF, Yang KL (2013) Hematopoietic stem cell donation. Int J Hematol 97:446–455. https://doi.org/10.1007/s12185-013-1298-8

Cheng H, Zheng Z, Cheng T (2020) New paradigms on hematopoietic stem cell differentiation. Protein Cell 11:34–44. https://doi.org/10.1007/s13238-019-0633-0

Chicha L, Feki A, Boni A, Irion O, Hovatta O, Jaconi M (2011) Human pluripotent stem cells differentiated in fully defined medium generate hematopoietic CD34+ and CD34− progenitors with distinct characteristics. PLoS One 6:e14733. https://doi.org/10.1371/journal.pone.0014733

Ciurea SO et al (2017) Phase 1 clinical trial using mbIL21 ex vivo–expanded donor-derived NK cells after haploidentical transplantation. Blood 130:1857–1868. https://doi.org/10.1182/blood-2017-05-785659

Cohen S et al (2020) Hematopoietic stem cell transplantation using single UM171-expanded cord blood: a single-arm, phase 1–2 safety and feasibility study. Lancet Haematol 7:134–145. https://doi.org/10.1016/S2352-3026(19)30202-9

De Smedt M, Leclercq G, Vandekerckhove B, Kerre T, Taghon T, Plum J (2011) T-lymphoid differentiation potential measured in vitro is higher in CD34+ CD38−/lo hematopoietic stem cells from umbilical cord blood than from bone marrow and is an intrinsic property of the cells. Haematologica 96:646–654. https://doi.org/10.3324/haematol.2010.036343

Demirci S et al (2020) Definitive hematopoietic stem/progenitor cells from human embryonic stem cells through serum/feeder-free organoid-induced differentiation. Stem Cell Res Ther 11:1–14. https://doi.org/10.1186/s13287-020-02019-5

Devine H, Tierney DK, Schmit-Pokorny K, McDermott K (2010) Mobilization of hematopoietic stem cells for use in autologous transplantation. Clin J Oncol Nurs 14:2. https://doi.org/10.1188/10.CJON.212-222

Dircio-Maldonado R et al (2018) Functional integrity and gene expression profiles of human cord blood-derived hematopoietic stem and progenitor cells generated in vitro. Stem Cells Transl Med 7:602–614. https://doi.org/10.1002/sctm.18-0013

Domen J, Wagers A, Weissman IL (2006) Bone marrow (hematopoietic) stem cells. Regen Med:13–34. https://www.uv.es/~elanuza/Dinamica/Regenerative_Medicine_2006.pdf#page=17

Domogala A, Madrigal JA, Saudemont A (2016) Cryopreservation has no effect on function of natural killer cells differentiated in vitro from umbilical cord blood CD34+ cells. Cytotherapy 18:754–759. https://doi.org/10.1016/j.jcyt.2016.02.008

Domogala A, Blundell M, Thrasher A, Lowdell MW, Madrigal JA, Saudemont A (2017) Natural killer cells differentiated in vitro from cord blood CD34+ cells are more advantageous for use as an immunotherapy than peripheral blood and cord blood natural killer cells. Cytotherapy 19:710–720. https://doi.org/10.1016/j.jcyt.2017.03.068

Elfeky R et al (2019) New graft manipulation strategies improve the outcome of mismatched stem cell transplantation in children with primary immunodeficiencies. J Allergy Clin Immunol 144:280–293. https://doi.org/10.1016/j.jaci.2019.01.030

Fan H et al (2012) Comparative study of regulatory T cells expanded ex vivo from cord blood and adult peripheral blood. Immunology 136:218–230. https://doi.org/10.1111/j.1365-2567.2012.03573.x

Fan X et al (2014) Low-dose insulin-like growth factor binding proteins 1 and 2 and angiopoietin-like protein 3 coordinately stimulate ex vivo expansion of human umbilical cord blood hematopoietic stem cells as assayed in NOD/SCID gamma null mice. Stem Cell Res Ther 5:1–9. https://doi.org/10.1186/scrt460

Farahbakhshian E et al (2014) Angiopoietin-like protein 3 promotes preservation of stemness during ex vivo expansion of murine hematopoietic stem cells. PLoS One 9:e105642. https://doi.org/10.1371/journal.pone.0105642

Fares IJ et al (2014) Cord blood expansion. Pyrimidoindole derivatives are agonists of human hematopoietic stem cell self-renewal. Science 345:1509–1512. https://doi.org/10.1126/science.1256337

Feng Q et al (2014) Scalable generation of universal platelets from human induced pluripotent stem cells. Stem Cell Rep 3:817–831. https://doi.org/10.1016/j.stemcr.2014.09.010

Ferreira MSV et al (2012) Cord blood-hematopoietic stem cell expansion in 3D fibrin scaffolds with stromal support. Biomaterials 33:6987–6997. https://doi.org/10.1016/j.biomaterials.2012.06.029

Flores-Guzmán P, Fernández-Sánchez V, Mayani H (2013) Concise review: ex vivo expansion of cord blood-derived hematopoietic stem and progenitor cells: basic principles, experimental approaches, and impact in regenerative medicine. Stem Cells Transl Med 2:830–838. https://doi.org/10.5966/sctm.2013-0071

Frassoni F et al (2003) Cord blood transplantation provides better reconstitution of hematopoietic reservoir compared with bone marrow transplantation. Blood 102:1138–1141. https://doi.org/10.1182/blood-2003-03-0720

Fry LJ, Querol S, Gomez SG, McArdle S, Rees R, Madrigal JA (2015) Assessing the toxic effects of DMSO on cord blood to determine exposure time limits and the optimum concentration for cryopreservation. Vox Sang 109:181–190. https://doi.org/10.1111/vox.12267

Guan X, Wang L, Wang H, Wang H, Dai W, Jiang Y (2020) Good manufacturing practice-grade of megakaryocytes produced by a novel ex vivo culturing platform. Clin Transl Sci 13:1115–1126. https://doi.org/10.1111/cts.12788

Harrison DE, Zhong RK (1992) The same exhaustible multilineage precursor produces both myeloid and lymphoid cells as early as 3-4 weeks after marrow transplantation. Proc Natl Acad Sci U S A 89:10134–10138. https://doi.org/10.1073/pnas.89.21.10134

Hayakawa J et al (2010) 5% dimethyl sulfoxide (DMSO) and pentastarch improves cryopreservation of cord blood cells over 10% DMSO. Transfusion 50:2158–2166. https://doi.org/10.1111/j.1537-2995.2010.02684.x

Heinz N, Ehrnström B, Schambach A, Schwarzer A, Modlich U, Schiedlmeier B (2015) Comparison of different cytokine conditions reveals resveratrol as a new molecule for ex vivo cultivation of cord blood-derived hematopoietic stem cells. Stem Cells Transl Med 4:1064–1072. https://doi.org/10.5966/sctm.2014-0284

Heshusius S et al (2019) Large-scale in vitro production of red blood cells from human peripheral blood mononuclear cells. Blood Adv 3:3337–3350. https://doi.org/10.1182/bloodadvances.2019000689

Hordyjewska A, Popiołek Ł, Horecka A (2015) Characteristics of hematopoietic stem cells of umbilical cord blood. Cytotechnology 67:387–396. https://doi.org/10.1007/s10616-014-9796-y

Hornberger K, Yu G, McKenna D, Hubel A (2019) Cryopreservation of hematopoietic stem cells: emerging assays, cryoprotectant agents, and technology to improve outcomes. Transfus Med Hemother 46:188–196. https://doi.org/10.1159/000496068

Horwitz ME et al (2019) Phase I/II study of stem-cell transplantation using a single cord blood unit expanded ex vivo with nicotinamide. J Clin Oncol 37:367–374. https://doi.org/10.1200/JCO.18.00053

Huang X, Guo B, Capitano M, Broxmeyer HE (2019) Past, present, and future efforts to enhance the efficacy of cord blood hematopoietic cell transplantation. F1000Research 8. https://doi.org/10.12688/f1000research.20002.1

Ichii M, Oritani K, Kanakura Y (2014) Early B lymphocyte development: similarities and differences in human and mouse. World J Stem Cells 6:421–431. https://doi.org/10.4252/wjsc.v6.i4.421

Ikonomi N, Kühlwein SD, Schwab JD, Kestler HA (2020) Awakening the HSC: dynamic modeling of HSC maintenance unravels regulation of the TP53 pathway and quiescence. Front Physiol 11:848. https://doi.org/10.3389/fphys.2020.00848

Jahan S, Adam MK, Manesia JK, Doxtator E, Ben RN, Pineault N (2020) Inhibition of ice recrystallization during cryopreservation of cord blood grafts improves platelet engraftment. Transfusion 60:769–778. https://doi.org/10.1111/trf.15759

Jaing TH, Chen SH, Wen YC, Chang TY, Yang YC, Tsay PK (2018) Effects of cryopreservation duration on the outcome of single-unit cord blood transplantation. Cell Transplant 27:515–519. https://doi.org/10.1177/0963689717753187

Kadekar D, Kale V, Limaye L (2015) Differential ability of MSCs isolated from placenta and cord as feeders for supporting ex vivo expansion of umbilical cord blood derived CD34+ cells. Stem Cell Res Ther 6:201. https://doi.org/10.1186/s13287-015-0194-y

Kiernan J et al (2017) Clinical studies of ex vivo expansion to accelerate engraftment after umbilical cord blood transplantation: a systematic review. Transfus Med Rev 31:173–182. https://doi.org/10.1016/j.tmrv.2016.12.004

Kim SK, Ghil HY, Song SU, Choi JW, Park SK (2005) Ex vivo expansion and clonal maintenance of CD34+ selected cells from cord blood and peripheral blood. Korean J Pediatr 48:894–900

Kline RM, Bertolone SJ (1998) Umbilical cord blood transplantation: providing a donor for everyone needing a bone marrow transplant? South Med J 91:821–828

Kobayashi H et al (2019) Environmental optimization enables maintenance of quiescent hematopoietic stem cells ex vivo. Cell Rep 28:145–158. https://doi.org/10.1016/j.celrep.2019.06.008

Kuhikar R, Khan N, Philip J, Melinkeri S, Kale V, Limaye L (2020) Transforming growth factor β1 accelerates and enhances in vitro red blood cell formation from hematopoietic stem cells by stimulating mitophagy. Stem Cell Res Ther 11:1–15. https://doi.org/10.1186/s13287-020-01603-z

Kumar S, Geiger H (2017) HSC niche biology and HSC expansion ex vivo. Trends Mol Med 23:799–819. https://doi.org/10.1016/j.molmed.2017.07.003

Kumar A, Verfaillie C (2012) Basic principles of multipotent stem cells. In: Progenitor and stem cell technologies and therapies. Woodhead Publishing Series in Biomaterials, pp 100–117. https://doi.org/10.1533/9780857096074.1.100

Kumar J, Kale V, Limaye L (2015) Umbilical cord blood-derived CD11c+ dendritic cells could serve as an alternative allogeneic source of dendritic cells for cancer immunotherapy. Stem Cell Res Ther 6:1–15. https://doi.org/10.1186/s13287-015-0160-8

Li R, Johnson R, Yu G, McKenna DH, Hubel A (2019) Preservation of cell-based immunotherapies for clinical trials. Cytotherapy 21:943–957. https://doi.org/10.1016/j.jcyt.2019.07.004

Lima D et al (2008) Transplantation of ex vivo expanded cord blood cells using the copper chelator tetraethylenepentamine: a phase I/II clinical trial. Bone Marrow Transplant 41:771–778. https://doi.org/10.1038/sj.bmt.1705979

Limaye LS, Kale VP (2001) Cryopreservation of human hematopoietic cells with membrane stabilizers and bioantioxidants as additives in the conventional freezing medium. J Hematother Stem Cell Res 10:709–718. https://doi.org/10.1089/152581601753193931

Luevano M et al (2014) Frozen cord blood hematopoietic stem cells differentiate into higher numbers of functional natural killer cells in vitro than mobilized hematopoietic stem cells or freshly isolated cord blood hematopoietic stem cells. PloS One 9:1. https://doi.org/10.1371/journal.pone.0087086

Mahony CB, Bertrand JY (2019) How HSCs colonize and expand in the fetal niche of the vertebrate embryo: an evolutionary perspective. Front Cell Dev Biol 7:34. https://doi.org/10.3389/fcell.2019.00034

Martinez AF, Miller WM (2019) Enabling large-scale ex vivo production of megakaryocytes from CD34+ cells using gas-permeable surfaces. Stem Cells Transl Med 8:658–670. https://doi.org/10.1002/sctm.18-0160

Matsunaga T et al (2006) Ex vivo large-scale generation of human platelets from cord blood CD34+ cells. Stem Cells 24:2877–2887. https://doi.org/10.1634/stemcells.2006-0309

Mayani H (2019) Human hematopoietic stem cells: concepts and perspectives on the biology and use of fresh versus in vitro–generated cells for therapeutic applications. Curr Stem Cell Rep 5:115–124. https://doi.org/10.1007/s40778-019-00162-1

Mayani H, Wagner JE, Broxmeyer HE (2020) Cord blood research, banking, and transplantation: achievements, challenges, and perspectives. Bone Marrow Transplant 55:48–61. https://doi.org/10.1038/s41409-019-0546-9

McKee C, Chaudhry GR (2017) Advances and challenges in stem cell culture. Colloids Surf B: Biointerfaces 159:62–77. https://doi.org/10.1016/j.colsurfb.2017.07.051

Mcmanus LM, Mitchell RN (2014) Pathobiology of human disease: a dynamic encyclopedia of disease mechanisms. Elsevier, pp 1800–1808

Meader E et al (2018) Pluripotent stem cell-derived hematopoietic progenitors are unable to downregulate key epithelial-mesenchymal transition-associated miRNAs. Stem Cells 36:55–64. https://doi.org/10.1002/stem.2724

Mikkola HK, Orkin SH (2006) The journey of developing hematopoietic stem cells. Development 133:3733–3744. https://doi.org/10.1242/dev.02568

Mitchell R et al (2015) Impact of long-term cryopreservation on single umbilical cord blood transplantation outcomes. Biol Blood Marrow Transplant 21:50–54. https://doi.org/10.1016/j.bbmt.2014.09.002

Niwa A et al (2011) A novel serum-free monolayer culture for orderly hematopoietic differentiation of human pluripotent cells via mesodermal progenitors. PLoS One 6:e22261. https://doi.org/10.1371/journal.pone.0022261

Norbnop P, Ingrungruanglert P, Israsena N, Suphapeetiporn K, Shotelersuk V (2020) Generation and characterization of HLA-universal platelets derived from induced pluripotent stem cells. Sci Rep 10:1–9. https://doi.org/10.1038/s41598-020-65577-x

Olbrich H et al (2020) Adult and cord blood-derived high-affinity gB-CAR-T cells effectively react against human Cytomegalovirus infections. Hum Gene Ther 31:423–439. https://doi.org/10.1089/hum.2019.149

Osawa M, Hanada KI, Hamada H, Nakauchi H (1996) Long-term lymphohematopoietic reconstitution by a single CD34-low/negative hematopoietic stem cell. Science 273:242–245. https://doi.org/10.1126/science.273.5272.242

Papa L, Djedaini M, Hoffman R (2020) Ex vivo HSC expansion challenges the paradigm of unidirectional human hematopoiesis. Ann N Y Acad Sci 1466:39. https://doi.org/10.1111/nyas.14133

Patel A et al (2019) Pre-clinical development of a cryopreservable megakaryocytic cell product capable of sustained platelet production in mice. Transfusion 59:3698–3713. https://doi.org/10.1111/trf.15546

Peled T et al (2012) Nicotinamide, a SIRT1 inhibitor, inhibits differentiation and facilitates expansion of hematopoietic progenitor cells with enhanced bone marrow homing and engraftment. Exp Hematol 40:342–355. https://doi.org/10.1016/j.exphem.2011.12.005

Perucca S et al (2017) Mesenchymal stromal cells (MSCs) induce ex vivo proliferation and erythroid commitment of cord blood haematopoietic stem cells (CB-CD34+ cells). PLoS One 12:e0172430. https://doi.org/10.1371/journal.pone.0172430

Pinho S, Frenette PS (2019) Haematopoietic stem cell activity and interactions with the niche. Nat Rev Mol

Cell Biol 20:303–320. https://doi.org/10.1038/s41580-019-0103-9
Plantinga M et al (2019) Cord-blood-stem-cell-derived conventional dendritic cells specifically originate from CD115-expressing precursors. Cancers 11:181. https://doi.org/10.3390/cancers11020181
Plantinga M et al (2020) Clinical grade production of Wilms' Tumor-1 loaded cord blood-derived dendritic cells to prevent relapse in pediatric AML after cord blood transplantation. Front Immunol 11:2559. https://doi.org/10.3389/fimmu.2020.559152
Rallapalli S, Guhathakurta S, Narayan S, Bishi DK, Balasubramanian V, Korrapati PS (2019) Generation of clinical-grade red blood cells from human umbilical cord blood mononuclear cells. Cell Tissue Res 375 (2):437–449. https://doi.org/10.1007/s00441-018-2919-6
Ramos-Mejía V et al (2012) Residual expression of the reprogramming factors prevents differentiation of iPSC generated from human fibroblasts and cord blood CD34+ progenitors. PLoS One 7:e35824. https://doi.org/10.1371/journal.pone.0035824
Ran D et al (2013) RUNX1a enhances hematopoietic lineage commitment from human embryonic stem cells and inducible pluripotent stem cells. Blood 121:2882–2890. https://doi.org/10.1182/blood-2012-08-451641
Risueño RM et al (2012) Inability of human induced pluripotent stem cell-hematopoietic derivatives to downregulate microRNAs in vivo reveals a block in xenograft hematopoietic regeneration. Stem Cells 30:131–139. https://doi.org/10.1002/stem.1684
Roy A et al (2012) Perturbation of fetal liver hematopoietic stem and progenitor cell development by trisomy 21. Proc Natl Acad Sci U S A 109:17579–17584. https://doi.org/10.1073/pnas.1211405109
Sangeetha VM, Kale VP, Limaye LS (2010) Expansion of cord blood CD34+ cells in presence of zVADfmk and zLLYfmk improved their in vitro functionality and in vivo engraftment in NOD/SCID mouse. PLoS One 5:e12221. https://doi.org/10.1371/journal.pone.0012221
Sangeetha VM, Kadekar D, Kale VP, Limaye LS (2012) Pharmacological inhibition of caspase and calpain proteases: a novel strategy to enhance the homing responses of cord blood HSPCs during expansion. PLoS One 7:e29383. https://doi.org/10.1371/journal.pone.0029383
Sasnoor LM, Kale VP, Limaye LS (2005) A combination of catalase and trehalose as additives to conventional freezing medium results in improved cryoprotection of human hematopoietic cells with reference to in vitro migration and adhesion properties. Transfusion 45:622–633. https://doi.org/10.1111/j.0041-1132.2005.04288.x
Schönberger S et al (2004) Transplantation of hematopoietic stem cells derived from cord blood, bone marrow or peripheral blood: a single centre matched-pair analysis in a heterogeneous risk population. Klin Padiatr 216:356–363. https://doi.org/10.1055/s-2004-832357
Seita J, Weissman IL (2010) Hematopoietic stem cell: self-renewal versus differentiation. Wiley Interdiscip Rev Syst Biol Med 2:640–653. https://doi.org/10.1002/wsbm.86
Sharma MB, Limaye LS, Kale VP (2012) Mimicking the functional hematopoietic stem cell niche in vitro: recapitulation of marrow physiology by hydrogel-based three-dimensional cultures of mesenchymal stromal cells. Haematologica 97:651–660. https://doi.org/10.3324/haematol.2011.050500
Shinde P, Melinkeri S, Santra MK, Kale V, Limaye L (2019) Autologous hematopoietic stem cells are a preferred source to generate dendritic cells for immunotherapy in multiple myeloma patients. Front Immunol 10:1079. https://doi.org/10.3389/fimmu.2019.01079
Shukla S et al (2017) Progenitor T-cell differentiation from hematopoietic stem cells using Delta-like-4 and VCAM-1. Nat Methods 14:531–538. https://doi.org/10.1038/nmeth.4258
Singh J, Chen EL, Xing Y, Stefanski HE, Blazar BR, Zúñiga-Pflücker JC (2019) Generation and function of progenitor T cells from StemRegenin-1–expanded CD34+ human hematopoietic progenitor cells. Blood Adv 3:2934–2948. https://doi.org/10.1182/bloodadvances.2018026575
Spanholtz J et al (2011) Clinical-grade generation of active NK cells from cord blood hematopoietic progenitor cells for immunotherapy using a closed-system culture process. PLoS One 6:e20740. https://doi.org/10.1371/journal.pone.0020740
Sugimura R et al (2017) Haematopoietic stem and progenitor cells from human pluripotent stem cells. Nature 545:432–438. https://doi.org/10.1038/nature22370
Sumide K et al (2018) A revised road map for the commitment of human cord blood CD34-negative hematopoietic stem cells. Nat Commun 9:1–17. https://doi.org/10.1038/s41467-018-04441-z
Svalgaard JD et al (2016) Low-molecular-weight carbohydrate Pentaisomaltose may replace dimethyl sulfoxide as a safer cryoprotectant for cryopreservation of peripheral blood stem cells. Transfusion 56:1088–1095. https://doi.org/10.1111/trf.13543
Tajer P, Pike-Overzet K, Arias S, Havenga M, Staal FJ (2019) Ex vivo expansion of hematopoietic stem cells for therapeutic purposes: lessons from development and the niche. Cell 8:169. https://doi.org/10.3390/cells8020169
Tal R et al (2019) Adult bone marrow progenitors become decidual cells and contribute to embryo implantation and pregnancy. PLoS Biol 17:e3000421. https://doi.org/10.1371/journal.pbio.3000421
Tao H, Gaudry L, Rice A, Chong B (1999) Cord blood is better than bone marrow for generating megakaryocytic progenitor cells. Exp Hematol 27:293–301. https://doi.org/10.1016/s0301-472x(98)00050-2

Thordardottir S et al (2014) The aryl hydrocarbon receptor antagonist StemRegenin 1 promotes human plasmacytoid and myeloid dendritic cell development from CD34+ hematopoietic progenitor cells. Stem Cells Dev 23:955–967. https://doi.org/10.1089/scd.2013.0521

Till JE, McCulloch EA (1961) A direct measurement of the radiation sensitivity of normal mouse bone marrow cells. Radiat Res 14:213–222. https://doi.org/10.1667/rrav01.1

Tunstall-Pedoe O et al (2008) Abnormalities in the myeloid progenitor compartment in down syndrome fetal liver precede acquisition of GATA1 mutations. Blood 112:4507–4511. https://doi.org/10.1182/blood-2008-04-152967

Valipour B et al (2020) Cord blood stem cell derived CD16+ NK cells eradicated acute lymphoblastic leukemia cells using with anti-CD47 antibody. Life Sci 242:117223. https://doi.org/10.1016/j.lfs.2019.117223

Wagner JE Jr et al (2016) Phase I/II trial of StemRegenin-1 expanded umbilical cord blood hematopoietic stem cells supports testing as a stand-alone graft. Cell Stem Cell 18:144–155. https://doi.org/10.1016/j.stem.2015.10.004

Watt SM, Austin E, Armitage S (2007) Cryopreservation of hematopoietic stem/progenitor cells for therapeutic use. In: Cryopreservation and freeze-drying protocols. Humana Press, pp 237–259. https://doi.org/10.1007/978-1-59745-362-2_17

Wilson A et al (2004) c-Myc controls the balance between hematopoietic stem cell self-renewal and differentiation. Genes Dev 18:2747–2763. https://doi.org/10.1101/gad.313104

Xue E, Milano F (2020) Are we underutilizing bone marrow and cord blood? Review of their role and potential in the era of cellular therapies. F1000Research 9:26. https://doi.org/10.12688/f1000research.20605.1

Yu J, Chen LXS, Zhang J, Guo G, Chen B (2016) The effects of a simple method for cryopreservation and thawing procedures on cord blood derived dc-based esophageal carcinoma vaccine. Cryo Letters 37:272–283

Zhang CC, Kaba M, Iizuka S, Huynh H, Lodish HF (2008) Angiopoietin-like 5 and IGFBP2 stimulate ex vivo expansion of human cord blood hematopoietic stem cells as assayed by NOD/SCID transplantation. Blood 111:3415–3423. https://doi.org/10.1182/blood-2007-11-122119

Zhang Y et al (2017) Large-scale ex vivo generation of human red blood cells from cord blood CD34+ cells. Stem Cells Transl Med 6:1698–1709. https://doi.org/10.1002/sctm.17-0057

Adv Exp Med Biol - Cell Biology and Translational Medicine (2021) 14: 45–64
https://doi.org/10.1007/5584_2021_647

Published online: 24 June 2021

Organoids in Tissue Transplantation

Derya Sağraç, Hatice Burcu Şişli, Selinay Şenkal,
Taha Bartu Hayal, Fikrettin Şahin, and Ayşegül Doğan

Abstract

Improvements in stem cell-based research and genetic modification tools enable stem cell-based tissue regeneration applications in clinical therapies. Although inadequate cell numbers in culture, invasive isolation procedures, and poor survival rates after transplantation remain as major challenges, cell-based therapies are useful tools for tissue regeneration.

Organoids hold a great promise for tissue regeneration, organ and disease modeling, drug testing, development, and genetic profiling studies. Establishment of 3D cell culture systems eliminates the disadvantages of 2D models in terms of cell adaptation and tissue structure and function. Organoids possess the capacity to mimic the specific features of tissue architecture, cell-type composition, and the functionality of real organs while preserving the advantages of simplified and easily accessible cell culture models. Thus, organoid technology might emerge as an alternative to cell and tissue transplantation. Although transplantation of various organoids in animal models has been demonstrated, liöitations related to vascularized structure formation, cell viability and functionality remain as obstacles in organoid-based transplantation therapies. Clinical applications of organoid-based transplantations might be possible in the near future, when limitations related to cell viability and tissue integration are solved. In this review, the literature was analyzed and discussed to explore the current status of organoid-based transplantation studies.

Keywords

3D cell culture · Organoid · Tissue regeneration · Transplantation · Vascularization

Abbreviations

2D	Two Dimensional
3D	Three Dimensional
APC	Adenomatous Polyposis Coli
ASC	Adult Stem Cell
ATG	Anti-thymocyte Globulin
Cas9	CRISPR Associated Protein 9
CNS	Central Nervous System
COMMD1	Copper Metabolism Domain Containing 1
CRISPR	Clustered Regularly Interspaced Short Palindromic Repeats

Derya Sağraç and Hatice Burcu Şişli have equally contributed to this chapter.

D. Sağraç, H. B. Şişli, S. Şenkal, T. B. Hayal, F. Şahin, and A. Doğan (✉)
Department of Genetics and Bioengineering, Faculty of Engineering, Yeditepe University, Istanbul, Turkey
e-mail: aysegul.dogan@yeditepe.edu.tr

CFTR	Cystic Fibrosis Transmembrane Conductance Regulator
EGF	Epidermal Growth Factor
ESC	Embryonic Stem Cell
FOXA3	Forkhead Box A3
GMP	Good Manufacturing Laboratory
GVHD	Graft-Versus-Host Disease
HLA	Human Leukocyte Antigen
HNF1A	Hepatocyte Nuclear Factor 1 Alpha
HNF4A	Hepatocyte Nuclear Factor 4 Alpha
HUN	Human Nuclear Protein
IL2-RA	Interleukin-2 Receptor Antibody
iPSC	Induced Pluripotent Stem Cell
Lgr5	Leucine-Rich Repeat-Containing G-Protein-Coupled Receptor
MBP	Myelin Basic Protein
MHC	Major Histocompatibility Complex
NOD	Non-obese Diabetic
NSG	NOD/SCID Gamma
NTHL1	nth-Like DNA Glycosylase-1
ONL	Outer Nuclear Layer
PDGFRα	Platelet-Derived Growth Factor Receptor Alpha
PLGA	Poly(Lactide-co-Glycolide)
PSC	Pluripotent Stem Cell
Salisphere	Salivary Gland Organoid
SCID	Severe Combined Immune Deficient
SMA	Smooth Muscle Actin
TP53	Tumor Protein p53
VEGF	Vascular Endothelial Growth Factor

1 Introduction

The first use of the word “organoid” in the literature was in a case study about cystic teratoma (Smith and Cochrane 1946) which defined the histological properties such as glandular organization observed in tumors. After development of the first intestinal organoids (Sato et al. 2009), the term has become more specific to self-organizing organ-like in vitro structures. Afterward, the definition of the term “organoid” has evolved into structures which resemble an organ (Huch and Koo 2015; Clevers 2016). Some criteria need to be presented in these structures for the characterization of an organoid (Lancaster and Huch 2019).

These characteristics include:

(i) A three-dimensional (3D) structure, which consists of cells to create or maintain the identity of the modeled organ
(ii) Multiple functional cell types of the desired organ
(iii) Self-organization inside the complex structure

Because organoids represent organ features in vitro, they are crucial tools to investigate organogenesis, homeostasis, adult organ repair, and disease mechanisms (Lancaster and Huch 2019) (Fig. 1). Organoids can be generated from pluripotent stem cells (PSCs), embryonic stem cells (ESCs), tissue-resident adult stem cells (ASCs) (Huch and Koo 2015), and cancer cells (Kim et al. 2019a). Human and mouse PSCs were used to generate brain (Mansour et al. 2018), retina (Volkner et al. 2016; Eiraku et al. 2011), intestine (Spence et al. 2011), inner ear (Longworth-Mills et al. 2016), stomach (McCracken et al. 2014), thyroid (Kurmann et al. 2015), liver (Takebe et al. 2013), lung (Dye et al. 2015), and kidney (Song et al. 2013) organoids. PSCs have been used for in vitro studies for a long period of time due to their limitless proliferation and differentiation potential. Although ASCs have limited proliferation and differentiation potential, they became popular in research and therapy thanks to easy isolation procedures and less ethical concerns (Lo and Parham 2009; Sridhar and Miller 2012). Utility of leucine-rich repeat-containing G-protein-coupled receptor expressing ($Lgr5^+$) stem cells of murine intestine is a leap in ASC-based organoid generation (Sato et al. 2009; Barker et al. 2007), and this achievement has led to development of the technology and brought a requirement for modified growth culture conditions. Continuous organoid cultures of adult tissues displaying structural and functional similarities to original organ tissue have been developed by simulating in vivo niche environment. Generation of intestinal organoids from $Lgr5^+$ intestinal stem cells was performed in Matrigel dome surrounded by organoid media, supplemented with epidermal growth factor (EGF), Wnt-3, R-spondin, and Noggin (Barker et al. 2007). The achievement of ASC-derived intestinal organoid development paved the way for the generation of various human and murine ASC-derived organoids such as stomach (Barker et al. 2010), colon (Sato et al. 2011), liver (Huch et al. 2015; Huch et al. 2013a), lung (Frank et al. 2016), prostate (Drost et al. 2016; Karthaus et al. 2014), pancreas (Huch et al. 2013b), ovaries (Kessler et al. 2015), endometrium (Boretto et al. 2017), mammary gland (Sachs et al. 2018), taste buds (Ren et al. 2014), and epithelium (Rock et al. 2009; Hisha et al. 2013) through the modifications of culture conditions. These achievements enhanced the potential of organoid-based transplantation therapy in clinics to improve personalized medicine and advance regenerative medicine applications in the near future.

2 Requirement of Organoids

Human disease and developmental processes are regulated by complex molecular and cellular events that are difficult to understand and directly implement (Jackson and Lu 2016). In vitro cellular systems that are capable of replacing model organisms are of interest in recent years. Because disease pathophysiology is controlled by several components at the cellular, tissue, and systemic level, more reliable research tools need to be established to support the development of preventive and therapeutic strategies (Mallo et al. 2010).

Development in stem cell culture technology has enabled the establishment of organoid culture systems as a multicellular and functional 3D tissue-like structures (Eiraku et al. 2011). Latest researches have shown that organoids are not only used to model organ and disease development but also allow a wide range of applications in basic research (Sato et al. 2009), drug discovery (Zhou et al. 2017), regenerative medicine (Shirai et al. 2016), and gene therapy in personalized medicine (Walsh et al. 2016) (Fig. 1). Although animal models are employed in research to mimic human physiology, completely similar disease pathology between animal and human is required for identification of the true response to treatments and molecular mechanism. In addition, access to high number of

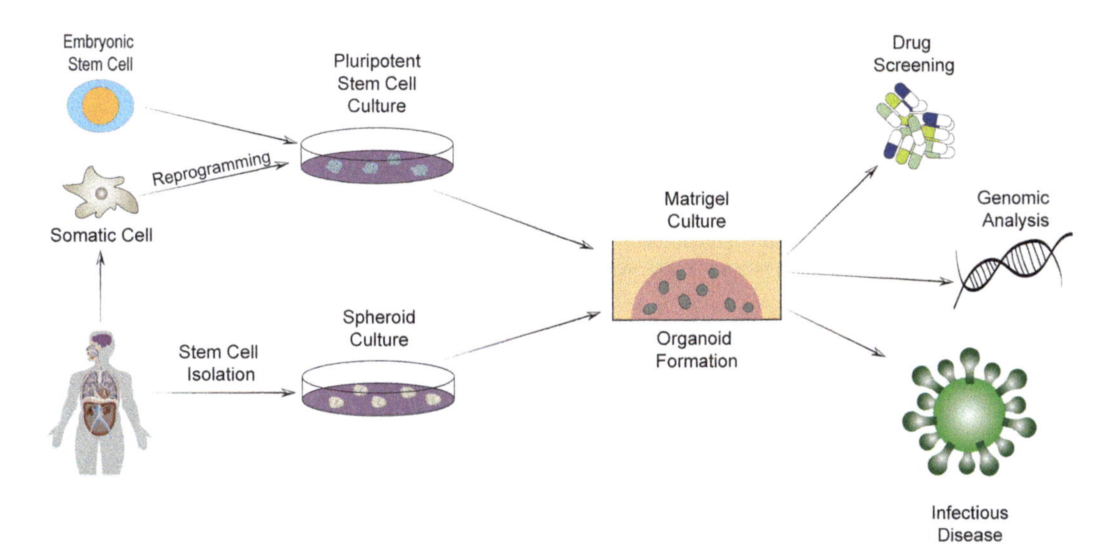

Fig. 1 A scheme of organoid formation and application. In vitro organoid generation is performed by using embryonic or patient-derived stem cells. Functional and mature organoids can be utilized for drug screening and investigation of novel mutations or infectious diseases

animals and ethical problems are important considerations for in vivo research. Therefore, organoids as an emerging technology might be preferred to replace animal models in practical drug discovery process (Miranda et al. 2018). A close resemblance of organoids to human tissues and organs is the major advantage for cell-based therapy and transplantation research. Successful expansion of various organoids such as intestine (Spence et al. 2011; Jung et al. 2011), prostate (Gao et al. 2014), and liver (Takebe et al. 2013) shows promise for the use of biologically similar organoid groups in autologous and allogeneic cell therapy approaches (Van De Wetering et al. 2015). Furthermore, combining organoid research with new gene editing tools such as clustered regularly interspaced short palindromic repeats (CRISPR)/CRISPR associated protein 9 (Cas9) might facilitate the easy genetic manipulation before autologous transplantation. Although organoids hold great promise in cell transplantation field, problems related to efficacy, safety, and immunogenicity still exist (Daviaud et al. 2018). Ex vivo expanded ASC-derived organoids can differentiate into almost all cellular lineages found in the relevant organ and maintain their organ identity as well as the genome stability. This feature might ease the generation of protocols for allogeneic cell therapies as an unlimited source for tissue replacement (Yang et al. 2017).

While simple models like 2D-monolayer cultures have advantages, lack of cell-cell and cell-matrix interactions, which are necessary to form tissue phenotypes, is a major restriction to mimic the cellular functions and signal pathways (Kim et al. 2019b) (Table 1). The most significant disadvantage of existing model systems is the large gap between cellular and tissue models (Yin et al. 2016). Organoids, representing a wide variety of 3D tissue models, have been used in filling this gap between cell culture and in vivo experiments over the last decade (Jee et al. 2019a). Certain features of 3D architecture such as cell-type composition and function can be applicable in organoid models (Hyun 2017) and,

Table 1 Comparison of 2D and 3D cell culture systems

2D cell culture	3D cell culture (model organism)
Advantages	**Disadvantages**
The cell culture period takes a short time (Chen et al. 2012)	The cell culture period takes a few days (Chen et al. 2012)
The cell culture protocols are easy to elucidate experimentally (Hickman et al. 2014)	The cell culture protocols are difficult to elucidate experimentally (Hickman et al. 2014)
Basic cell culture protocols are easy to establish (Hickman et al. 2014)	The cell culture protocols and organoid formation steps are difficult to establish (Hickman et al. 2014)
There is a high reproducibility in cell culture setup (Hickman et al. 2014)	Reproducibility of culture setup is low (Hickman et al. 2014)
Cells have unlimited access to essential products such as oxygen, nutrient, and signaling molecules (Frieboes et al. 2006).	Cells have variable access to essential products such as oxygen, nutrient, and signaling molecules (Frieboes et al. 2006)
Cell culture materials are inexpensive (Krishnamurthy and Nör 2013)	Cell culture materials are expensive (Krishnamurthy and Nör 2013)
The culture process has commercially available tests for validation (Krishnamurthy and Nör 2013; Weiswald and Bellet 2015)	The culture process has fewer commercially available tests for validation (Krishnamurthy and Nör 2013; Weiswald and Bellet 2015)
Disadvantages	**Advantages**
The cell culture does not simulate natural structure of tissues (Bokhari et al. 2007)	The cell culture simulates the in vivo 3D form of tissues and organs (Bokhari et al. 2007)
The protocols lack the cell-to-cell and cell-to-matrix interactions (Gilbert et al. 2010)	The protocols have convenient interactions of the cell to cell and cell to matrix (Gilbert et al. 2010)
Only in vitro utilization of molecular mechanism, such as gene expression and mRNA splicing, is available (Li et al. 2006; Gómez-Lechón et al. 1998)	Utilization of in vivo gene expressions, splicing, and biochemistry of cells is also possible (Li et al. 2006; Gómez-Lechón et al. 1998)

therefore, offer an opportunity for a variety of biological and biomedical applications.

3 Organoid Models for Tissue Transplantation and Applications

3D cell culture systems have been popularized in recent years due to the disparities of 2D models in terms of tissue modeling and architecture (Table 1). Organoids, which possess the structural and functional properties of the original organs in vitro, have shed light to principles of organ biology and grant researchers an accessible system to study on the processes of organogenesis and morphogenesis. Organoid technology has been evolved and put in use in transplantation and regenerative medicine (Table 2). The first organoid model generated by using ASCs was intestinal organoids created by Sato et al. in 2009 with $Lgr5^{+}$ mouse intestinal stem cells (Sato et al. 2009). Numerous human and murine organoid models were derived from ASCs, ESCs, and induced pluripotent stem cells (iPSCs) so far (Khalil et al. 2019) (Table 2). In this section, organoid models for disease research and transplantation studies were reviewed based on the literature.

Organoid systems can reflect various characteristics of tissues and organs both in vitro (Huch and Koo 2015; Garcez et al. 2016; Sun et al. 2019; Xie et al. 2018; Ren et al. 2020) and in vivo (Mansour et al. 2018; Takebe et al. 2013; Yui et al. 2012; Varzideh et al. 2019) as a model system. Successful transplantation of such model systems into living organisms might enhance insights for disease pathology and organ regeneration and replacement. Animal model experiments to assess the clinical utility and potency of organoid transplantation demonstrated that organoids are promising sources for cell therapy and regenerative medicine applications. Assawachananont et al. reported that when progressive retinal degradation mouse model, lacking outer nuclear layer (ONL), was transplanted with retinal sheet organoids, derived from mouse ESCs and mouse iPSCs, ONL structure with mature photoreceptors was formed from the graft and integration of the photoreceptors to host bipolar cells via synaptogenesis was observed (Eiraku et al. 2011; Assawachananont et al. 2014). Furthermore, retinal organoids derived from human ESCs were engrafted into retinal degradation models of immunodeficient rats and monkeys and resulted in the generation of various retinal cell types and ONL with mature photoreceptors. Contact between the graft photoreceptor with the host bipolar cells, possibly with synaptic connections, was also observed after transplantations (Shirai et al. 2016). Moreover, experiments for functional evaluation of mouse iPSCs-derived retinal sheet transplantation in advanced retinal degradation mouse model indicated that dendrites of bipolar cells of the host reached into the graft and these mice responded to light (Mandai et al. 2017). Aside from these studies, transplantation of several other organoids, including intestine (Khalil et al. 2019), kidney (Nam et al. 2019), and liver (Huch et al. 2013a), into the animal models was also reported.

Animal models and transplantation area are two important factors for efficient transplantation. Different animal hosts have been used in organoid transplantation studies. Non-obese diabetic/severe combined immune-deficient (NOD/SCID) mice (Takebe et al. 2013; Sun et al. 2019), NSG (NOD/SCID gamma) mice (Tan et al. 2017), Sprague-Dawley rats (Song et al. 2013), and C57Bl/6 mice (Boj et al. 2015; O'Rourke et al. 2017) have been used as transplantation host for organoid research. In addition, specialized mouse models, including $Rag^{2-/-}$ mice (Yui et al. 2012) to avoid immune rejection, male athymic nude rats (Xinaris et al. 2012) which have less T cells in the periphery, type 1 fulminant diabetic mouse model (Takebe et al. 2015), and Athymic Nude-$Foxn1^{nu}$ mice (Santos et al. 2019) which do not have T cells and graft-versus-host response, have been used for specific organoid transplantations.

Transportation of nutrients and oxygen to the tissue and removal of waste materials are required for the proliferation of cells indide the tissue. The

Table 2 Organoid transplantation models for different tissues

Tissue	Source	Models	References
Brain	Human ESCs Human iPSCs Patient skin fibroblasts	Cancer Infectious disease Autism Microcephaly	Mansour et al. (2018), Daviaud et al. (2018), Ogawa et al. (2018), Pollen et al. (2019), Mariani et al. (2012), Mariani et al. (2015), Lancaster et al. (2013)
Lung	Human ESCs Human iPSCs Mouse fetal pulmonary cells Mouse basal cells	Cancer Infectious disease Cystic fibrosis Regeneration Fibrotic pulmonary disease	Clevers (2016), Kim et al. (2019a), Dye et al. (2015), Rock et al. (2009), Tan et al. (2017), Sachs et al. (2019), Takahashi et al. (2019), Chen et al. (2017)
Stomach	Mouse/human Adult tissue Mouse/human ESCs Human iPSCs	Cancer Infectious disease	McCracken et al. (2014), Barker et al. (2010), Nanki et al. (2018), Wroblewski et al. (2009), Yan et al. (2018), Bartfeld et al. (2015)
Pancreas	Mouse/human Adult tissue	Cancer Diabetes mellitus	Huch et al. (2013a), Boj et al. (2015), Huang et al. (2015), Raimondi et al. (2020), Dorrell et al. (2014), Loomans et al. (2018), Hindley et al. (2016)
Liver	Human iPSCs Mouse adult tissue Human adult tissue	Cancer Rare disease, diagnosis Hepatitis	Takebe et al. (2013), Huch et al. (2013a), Boj et al. (2015), Broutier et al. (2016), Broutier et al. (2017), Elbadawy et al. (2020)
Intestine	Adult tissue Mouse/human ESCs Human iPSCs	Cancer Cystic fibrosis Infectious disease Regeneration Alagille syndrome Steatosis Alcohol-related liver disease	Sato et al. (2009), Spence et al. (2011), Barker et al. (2007), Huch et al. (2015), Cortez et al. (2018), Finkbeiner et al. (2015), Schwank et al. (2013), Fatehullah et al. (2016), Hindley et al. (2016), Finkbeiner et al. (2012), Cao et al. (2015), Dekkers et al. (2013), McCracken et al. (2011), Mustata et al. (2013), Wang et al. (2019)
Colon	Adult tissue Mouse/human iPSCs	Cancer Bowel disease Regeneration	Clevers (2016), Barker et al. (2007), Sato et al. (2011), Van De Wetering et al. (2015), Yui et al. (2012), Matano et al. (2015), Michels et al. (2020)
Kidney	Human iPSCs Human adult tissue	Cystic fibrosis Regeneration Cancer Renal disease Congenital nephrotic syndrome (NPHS1)	Takasato et al. (2015), van den Berg et al. (2018), Freedman et al. (2015), Jun et al. (2018), Schutgens et al. (2019), Hale et al. (2018), Peters and Breuning (2001)
Bladder	Mouse/human adult tissue	Cancer	Santos et al. (2019), Mullenders et al. (2019), Vasyutin et al. (2019), Lee et al. (2018)
Fallopian tube	Human adult tissue	Cancer Infectious disease	Kessler et al. (2015), Kopper et al. (2019), Maenhoudt et al. (2020)
Breast	Human adult tissue	Cancer	Sachs et al. (2018), Rosenbluth et al. (2020)
Prostate	Mouse and human Adult tissue Fetal pancreatic tissue	Cancer Pancreatic ductal adenocarcinoma (PDAC)	Drost et al. (2016), Karthaus et al. (2014), Gao et al. (2014), Chua et al. (2014), Balak et al. (2019)
Retinal	Mouse ESCs Mouse iPSCs	Retinal development Leber congenital amaurosis (LCA) X-linked juvenile retinoschisis	Singh et al. (2015), Shimada et al. (2017), Gao et al. (2020), Huang et al. (2019)

(continued)

Table 2 (continued)

Tissue	Source	Models	References
		Retinitis pigmentosa	
Salivary gland	Primary mouse cell	Hyposalivation	Nanduri et al. (2014)
Cortical	Human ESCs, human iPSCs, human endothelial cells	Development Microcephaly Miller-Dieker syndrome	Shi et al. (2020), Cakir et al. (2019), Iefremova et al. (2017), Lancaster et al. (2013)
Esophagus	Mouse adult tissue Human adult tissue	Barrett's esophagus Esophageal atresia Cancer	Sato et al. (2011), DeWard et al. (2014), Li et al. (2018), Fantes et al. (2003)
Human adult tissue	Adult tissue-specific stem cells	Cystic fibrosis Host-pathogen interactions Chronic helicobacter infection	Dekkers et al. (2013), Leslie and Young (2016), Salama et al. (2013)

content of the culture medium, transport of this content to the cells, and the amount of the oxygen in the incubator environment are important factors for 2D and 3D cell culture in vitro. From this perspective, transplanted organoids must be supported by the host and newly generated vessels. The host system must provide signals for survival, maturation, differentiation, function (Watson et al. 2014), and/or cellular polarity (Vargas-Valderrama et al. 2020) of organoids at the transplantation site (Dye et al. 2016). Therefore, being disconnected to the vascular circulation has been one of the major limitations of organoid transplantation (Shi et al. 2020). It has been reported that transplantation of organoids into vascularized areas, especially under the kidney capsule (Nam et al. 2019; Xinaris et al. 2012) or the frontoparietal cortex (Daviaud et al. 2018), facilitates the spread of vascular structures to newly engrafted organoids. Two different organoid engraftment types have been used: orthotopically, indicating the transplantation into a similar field with the native environment (Song et al. 2013; Boj et al. 2015), and ectopically, indicating the transplantation into different fields from the native environment (Tan et al. 2017; Takebe et al. 2015). In both cases, viability after transplantation is the serious issue for replacing the tissue function by organoids. Insufficiency or absence of vascularization causes impaired transfer of nutrients and subsequent death of organoids (Vargas-Valderrama et al. 2020; Shi et al. 2020; Cakir et al. 2019). It was reported that organoids are exposed to hypoxia due to the lack of vascularization and results in necrosis which prevents the development of cellular organization (Giandomenico and Lancaster 2017). Although the host system provides the vascularization mainly after organoid transplantation, integration of endothelial cells into organoid models increases the survival of organoids which was shown in a previous study. Takebe et al. showed that human vascular structure inside the organoids increased the maturation of liver buds via the connection with host vasculature (Takebe et al. 2013). Because organoids are composed of a large number of cells, accessibility of the required materials from external blood vessels might be limited. Takebe's transplantation model which is called organ-bud transplantation (Takebe et al. 2013) overcomes this problem. In this technique, the liver organoids have been generated within a system that contains human umbilical vein endothelial cells, human mesenchymal stem cells, and human iPSCs-derived hepatic endoderm cells to facilitate the establishment of vascular structure within the host tissue and enhance the survival and penetrance of engraftments. Endothelial cell proliferation and development of vasculature have been accomplished in many studies (Takebe et al. 2013; Varzideh et al. 2019; Tan et al. 2017; Sharmin et al. 2016). Apart from endothelial cell incorporation into the organoids, direct differentiation of PSCs into the desired organoid might generate vascular like structures as a result of their pluripotent nature. Takasato and colleagues demonstrated that differentiation of iPSCs to the kidney organoids resulted in endogenous endothelial cell formation (Takasato et al. 2015). In another study, 28 days after transplantation under the renal capsule of mice, kidney organoids were vascularized progressively and matured into a glomerular form (van den Berg et al. 2018).

Cortez et al. showed that human intestinal organoids that are transplanted into the mesentery of mice are maintained by the splanchnic circulation through the mesenteric vessels of the host (Cortez et al. 2018). In another study, vascular cortical organoids were transplanted into the S1 cortex of NOD-SCID mice. Presynaptic and postsynaptic proteins SYB2 and PSD95, respectively, were positively expressed at the host-graft border within 60 days after transplantation. Colocalization of markers indicated a synaptic connection between the graft and the host. The distinction of the organoids from host cells has been demonstrated by staining of human nuclear protein (HUN) and laminin. Two months after implantation, HUVEC-derived HUN^+ endothelial cells and mouse HUN^- endothelial cells coexisted in the blood vessels of organoid graft. Myelin basic protein (MBP) and HUN staining have shown myelinization at the graft-host borders. It has also been reported that the myelinated fibers observed in the organoid graft were derived primarily from the host's brain and are located only at the implantation border (Shi et al. 2020). Similarly, Mansour et al. transplanted GFP + organoids into the mice brain, and GFP+ neurites were spread from organoids inside to the host brain on the 50th day after transplantation. Specifically, axonal growth was found not only in the grafted area but also in the rostral region of the host. These results showed that transplanted organoids can generate long axonal projections to distant targets in the host brain. In addition, the relationship between synaptic human axon and host brain demonstrated that synaptophysin+ structures were colocalized with GFP+ fibers in the host cortex 90 days after transplantation. Host-derived vascular network was demonstrated by the presence of mouse-specific CD31 staining. Collectively, these results showed the importance of organoid and the host-tissue relationship after transplantation and the importance of new vessel formation in the transplantation area, which is one of the most important reasons for the maturation and functionality of the organoids (Mansour et al. 2018).

As mentioned above, although organoid transplantation models showed promising results, there are reports demonstrating the insufficient tissue integration of organoids and lack of vascularization and dysfunction after transplantation. Nam and colleagues showed that human iPSCs-derived kidney organoids were transplanted under the kidney capsule of NOD/SCID mice and survived for up to 42 days. Even though kidney organoids managed to survive, they remained as small-sized, partially vascularized, and lacked nephron-like tubule formation (Nam et al. 2019). Apart from unsuccessful models, kidney organoids generated from both human ESCs and human iPSCs without adding any growth factor survived in the sub-renal capsule of NOD/SCID mice, formed glomerular and tubular structures, and induced glomerular vascularization by secreting organoid-derived vascular endothelial growth factor (VEGF) (van den Berg et al. 2018). This organoid transplantation model can therefore be preferred for understanding the mechanism of glomerulus-derived kidney diseases and screening nephrotoxicity of drugs.

In a previous study, liver organoids generated from human iPSCs became functional by connection of the host's vascular structure within 48 h after ectopic transplantation into the mesentery of mice. Functional liver-like tissue exhibiting the maturation signs, including production of human albumin, whose secretion into the bloodstream of mice was observed approximately 10 days after the transplantation and recorded up to day 45, and control of the drug metabolism, was achieved (Takebe et al. 2017). This study is an important report for the formation of a functional human organoid using human PSCs which created vessels in the host tissue.

The functionality of the transplanted organoids is one of the most important criteria for the future regenerative medicine applications. In another study, salivary gland organoids (salispheres), derived from human submandibular gland stem cells, were placed into the neck of mice 1 month after the irradiation, which mimics the condition of hyposalivation. In the irradiated environment, α-amylase, aquaporin 5, and cytokeratin, which are markers of salivary gland, were expressed on the surface of salispheres, and saliva production was restored in the organoid transplanted group compared to the control group (Pringle et al. 2016). These promising results have a potential for the treatment of dry mouth due to diabetes, heart attack, and chronic xerostomia. Two different organoid transplantation models related to lung regeneration have been reported. Human lung organoids, produced from human ESCs, in the poly(lactide-co-glycolide) (PLGA) scaffolds were injected into the NSG mice's epididymal fat pad which is highly vascularized and suitable for large transplantations. After 8 weeks of transplantation, human lung organoids expressed lung epithelium marker, Nkx2.1, produced mucin, and showed the alveolar-like branching characteristics as well as similar gene expression pattern with the mature lung. Furthermore, human lung organoids created adult airway-like epithelium and differentiated into basal, goblet, ciliated, and club cells. Besides, the transplantation area was highly vascularized and surrounded by smooth muscle actin-positive (SMA^+) myofibroblasts and mesenchymal cells, similar to the adult airway (Dye et al. 2016). In the second model, bud-type progenitor organoids were differentiated from lung progenitor cells in 22 days and were transplanted into mice after naphthalene-induced airway injury. The injured airway of mice began to heal and produced mucin within 6 weeks after ectopic transplantation (Miller et al. 2018). Both studies have remarkable regenerative outcomes for lung diseases or people who suffered from airway damage such as mining workers.

Organoid transplantation systems are often used in cancer modeling experiments which is one of the most important fields for molecular mechanism analysis and drug discovery. Organoids generated from healthy tissues are utilized to model cancer by inducing cancer-causing gene mutations such as nth-like DNA glycosylase-1 (NTHL1) (Weren et al. 2015), KRAS, and adenomatous polyposis coli (APC) (Drost et al. 2015; Matano et al. 2015). The first example for organoid-based cancer modeling through transplantation is intestinal organoid systems to mimic colorectal cancer. This model was created by altering the expression of APC, SMAD4, tumor protein p53 (TP53), and KRAS genes in intestinal stem cells using CRISPR/Cas9 technology. Knockout of aforementioned genes resulted in adenoma formation in the organoid systems after transplantation under the kidney subcapsule. These tumoroids exert colorectal cancer characteristics, form micro-metastases after injected into the mouse spleen, and showed cancer cell markers like aneuploidy and chromosomal instability (Drost et al. 2015; Matano et al. 2015).

Following the direct reprogramming of fibroblasts by ectopic expression of forkhead box A3 (FOXA3), hepatocyte nuclear factor 1 alpha (HNF1A), hepatocyte nuclear factor 4 alpha (HNF4A) genes, and liver organoids were formed in vitro (Huang et al. 2014). Cancer was induced by Ras and c-Myc genes in order to observe the initial stages of intrahepatic cholangiocarcinoma and hepatocellular carcinoma, respectively. In vitro conversion of liver organoids into both intrahepatic cholangiocarcinoma and hepatocellular carcinoma organoids resulted in the formation of liver cancer in NOD/SCID mice, when transplanted orthotopically into the liver capsule. In this study, Ras- and c-Myc-induced cancer modeling represented an organoid system which is suitable for genetic manipulation-based cancer models (Sun et al. 2019).

Apart from cancer models, organoid systems are useful tools for generation of organogenesis models and their use for regenerative medicine purposes. Organoids are useful tools to study specific organ development and their regenerative potential, disease modeling, and drug testing.

Finkbeiner and colleagues demonstrated that immature fetal intestinal organoids, transplanted under kidney capsule, exerted fetal to adult development process in vivo. Briefly, human PSCs-derived intestinal organoids expressed fetal intestine markers initially and produce adult intestinal enzymes such as trehalase, lactase, and maltase glucoamylase followed by transplantation under the kidney capsule of mice (Finkbeiner et al. 2015). Similar to the adult intestine, organization of the cells expressing SMA and platelet-derived growth factor receptor alpha (PDGFRα) was observed in the transplanted area. Thus, it could be assumed that unknown adult signals can trigger the fetal to adult conversation of transplanted organoids inside the body. Another study has shown that colon organoids can be formed from a single stem cell in vitro, the epithelial barrier of the colon was completely restored, and functional crypts were formed after transplantation into the damaged colonic lumen of mice (Yui et al. 2012). These two examples revealed that organoid transplantation might have recovery ability, which might make them a useful tool in the tissue regeneration field such as the treatment of damaged or dysfunctional organs, deep burns characterized by loss of skin and muscles, and autoimmune disorders.

One of the most important models for organoid research is brain organoids, and their transplantation for tissue regeneration might be important for treatment of neurodegenerative disorders. Transplantation of cerebral organoids generated from human PSCs is successfully integrated into the host tissue and differentiated into the neuronal lineages. As a result, transplantation into the lesioned frontoparietal cortex increased survival and showed robust vascularization, indicating the hope in replacement therapy for progressive central nervous system (CNS) and neurodegenerative diseases (Daviaud et al. 2018). Similarly, Mansour and colleagues reported that intracerebral engraftment of human PSCs-derived cerebral organoids into the mouse brain demonstrated successful vascularization, neuronal differentiation and maturation, integration into the host tissue by synapses, and functional neuronal activity in the graft in response to stimuli (Mansour et al. 2018).

The ability of organoids to mimic a real tissue or a disease was enhanced with the help of developing technology. Assembloids can be defined as a new generation 3D brain organoids, which consist of multiple cell lineages, and are useful to study various cellular interactions or even neurological disorders such as epilepsy or autism (Pasca 2019; Marton and Paşca 2020; Birey et al. 2017). In a previous study, successful combination of striatum and midbrain organoids for reliable screening of glutamatergic and dopaminergic neurons was conducted (Pasca 2019). Wörsdörfer et al. developed a technique based on the co-culture of neural and mesodermal progenitor cell organoids to visualize the vascularization of the brain (Wörsdörfer et al. 2020). Furthermore, retinal and brain assembloids were used for development of better drug strategies for numerous retinal diseases such as glaucoma and macular degeneration (Gopalakrishnan 2019; Singh and Nasonkin 2020).

Sweat-gland organoids were transplanted on the injured dorsal thoraco-lumber region of mice and contributed to the regeneration of the epidermis and sweat gland, indicating the skin regeneration potential (Diao et al. 2019). Elbadawi et al. indicated that human airway organoids can be used as a specific tool to develop new therapeutics against SARS-CoV-2. They cultured airway organoids with SARS-CoV-2 and co-cultured with macrophages in order to understand the interactions of the viruses with the immune system on a small scale (Elbadawi and Efferth 2020). Cytokines secreted from macrophages in response to SARS-CoV-2 infection would be a hope for patients with cytokine release syndrome. Miller et al. showed that lung organoids and bud-type organoids, generated from human pluripotent stem cells in 3D culture model, were used to understand epithelial cell fate decision and epithelial mesenchymal crosstalk, respectively. Moreover, therapeutic properties of these organoid models were shown against ciliopathies, cystic fibrosis, or the process of goblet cell hyperplasia (Miller et al. 2019). Therefore, utilization of organoids which have distinctive features provides suitable models in the tissue transplantation in the future.

A cardiac organoid transplantation model was established using human ESCs with organ-bud technique within polylactic acid scaffolds. Varzideh and colleagues heterotopically transplanted the cardiac organoids into the peritoneal cavity of mice. Histological findings and function analysis showed that cardiac organoids triggered new vessel formation and maturation after 28 days. Cardiomyocyte fibril ultrastructure with high expression of genes related to vascularization and contraction was observed (Varzideh et al. 2019). Therefore, these organoid transplantation models might be a hope for treatment of cardiovascular diseases.

Furthermore, genetic modifications of organoids serve an opportunity to study genetic disorders, cancer, and gene therapy approaches. Organoid transplantation models are useful in the diagnosis and treatment of progressive and chronic rare genetic disorders (Huch and Koo 2015; Yui et al. 2012; Iefremova et al. 2017). In a previous research, CRISPR-Cas9 system was utilized for intestinal organoids from patients with cystic fibrosis to treat the disease by gene correction. Adult intestinal stem cells, harboring the ΔF508 mutation in the cystic fibrosis transmembrane conductance regulator (CTFR) locus, were obtained from the patients with cystic fibrosis. After collecting intestine samples, Lgr5-expressing stem cells were sorted and the mutation in the CTFR locus was corrected with CRISPR-Cas9 technology. Organoids obtained from both small and large intestines were developed from adult intestinal stem cells with an induction cocktail. The forskolin-induced swelling assay exhibited that organoids with the restored CTFR locus successfully acquired receptor function in vitro compared to untreated organoids (Schwank et al. 2013). Although promising results have been obtained after successful transplantation into the colon epithelium of mice (Yui et al. 2012), more in vivo studies are required to verify the effect of CRISPR-Cas9 in cystic fibrosis. Liver organoids from patients with Alagille syndrome and alpha 1-antitrypsin deficiency displayed the pathophysiology of the related disease in vitro. However, in vivo research is required to understand whether transplantation of genetically modified or healthy organoids can replace the function (Huch et al. 2015). In a study published in 2017, iPSCs derived from skin fibroblast cells of Miller-Dieker syndrome patients were used to generate forebrain organoids. These organoids were smaller compared to healthy organoids due to asymmetric cell division of ventricular zone radial glial cells, indicating the importance of organoids in disease modeling (Iefremova et al. 2017). This study contributes to further understanding of the developmental mechanisms and changes associated with the mutation in a single gene within complex tissue environment.

Research on infectious diseases is another area in which organoid transplantation models are used. Brain organoids have been generated to examine the development and effects of the Zika virus infection. When human iPSCs-derived brain organoids were infected by the virus, the growth rate of these organoids decreased (Garcez et al. 2016). Gastric organoids generated from human PSCs have been reported as the first model in which both the developmental process of the stomach and the histopathology of *Helicobacter pylori* infection were examined (McCracken et al. 2014).

As in organ transplantation procedures, organoid transplantations also have immunity concerns and rejection risks depending on several factors such as human leukocyte antigen (HLA) system of organoids. HLA, which is important in allogeneic transplantation, is a group of proteins encoded by the major histocompatibility complex (MHC) genes in humans (Agarwal et al. 2017). The HLA system, which consists of three different classes, is responsible for the acceptance or rejection of an organ/organoid by the host, according to the graft-host compatibility. Transplanted organoids, which do not have the similar HLA pattern with the host, showed an increased rejection risk. Therefore, a severe disease called graft-versus-host disease (GVHD) or even transplant-related death occurs, and the transplant is recognized as invasive by the host's immune system and causes graft rejection (Yu et al. 2019). To overcome this problem, four main approaches are used. Firstly, immune-deficient animals such as Balb/c strain or NOD/SCID mice are preferred for avoiding from host's immune system. Likewise, using the host's immune cells while preparing organoids is the second method used in the prevention of immune rejection (Min et al. 2020). Although suppressing the host's immune system is one of the feasible ways to avoid immune activation, another approach is engrafting organoids into the specialized area. The liver is one of such transplant sites for islet transplants. Cantarelli et al. showed that when islets are transplanted into the liver, graft-associated T cell responses are reduced compared to the one at the bone marrow transplantation site (Cantarelli et al. 2017). This strategy might be a good alternative for organoid tissue transplantation models. Autologous organoid transplantation models have a huge promise. Kruitwagen et al. reported that genetically corrected organoids were transplanted autologously into the liver of copper metabolism domain containing 1 (COMMD1)-deficient canine and lived up to 2 years without postoperative immune complications (Kruitwagen et al. 2020). Eventually, the last method is using immunosuppressive drugs such as interleukin-2-receptor antibodies (IL2-RA) basiliximab, anti-thymocyte globulin (ATG), or mTOR inhibitors to inhibit the immune rejection (Singh et al. 2019). The study conducted by Xian et al. demonstrated that using dexamethasone as an immunosuppressive agent protects the retinal organoids against immune attack (Xian et al. 2019). Similarly, Singh et al. showed that transplantation of retinal organoids into the eyes of cats who were administered with immunosuppressed drugs showed better survival rates and low immune response in comparison with the control group (Singh et al. 2019).

HLA-matched tissue-specific organoids can be generated for human transplantation studies by iPSC technology and can be used to generate clinically relevant organoid transplants in the future. Therefore, it is anticipated that real-sized organ-based organoids might be produced under good manufacturing laboratory (GMP) compliance in the further future and this might be a hope for patients waiting for transplants.

4 Conclusion

Organoids have been popularized as 3D tissue culture systems in recent years and used to model complex mechanisms of mammalian development, organogenesis, and disease pathology. Organoid technology is a useful tool for understanding organ development, tumor formation, personalized drug screening, regenerative medicine, and vaccine discovery (Fig. 1). Advantages over in vitro and in vivo models in terms of easy procedures and similarities to organ structure and function made organoids indispensable tools for molecular biology and stem cell research. Application of organoid technology in transplantation therapies might solve problems related to cell and tissue sources, rejection problems, and ethical considerations. Considering the success, efficiency, and safety of in vivo mouse (Yui et al. 2012; Jee et al. 2019b) and human (Okamoto et al. 2020; Sugimoto et al. 2018) organoid engraftments as well as pre-clinical trials (Takebe et al. 2018) in intestinal regeneration, the prognosis of organoid-based therapy seems quite encouraging. For refractory cases of mucosal healing in inflammatory gastrointestinal diseases, such as Crohn's disease and ulcerative colitis, organoid-based therapy may be used as a primary treatment to promote regeneration of the intestinal tissue properly (Okamoto et al. 2020). Although organoid technology is quite promising for the future regenerative medicine applications, efficacy and continuity still need to be searched.

Identification of an effective, easy, and inexpensive method for tissue and organ replacement is crucial. Although organoid technology might be promising for future organ replacement therapies, autologous tissue/organ transplantation is preferred over autologous organoid transplantation. The application time, cost, integration to the host tissue, survival, immune rejections, and function can be considered as the main limitations of autologous/allogeneic organoid transplantation (Table 3). Although patient-derived iPSC technology is a remarkable tool in cell therapy field, the complete methodology of both iPSC

Table 3 Comparison of organoid and tissue transplantation

Organoid transplantation	Tissue transplantation
There is an obstacle about how much organoids can survive after transplantation without cell death	Cell types in tissue transplantation may vary in their capability to proliferate in the body
There is no rejection risk for autologous cell-derived organoids	There is not immune rejection risk for partial autologous tissue transplantation but there are immune rejection and responses for organ transplantations
It lacks vasculature after transplantation	There is promising in vivo and ex vivo experiments for vascularization after transplantation
It might lack certain organ-specific cell types and function after transplantation	There are solid-organ transplantation applications in clinics which replace the whole organ function

generation and quality control of generated iPSCs is highly expensive and takes a very long time. Therefore, routine utility of patient-derived iPSCs for organoid transplantation could not be widespread in the near future. Furthermore, the epigenetic variations of iPSCs (Liang and Zhang 2013), generated from different patients, are not considered suitable for organoid transplantation since they may increase the possibility of unsuccessful differentiation (Singh et al. 2018). Despite all of these limitations, iPSCs-derived autologous organoid transplantation might be a powerful candidate in the further future for high clinical efficacy with no immune rejection (Fatehullah et al. 2016).

Adaptation of transplanted organoids to the tissue environment could only be possible by a functional vascular structure which provides bloodstream to the organoid and increase cell viability inside the 3D structure. Establishment of vascularization inside and in the surrounding environment of the transplanted organoid is one of the challenges in clinical applications. In addition, organization of cells inside the 3D structure and production of functional tissue-specific enzymes, hormones, and cytokines are required after transplantation. As a conclusion, tissue/organ transplantation still is the most popular solution for damaged tissue/organ regeneration in terms of availability, efficacy, and reproducibility. Organoid technology has brought a new perspective to cellular therapy and regenerative medicine area as a novel emerging method. Improvement of organoid technology and research might enable generation of functional organ-like transplantable in vitro structures in the near future.

References

Agarwal RK, Kumari A, Sedai A, Parmar L, Dhanya R, Faulkner L (2017) The case for high resolution extended 6-loci HLA typing for identifying related donors in the Indian subcontinent. Biol Blood Marrow Transplant 23:1592–1596. https://doi.org/10.1016/j.bbmt.2017.05.030

Assawachananont J, Mandai M, Okamoto S, Yamada C, Eiraku M, Yonemura S et al (2014) Transplantation of embryonic and induced pluripotent stem cell-derived 3D retinal sheets into retinal degenerative mice. Stem Cell Reports 2:662–674. https://doi.org/10.1016/j.stemcr.2014.03.011

Balak JRA, Juksar J, Carlotti F, Lo Nigro A, de Koning EJP (2019) Organoids from the human fetal and adult pancreas. Curr Diab Rep. https://doi.org/10.1007/s11892-019-1261-z

Barker N, van Es JH, Kuipers J, Kujala P, van den Born M, Cozijnsen M et al (2007) Identification of stem cells in small intestine and colon by marker gene Lgr5. Nature 449:1003–1007. https://doi.org/10.1038/nature06196

Barker N, Huch M, Kujala P, van de Wetering M, Snippert HJ, van Es JH et al (2010) Lgr5(+ve) stem cells drive self-renewal in the stomach and build long-lived gastric units in vitro. Cell Stem Cell 6:25–36. https://doi.org/10.1016/j.stem.2009.11.013

Bartfeld S, Bayram T, Van De Wetering M, Huch M, Begthel H, Kujala P et al (2015) In vitro expansion of human gastric epithelial stem cells and their responses to bacterial infection. Gastroenterology 148:126–136.e6. https://doi.org/10.1053/j.gastro.2014.09.042

Birey F, Andersen J, Makinson CD, Islam S, Wei W, Huber N et al (2017) Assembly of functionally integrated human forebrain spheroids. Nature 545:54–59. https://doi.org/10.1038/nature22330

Boj SF, Hwang C II, Baker LA, IIC C, Engle DD, Corbo V et al (2015) Organoid models of human and mouse

ductal pancreatic cancer. Cell 160:324–338. https://doi.org/10.1016/j.cell.2014.12.021
Bokhari M, Carnachan RJ, Cameron NR, Przyborski SA (2007) Novel cell culture device enabling three-dimensional cell growth and improved cell function. Biochem Biophys Res Commun 354:1095–1100. https://doi.org/10.1016/j.bbrc.2007.01.105
Boretto M, Cox B, Noben M, Hendriks N, Fassbender A, Roose H et al (2017) Development of organoids from mouse and human endometrium showing endometrial epithelium physiology and long-term expandability. Development 144:1775–1786. https://doi.org/10.1242/dev.148478
Broutier L, Andersson-Rolf A, Hindley CJ, Boj SF, Clevers H, Koo BK et al (2016) Culture and establishment of self-renewing human and mouse adult liver and pancreas 3D organoids and their genetic manipulation. Nat Protoc 11:1724–1743. https://doi.org/10.1038/nprot.2016.097
Broutier L, Mastrogiovanni G, Verstegen MMA, Francies HE, Gavarró LM, Bradshaw CR et al (2017) Human primary liver cancer-derived organoid cultures for disease modeling and drug screening. Nat Med 23:1424–1435. https://doi.org/10.1038/nm.4438
Cakir B, Xiang Y, Tanaka Y, Kural MH, Parent M, Kang YJ et al (2019) Engineering of human brain organoids with a functional vascular-like system. Nat Methods 16:1169–1175. https://doi.org/10.1038/s41592-019-0586-5
Cantarelli E, Citro A, Pellegrini S, Mercalli A, Melzi R, Dugnani E et al (2017) Transplant site influences the immune response after islet transplantation. Transplantation 101:1046–1055. https://doi.org/10.1097/TP.0000000000001462
Cao L, Kuratnik A, Xu W, Gibson JD, Kolling F, Falcone ER et al (2015) Development of intestinal organoids as tissue surrogates: cell composition and the epigenetic control of differentiation. Mol Carcinog 54:189–202. https://doi.org/10.1002/mc.22089
Chen S-F, Chang Y-C, Nieh S, Liu C-L, Yang C-Y, Lin Y-S (2012) Nonadhesive culture system as a model of rapid sphere formation with Cancer stem cell properties. PLoS One 7:e31864. https://doi.org/10.1371/journal.pone.0031864
Chen YW, Huang SX, De Carvalho ALRT, Ho SH, Islam MN, Volpi S et al (2017) A three-dimensional model of human lung development and disease from pluripotent stem cells. Nat Cell Biol. https://doi.org/10.1038/ncb3510
Chua CW, Shibata M, Lei M, Toivanen R, Barlow LJ, Bergren SK et al (2014) Single luminal epithelial progenitors can generate prostate organoids in culture. Nat Cell Biol 16:951–961. https://doi.org/10.1038/ncb3047
Clevers H (2016) Modeling development and disease with organoids. Cell 165:1586–1597. https://doi.org/10.1016/j.cell.2016.05.082
Cortez AR, Poling HM, Brown NE, Singh A, Mahe MM, Helmrath MA (2018) Transplantation of human intestinal organoids into the mouse mesentery: a more physiologic and anatomic engraftment site. Surg (United States) 164:643–650. https://doi.org/10.1016/j.surg.2018.04.048
Daviaud N, Friedel RH, Zou H (2018) Vascularization and engraftment of transplanted human cerebral organoids in mouse cortex. ENeuro 5. https://doi.org/10.1523/ENEURO.0219-18.2018
Dekkers JF, Wiegerinck CL, De Jonge HR, Bronsveld I, Janssens HM, De Winter-De Groot KM et al (2013) A functional CFTR assay using primary cystic fibrosis intestinal organoids. Nat Med 19:939–945. https://doi.org/10.1038/nm.3201
DeWard AD, Cramer J, Lagasse E (2014) Cellular heterogeneity in the mouse esophagus implicates the presence of a nonquiescent epithelial stem cell population. Cell Rep 9:701–711. https://doi.org/10.1016/j.celrep.2014.09.027
Diao J, Liu J, Wang S, Chang M, Wang X, Guo B et al (2019) Sweat gland organoids contribute to cutaneous wound healing and sweat gland regeneration. Cell Death Dis. https://doi.org/10.1038/s41419-019-1485-5
Dorrell C, Tarlow B, Wang Y, Canaday PS, Haft A, Schug J et al (2014) The organoid-initiating cells in mouse pancreas and liver are phenotypically and functionally similar. Stem Cell Res 13:275–283. https://doi.org/10.1016/j.scr.2014.07.006
Drost J, Van Jaarsveld RH, Ponsioen B, Zimberlin C, Van Boxtel R, Buijs A et al (2015) Sequential cancer mutations in cultured human intestinal stem cells. Nature 521:43–47. https://doi.org/10.1038/nature14415
Drost J, Karthaus WR, Gao D, Driehuis E, Sawyers CL, Chen Y et al (2016) Organoid culture systems for prostate epithelial and cancer tissue. Nat Protoc 11:347–358. https://doi.org/10.1038/nprot.2016.006
Dye BR, Hill DR, Ferguson MA, Tsai YH, Nagy MS, Dyal R et al (2015) In vitro generation of human pluripotent stem cell derived lung organoids. elife 4. https://doi.org/10.7554/eLife.05098
Dye BR, Dedhia PH, Miller AJ, Nagy MS, White ES, Shea LD et al (2016) A bioengineered niche promotes in vivo engraftment and maturation of pluripotent stem cell derived human lung organoids. elife 5. https://doi.org/10.7554/eLife.19732
Eiraku M, Takata N, Ishibashi H, Kawada M, Sakakura E, Okuda S et al (2011) Self-organizing optic-cup morphogenesis in three-dimensional culture. Nature 472:51–56. https://doi.org/10.1038/nature09941
Elbadawi M, Efferth T (2020) Organoids of human airways to study infectivity and cytopathy of SARS-CoV-2. Lancet Respir 8:e55–e56. https://doi.org/10.1016/S2213-2600(20)30238-1
Elbadawy M, Yamanaka M, Goto Y, Hayashi K, Tsunedomi R, Hazama S et al (2020) Efficacy of primary liver organoid culture from different stages of non-alcoholic steatohepatitis (NASH) mouse model. Biomaterials 237:119823. https://doi.org/10.1016/j.biomaterials.2020.119823

Fantes J, Ragge NK, Lynch SA, McGill NI, Collin JRO, Howard-Peebles PN et al (2003) Mutations in SOX2 cause anophthalmia. Nat Genet. https://doi.org/10.1038/ng1120

Fatehullah A, Tan SH, Barker N (2016) Organoids as an in vitro model of human development and disease. Nat Cell Biol 18:246–254. https://doi.org/10.1038/ncb3312

Finkbeiner SR, Zeng XL, Utama B, Atmar RL, Shroyer NF, Estesa MK (2012) Stem cell-derived human intestinal organoids as an infection model for rotaviruses. MBio 3. https://doi.org/10.1128/mBio.00159-12

Finkbeiner SR, Hill DR, Altheim CH, Dedhia PH, Taylor MJ, Tsai YH et al (2015) Transcriptome-wide analysis reveals hallmarks of human intestine development and maturation in vitro and in vivo. Stem Cell Reports 4:1140–1155. https://doi.org/10.1016/j.stemcr.2015.04.010

Frank DB, Peng T, Zepp JA, Snitow M, Vincent TL, Penkala IJ et al (2016) Emergence of a wave of Wnt signaling that regulates lung alveologenesis by controlling epithelial self-renewal and differentiation. Cell Rep 17:2312–2325. https://doi.org/10.1016/j.celrep.2016.11.001

Freedman BS, Brooks CR, Lam AQ, Fu H, Morizane R, Agrawal V et al (2015) Modelling kidney disease with CRISPR-mutant kidney organoids derived from human pluripotent epiblast spheroids. Nat Commun 6:1–13. https://doi.org/10.1038/ncomms9715

Frieboes HB, Zheng X, Sun CH, Tromberg B, Gatenby R, Cristini V (2006) An integrated computational/experimental model of tumor invasion. Cancer Res 66:1597–1604. https://doi.org/10.1158/0008-5472.CAN-05-3166

Gao D, Vela I, Sboner A, Iaquinta PJ, Karthaus WR, Gopalan A et al (2014) Organoid cultures derived from patients with advanced prostate cancer. Cell 159:176–187. https://doi.org/10.1016/j.cell.2014.08.016

Gao ML, Lei XL, Han F, He KW, Jin SQ, Zhang YY et al (2020) Patient-specific retinal organoids recapitulate disease features of late-onset retinitis Pigmentosa. Front Cell Dev Biol. https://doi.org/10.3389/fcell.2020.00128

Garcez PP, Loiola EC, Da Costa RM, Higa LM, Trindade P, Delvecchio R et al (2016) Zika virus: Zika virus impairs growth in human neurospheres and brain organoids. Science (80-) 352:816–818. https://doi.org/10.1126/science.aaf6116

Giandomenico SL, Lancaster MA (2017) Probing human brain evolution and development in organoids. Curr Opin Cell Biol 44:36–43. https://doi.org/10.1016/j.ceb.2017.01.001

Gilbert PM, Havenstrite KL, Magnusson KEG, Sacco A, Leonardi NA, Kraft P et al (2010) Substrate elasticity regulates skeletal muscle stem cell self-renewal in culture. Science (80-) 329:1078–1081. https://doi.org/10.1126/science.1191035

Gómez-Lechón MJ, Jover R, Donato T, Ponsoda X, Rodriguez C, Stenzel KG et al (1998) Long-term expression of differentiated functions in hepatocytes cultured in three-dimensional collagen matrix. J Cell Physiol:177. https://doi.org/10.1002/(SICI)1097-4652(199812)177:4<553::AID-JCP6>3.0.CO;2-F

Gopalakrishnan J (2019) The emergence of stem cell-based brain organoids: trends and challenges. BioEssays 41:1–10. https://doi.org/10.1002/bies.201900011

Hale LJ, Howden SE, Phipson B, Lonsdale A, Er PX, Ghobrial I et al (2018) 3D organoid-derived human glomeruli for personalised podocyte disease modelling and drug screening. Nat Commun. https://doi.org/10.1038/s41467-018-07594-z

Hickman JA, Graeser R, de Hoogt R, Vidic S, Brito C, Gutekunst M et al (2014) Three-dimensional models of cancer for pharmacology and cancer cell biology: capturing tumor complexity in vitro/ex vivo. Biotechnol J 9:1115–1128. https://doi.org/10.1002/biot.201300492

Hindley CJ, Cordero-Espinoza L, Huch M (2016) Organoids from adult liver and pancreas: stem cell biology and biomedical utility. Dev Biol. https://doi.org/10.1016/j.ydbio.2016.06.039

Hisha H, Tanaka T, Kanno S, Tokuyama Y, Komai Y, Ohe S et al (2013) Establishment of a novel lingual organoid culture system: generation of organoids having mature keratinized epithelium from adult epithelial stem cells. Sci Rep 3:3224. https://doi.org/10.1038/srep03224

Huang P, Zhang L, Gao Y, He Z, Yao D, Wu Z et al (2014) Direct reprogramming of human fibroblasts to functional and expandable hepatocytes. Cell Stem Cell 14:370–384. https://doi.org/10.1016/j.stem.2014.01.003

Huang L, Holtzinger A, Jagan I, Begora M, Lohse I, Ngai N et al (2015) Ductal pancreatic cancer modeling and drug screening using human pluripotent stem cell- and patient-derived tumor organoids. Nat Med 21:1364–1371. https://doi.org/10.1038/nm.3973

Huang KC, Wang ML, Chen SJ, Kuo JC, Wang WJ, Nhi Nguyen PN et al (2019) Morphological and molecular defects in human three-dimensional retinal organoid model of X-linked juvenile Retinoschisis. Stem Cell Rep. https://doi.org/10.1016/j.stemcr.2019.09.010

Huch M, Koo BK (2015) Modeling mouse and human development using organoid cultures. Development 142:3113–3125. https://doi.org/10.1242/dev.118570

Huch M, Dorrell C, Boj SF, van Es JH, Li VS, van de Wetering M et al (2013a) In vitro expansion of single Lgr5+ liver stem cells induced by Wnt-driven regeneration. Nature 494:247–250. https://doi.org/10.1038/nature11826

Huch M, Bonfanti P, Boj SF, Sato T, Loomans CJ, van de Wetering M et al (2013b) Unlimited in vitro expansion of adult bi-potent pancreas progenitors through the Lgr5/R-spondin axis. EMBO J 32:2708–2721. https://doi.org/10.1038/emboj.2013.204

Huch M, Gehart H, van Boxtel R, Hamer K, Blokzijl F, Verstegen MM et al (2015) Long-term culture of genome-stable bipotent stem cells from adult human liver. Cell 160:299–312. https://doi.org/10.1016/j.cell.2014.11.050

Hyun I (2017) Engineering ethics and self-organizing models of human development: opportunities and challenges. Cell Stem Cell 21:718–720. https://doi.org/10.1016/j.stem.2017.09.002

Iefremova V, Manikakis G, Krefft O, Jabali A, Weynans K, Wilkens R et al (2017) An organoid-based model of cortical development identifies non-cell-autonomous defects in Wnt signaling contributing to Miller-Dieker syndrome. Cell Rep 19:50–59. https://doi.org/10.1016/j.celrep.2017.03.047

Jackson EL, Lu H (2016) Three-dimensional models for studying development and disease: moving on from organisms to organs-on-a-chip and organoids. Integr Biol (United Kingdom) 8:672–683. https://doi.org/10.1039/c6ib00039h

Jee JH, Lee DH, Ko J, Hahn S, Jeong SY, Kim HK et al (2019a) Development of collagen-based 3D matrix for gastrointestinal tract-derived organoid culture. Stem Cells Int:2019. https://doi.org/10.1155/2019/8472712

Jee J, Jeong SY, Kim HK, Choi SY, Jeong S, Lee J et al (2019b) In vivo evaluation of scaffolds compatible for colonoid engraftments onto injured mouse colon epithelium. FASEB J. https://doi.org/10.1096/fj.201802692RR

Jun D-Y, Kim SY, Na JC, Ho Leeid H, Kim J, Yoon YE et al (2018) Tubular organotypic culture model of human kidney. PLoS One. https://doi.org/10.1371/journal.pone.0206447

Jung P, Sato T, Merlos-Suárez A, Barriga FM, Iglesias M, Rossell D et al (2011) Isolation and in vitro expansion of human colonic stem cells. Nat Med 17:1225–1227. https://doi.org/10.1038/nm.2470

Karthaus WR, Iaquinta PJ, Drost J, Gracanin A, van Boxtel R, Wongvipat J et al (2014) Identification of multipotent luminal progenitor cells in human prostate organoid cultures. Cell 159:163–175. https://doi.org/10.1016/j.cell.2014.08.017

Kessler M, Hoffmann K, Brinkmann V, Thieck O, Jackisch S, Toelle B et al (2015) The Notch and Wnt pathways regulate stemness and differentiation in human fallopian tube organoids. Nat Commun 6:8989. https://doi.org/10.1038/ncomms9989

Khalil HA, Hong SN, Rouch JD, Scott A, Cho Y, Wang J et al (2019) Intestinal epithelial replacement by transplantation of cultured murine and human cells into the small intestine. PLoS One 14. https://doi.org/10.1371/journal.pone.0216326

Kim M, Mun H, Sung CO, Cho EJ, Jeon HJ, Chun SM et al (2019a) Patient-derived lung cancer organoids as in vitro cancer models for therapeutic screening. Nat Commun 10:3991. https://doi.org/10.1038/s41467-019-11867-6

Kim S, Cho A-N, Min S, Kim S, Cho S-W (2019b) Organoids for advanced therapeutics and disease models. Adv Ther 2:1800087. https://doi.org/10.1002/adtp.201800087

Kopper O, de Witte CJ, Lõhmussaar K, Valle-Inclan JE, Hami N, Kester L et al (2019) An organoid platform for ovarian cancer captures intra- and interpatient heterogeneity. Nat Med 25:838–849. https://doi.org/10.1038/s41591-019-0422-6

Krishnamurthy S, Nör JE (2013) Orosphere assay: a method for propagation of head and neck cancer stem cells. Head Neck 35:1015–1021. https://doi.org/10.1002/hed.23076

Kruitwagen HS, Oosterhoff LA, van Wolferen ME, Chen C, Nantasanti Assawarachan S, Schneeberger K et al (2020) Long-term survival of transplanted autologous canine liver organoids in a COMMD1-deficient dog model of metabolic liver disease. Cell. https://doi.org/10.3390/cells9020410

Kurmann AA, Serra M, Hawkins F, Rankin SA, Mori M, Astapova I et al (2015) Regeneration of thyroid function by transplantation of differentiated pluripotent stem cells. Cell Stem Cell 17:527–542. https://doi.org/10.1016/j.stem.2015.09.004

Lancaster MA, Huch M (2019) Disease modelling in human organoids. Dis Model Mech 12. https://doi.org/10.1242/dmm.039347

Lancaster MA, Renner M, Martin CA, Wenzel D, Bicknell LS, Hurles ME et al (2013) Cerebral organoids model human brain development and microcephaly. Nature 501:373–379. https://doi.org/10.1038/nature12517

Lee SH, Hu W, Matulay JT, Silva MV, Owczarek TB, Kim K et al (2018) Tumor evolution and drug response in patient-derived organoid models of bladder Cancer. Cell 173:515–528.e17. https://doi.org/10.1016/j.cell.2018.03.017

Leslie JL, Young VB (2016) A whole new ball game: stem cell-derived epithelia in the study of host-microbe interactions. Anaerobe. https://doi.org/10.1016/j.anaerobe.2015.10.016

Li C, Kato M, Shiue L, Shively JE, Ares M, Lin RJ (2006) Cell type and culture condition-dependent alternative splicing in human breast cancer cells revealed by splicing-sensitive microarrays. Cancer Res 66:1990–1999. https://doi.org/10.1158/0008-5472.CAN-05-2593

Li X, Francies HE, Secrier M, Perner J, Miremadi A, Galeano-Dalmau N et al (2018) Organoid cultures recapitulate esophageal adenocarcinoma heterogeneity providing a model for clonality studies and precision therapeutics. Nat Commun 9:1–13. https://doi.org/10.1038/s41467-018-05190-9

Liang G, Zhang Y (2013) Embryonic stem cell and induced pluripotent stem cell: an epigenetic perspective. Cell Res 23:49–69. https://doi.org/10.1038/cr.2012.175

Lo B, Parham L (2009) Ethical issues in stem cell research. Endocr Rev 30:204–213. https://doi.org/10.1210/er.2008-0031

Longworth-Mills E, Koehler KR, Hashino E (2016) Generating inner ear organoids from mouse embryonic stem cells. Methods Mol Biol 1341:391–406. https://doi.org/10.1007/7651_2015_215

Loomans CJM, Williams Giuliani N, Balak J, Ringnalda F, van Gurp L, Huch M et al (2018) Expansion of adult human pancreatic tissue yields organoids harboring progenitor cells with endocrine differentiation potential. Stem Cell Reports 10:712–724. https://doi.org/10.1016/j.stemcr.2018.02.005

Maenhoudt N, Defraye C, Boretto M, Jan Z, Heremans R, Boeckx B et al (2020) Developing organoids from ovarian Cancer as experimental and preclinical models. Stem Cell Reports 14:717–729. https://doi.org/10.1016/j.stemcr.2020.03.004

Mallo M, Wellik DM, Deschamps J (2010) Hox genes and regional patterning of the vertebrate body plan. Dev Biol 344:7–15. https://doi.org/10.1016/j.ydbio.2010.04.024

Mandai M, Fujii M, Hashiguchi T, Sunagawa GA, Ito S, Sun J et al (2017) iPSC-derived retina transplants improve vision in rd1 end-stage retinal-degeneration mice. Stem Cell Reports 8:69–83. https://doi.org/10.1016/j.stemcr.2016.12.008

Mansour AA, Gonçalves JT, Bloyd CW, Li H, Fernandes S, Quang D et al (2018) An in vivo model of functional and vascularized human brain organoids. Nat Biotechnol 36:432–441. https://doi.org/10.1038/nbt.4127

Mariani J, Simonini MV, Palejev D, Tomasini L, Coppola G, Szekely AM et al (2012) Modeling human cortical development in vitro using induced pluripotent stem cells. Proc Natl Acad Sci U S A 109:12770–12775. https://doi.org/10.1073/pnas.1202944109

Mariani J, Coppola G, Zhang P, Abyzov A, Provini L, Tomasini L et al (2015) FOXG1-dependent dysregulation of GABA/glutamate neuron differentiation in autism Spectrum disorders. Cell 162:375–390. https://doi.org/10.1016/j.cell.2015.06.034

Marton RM, Paşca SP (2020) Organoid and Assembloid Technologies for Investigating Cellular Crosstalk in human brain development and disease. Trends Cell Biol 30:133–143. https://doi.org/10.1016/j.tcb.2019.11.004

Matano M, Date S, Shimokawa M, Takano A, Fujii M, Ohta Y et al (2015) Modeling colorectal cancer using CRISPR-Cas9-mediated engineering of human intestinal organoids. Nat Med 21:256–262. https://doi.org/10.1038/nm.3802

McCracken KW, Howell JC, Wells JM, Spence JR (2011) Generating human intestinal tissue from pluripotent stem cells in vitro. Nat Protoc 6:1920–1928. https://doi.org/10.1038/nprot.2011.410

McCracken KW, Catá EM, Crawford CM, Sinagoga KL, Schumacher M, Rockich BE et al (2014) Modelling human development and disease in pluripotent stem-cell-derived gastric organoids. Nature 516:400–404. https://doi.org/10.1038/nature13863

Michels BE, Mosa MH, Streibl BI, Zhan T, Menche C, Abou-El-Ardat K et al (2020) Pooled in vitro and in vivo CRISPR-Cas9 screening identifies tumor suppressors in human Colon organoids. Cell Stem Cell 26:782–792.e7. https://doi.org/10.1016/j.stem.2020.04.003

Miller AJ, Hill DR, Nagy MS, Aoki Y, Dye BR, Chin AM et al (2018) In vitro induction and in vivo engraftment of lung bud tip progenitor cells derived from human pluripotent stem cells. Stem Cell Reports 10:101–119. https://doi.org/10.1016/j.stemcr.2017.11.012

Miller AJ, Dye BR, Ferrer-Torres D, Hill DR, Overeem AW, Shea LD et al (2019) Generation of lung organoids from human pluripotent stem cells in vitro. Nat Protoc. https://doi.org/10.1038/s41596-018-0104-8

Min S, Kim S, Cho SW (2020) Gastrointestinal tract modeling using organoids engineered with cellular and microbiota niches. Exp Mol Med. https://doi.org/10.1038/s12276-020-0386-0

Miranda CC, Fernandes TG, Diogo MM, Cabral JMS (2018) Towards multi-organoid systems for drug screening applications. Bioengineering 5:49. https://doi.org/10.3390/bioengineering5030049

Mullenders J, de Jongh E, Brousali A, Roosen M, Blom JPA, Begthel H et al (2019) Mouse and human urothelial cancer organoids: a tool for bladder cancer research. Proc Natl Acad Sci U S A 116:4567–4574. https://doi.org/10.1073/pnas.1803595116

Mustata RC, Vasile G, Fernandez-Vallone V, Strollo S, Lefort A, Libert F et al (2013) Identification of Lgr5-independent spheroid-generating progenitors of the mouse fetal intestinal epithelium. Cell Rep 5:421–432. https://doi.org/10.1016/j.celrep.2013.09.005

Nam SA, Seo E, Kim JW, Kim HW, Kim HL, Kim K et al (2019) Graft immaturity and safety concerns in transplanted human kidney organoids. Exp Mol Med 51. https://doi.org/10.1038/s12276-019-0336-x

Nanduri LSY, Baanstra M, Faber H, Rocchi C, Zwart E, De Haan G et al (2014) Purification and ex vivo expansion of fully functional salivary gland stem cells. Stem Cell Reports 3:957–964. https://doi.org/10.1016/j.stemcr.2014.09.015

Nanki K, Toshimitsu K, Takano A, Fujii M, Shimokawa M, Ohta Y et al (2018) Divergent routes toward Wnt and R-spondin niche independency during human gastric carcinogenesis. Cell 174:856–869.e17. https://doi.org/10.1016/j.cell.2018.07.027

O'Rourke KP, Loizou E, Livshits G, Schatoff EM, Baslan T, Manchado E et al (2017) Transplantation of engineered organoids enables rapid generation of metastatic mouse models of colorectal cancer. Nat Biotechnol 35:577–582. https://doi.org/10.1038/nbt.3837

Ogawa J, Pao GM, Shokhirev MN, Verma IM (2018) Glioblastoma Model Using Human Cerebral Organoids. Cell Rep 23:1220–1229. https://doi.org/10.1016/j.celrep.2018.03.105

Okamoto R, Shimizu H, Suzuki K, Kawamoto A, Takahashi J, Kawai M et al (2020) Organoid-based

regenerative medicine for inflammatory bowel disease. Regen Ther 13:1–6. https://doi.org/10.1016/j.reth.2019.11.004
Pasca SP (2019) Assembling human brain organoids. Science (80-) 363:126–127. https://doi.org/10.1126/science.aau5729
Peters DJM, Breuning MH (2001) Autosomal dominant polycystic kidney disease. Mol Genet Basis Ren Dis 358:1439–1444. https://doi.org/10.1016/B978-1-4160-0252-9.50010-0
Pollen AA, Bhaduri A, Andrews MG, Nowakowski TJ, Meyerson OS, Mostajo-Radji MA et al (2019) Establishing cerebral organoids as models of human-specific brain evolution. Cell 176:743–756.e17. https://doi.org/10.1016/j.cell.2019.01.017
Pringle S, Maimets M, van der Zwaag M, Stokman MA, van Gosliga D, Zwart E et al (2016) Human salivary gland stem cells functionally restore radiation damaged salivary glands. Stem Cells 34:640–652. https://doi.org/10.1002/stem.2278
Raimondi G, Mato-Berciano A, Pascual-Sabater S, Rovira-Rigau M, Cuatrecasas M, Fondevila C et al (2020) Patient-derived pancreatic tumour organoids identify therapeutic responses to oncolytic adenoviruses. EBioMedicine 56:102786. https://doi.org/10.1016/j.ebiom.2020.102786
Ren W, Lewandowski BC, Watson J, Aihara E, Iwatsuki K, Bachmanov AA et al (2014) Single Lgr5- or Lgr6-expressing taste stem/progenitor cells generate taste bud cells ex vivo. Proc Natl Acad Sci U S A 111:16401–16406. https://doi.org/10.1073/pnas.1409064111
Ren W, Liu Q, Zhang X, Yu Y (2020) Age-related taste cell generation in circumvallate papillae organoids via regulation of multiple signaling pathways. Exp Cell Res 394:112150. https://doi.org/10.1016/j.yexcr.2020.112150
Rock JR, Onaitis MW, Rawlins EL, Lu Y, Clark CP, Xue Y et al (2009) Basal cells as stem cells of the mouse trachea and human airway epithelium. Proc Natl Acad Sci U S A 106:12771–12775. https://doi.org/10.1073/pnas.0906850106
Rosenbluth JM, Schackmann RCJ, Gray GK, Selfors LM, Li CMC, Boedicker M et al (2020) Organoid cultures from normal and cancer-prone human breast tissues preserve complex epithelial lineages. Nat Commun 11:1–14. https://doi.org/10.1038/s41467-020-15548-7
Sachs N, de Ligt J, Kopper O, Gogola E, Bounova G, Weeber F et al (2018) A living biobank of breast Cancer organoids captures disease heterogeneity. Cell 172:373–386 e10. https://doi.org/10.1016/j.cell.2017.11.010
Sachs N, Papaspyropoulos A, Zomer-van Ommen DD, Heo I, Böttinger L, Klay D et al (2019) Long-term expanding human airway organoids for disease modeling. EMBO J 38. https://doi.org/10.15252/embj.2018100300
Salama NR, Hartung ML, Müller A (2013) Life in the human stomach: persistence strategies of the bacterial pathogen helicobacter pylori. Nat Rev Microbiol. https://doi.org/10.1038/nrmicro3016
Santos CP, Lapi E, Martínez de Villarreal J, Álvaro-Espinosa L, Fernández-Barral A, Barbáchano A et al (2019) Urothelial organoids originating from Cd49fhigh mouse stem cells display notch-dependent differentiation capacity. Nat Commun 10:1–17. https://doi.org/10.1038/s41467-019-12307-1
Sato T, Vries RG, Snippert HJ, Van De Wetering M, Barker N, Stange DE et al (2009) Single Lgr5 stem cells build crypt-villus structures in vitro without a mesenchymal niche. Nature 459:262–265. https://doi.org/10.1038/nature07935
Sato T, Stange DE, Ferrante M, Vries RG, Van Es JH, Van den Brink S et al (2011) Long-term expansion of epithelial organoids from human colon, adenoma, adenocarcinoma, and Barrett's epithelium. Gastroenterology 141:1762–1772. https://doi.org/10.1053/j.gastro.2011.07.050
Schutgens F, Rookmaaker MB, Margaritis T, Rios A, Ammerlaan C, Jansen J et al (2019) Tubuloids derived from human adult kidney and urine for personalized disease modeling. Nat Biotechnol 37:303–313. https://doi.org/10.1038/s41587-019-0048-8
Schwank G, Koo BK, Sasselli V, Dekkers JF, Heo I, Demircan T et al (2013) Functional repair of CFTR by CRISPR/Cas9 in intestinal stem cell organoids of cystic fibrosis patients. Cell Stem Cell 13:653–658. https://doi.org/10.1016/j.stem.2013.11.002
Sharmin S, Taguchi A, Kaku Y, Yoshimura Y, Ohmori T, Sakuma T et al (2016) Human induced pluripotent stem cell-derived podocytes mature into vascularized glomeruli upon experimental transplantation. J Am Soc Nephrol 27:1778–1791. https://doi.org/10.1681/ASN.2015010096
Shi Y, Sun L, Wang M, Liu J, Zhong S, Li R et al (2020) Vascularized human cortical organoids (vOrganoids) model cortical development in vivo. PLoS Biol 18. https://doi.org/10.1371/journal.pbio.3000705
Shimada H, Lu Q, Insinna-Kettenhofen C, Nagashima K, English MA, Semler EM et al (2017) In vitro modeling using ciliopathy-patient-derived cells reveals distinct cilia dysfunctions caused by CEP290 mutations. Cell Rep. https://doi.org/10.1016/j.celrep.2017.06.045
Shirai H, Mandai M, Matsushita K, Kuwahara A, Yonemura S, Nakano T et al (2016) Transplantation of human embryonic stem cell-derived retinal tissue in two primate models of retinal degeneration. Proc Natl Acad Sci U S A 113:E81–E90. https://doi.org/10.1073/pnas.1512590113
Singh RK, Nasonkin IO (2020) Limitations and promise of retinal tissue from human pluripotent stem cells for developing therapies of blindness. Front Cell Neurosci 14:1–26. https://doi.org/10.3389/fncel.2020.00179
Singh RK, Mallela RK, Cornuet PK, Reifler AN, Chervenak AP, West MD et al (2015) Characterization

of three-dimensional retinal tissue derived from human embryonic stem cells in adherent monolayer cultures. Stem Cells Dev. https://doi.org/10.1089/scd.2015.0144

Singh R, Cuzzani O, Binette F, Sternberg H, West MD, Nasonkin IO (2018) Pluripotent stem cells for retinal tissue engineering: current status and future prospects. Stem Cell Rev Reports 14:463–483. https://doi.org/10.1007/s12015-018-9802-4

Singh RK, Occelli LM, Binette F, Petersen-Jones SM, Nasonkin IO (2019) Transplantation of human embryonic stem cell-derived retinal tissue in the subretinal space of the cat eye. Stem Cells Dev. https://doi.org/10.1089/scd.2019.0090

Smith E, Cochrane WJ (1946) Cystic organoid teratoma; report of a case. Can Med Assoc J 55:151

Song JJ, Guyette JP, Gilpin SE, Gonzalez G, Vacanti JP, Ott HC (2013) Regeneration and experimental orthotopic transplantation of a bioengineered kidney. Nat Med 19:646–651. https://doi.org/10.1038/nm.3154

Spence JR, Mayhew CN, Rankin SA, Kuhar MF, Vallance JE, Tolle K et al (2011) Directed differentiation of human pluripotent stem cells into intestinal tissue in vitro. Nature 470:105–109. https://doi.org/10.1038/nature09691

Sridhar KN, Miller DG (2012) Clinical relevance of tissue engineering. In: Ramalingam M, Pvurw JL (eds) Tissue engineering and regenerative medicine: a nano approach. CRC Press, pp 519–533

Sugimoto S, Ohta Y, Fujii M, Matano M, Shimokawa M, Nanki K et al (2018) Reconstruction of the human Colon epithelium in vivo. Cell Stem Cell. https://doi.org/10.1016/j.stem.2017.11.012

Sun L, Wang Y, Cen J, Ma X, Cui L, Qiu Z et al (2019) Modelling liver cancer initiation with organoids derived from directly reprogrammed human hepatocytes. Nat Cell Biol 21:1015–1026. https://doi.org/10.1038/s41556-019-0359-5

Takahashi N, Hoshi H, Higa A, Hiyama G, Tamura H, Ogawa M et al (2019) An in vitro system for evaluating molecular targeted drugs using lung patient-derived tumor organoids. Cell 8:481. https://doi.org/10.3390/cells8050481

Takasato M, Er PX, Chiu HS, Maier B, Baillie GJ, Ferguson C et al (2015) Kidney organoids from human iPS cells contain multiple lineages and model human nephrogenesis. Nature 526:564–568. https://doi.org/10.1038/nature15695

Takebe T, Sekine K, Enomura M, Koike H, Kimura M, Ogaeri T et al (2013) Vascularized and functional human liver from an iPSC-derived organ bud transplant. Nature 499:481–484. https://doi.org/10.1038/nature12271

Takebe T, Enomura M, Yoshizawa E, Kimura M, Koike H, Ueno Y et al (2015) Vascularized and complex organ buds from diverse tissues via mesenchymal cell-driven condensation. Cell Stem Cell 16:556–565. https://doi.org/10.1016/j.stem.2015.03.004

Takebe T, Sekine K, Kimura M, Yoshizawa E, Ayano S, Koido M et al (2017) Massive and reproducible production of liver buds entirely from human pluripotent stem cells. Cell Rep 21:2661–2670. https://doi.org/10.1016/j.celrep.2017.11.005

Takebe T, Wells JM, Helmrath MA, Zorn AM (2018) Organoid center strategies for accelerating clinical translation. Cell Stem Cell. https://doi.org/10.1016/j.stem.2018.05.008

Tan Q, Choi KM, Sicard D, Tschumperlin DJ (2017) Human airway organoid engineering as a step toward lung regeneration and disease modeling. Biomaterials 113:118–132. https://doi.org/10.1016/j.biomaterials.2016.10.046

Van De Wetering M, Francies HE, Francis JM, Bounova G, Iorio F, Pronk A et al (2015) Prospective derivation of a living organoid biobank of colorectal cancer patients. Cell 161:933–945. https://doi.org/10.1016/j.cell.2015.03.053

van den Berg CW, Ritsma L, Avramut MC, Wiersma LE, van den Berg BM, Leuning DG et al (2018) Renal subcapsular transplantation of PSC-derived kidney organoids induces neo-vasculogenesis and significant glomerular and tubular maturation in vivo. Stem Cell Reports 10:751–765. https://doi.org/10.1016/j.stemcr.2018.01.041

Vargas-Valderrama A, Messina A, Mitjavila-Garcia MT, Guenou H (2020) The endothelium, a key actor in organ development and hPSC-derived organoid vascularization. J Biomed Sci 27:67. https://doi.org/10.1186/s12929-020-00661-y

Varzideh F, Pahlavan S, Ansari H, Halvaei M, Kostin S, Feiz MS et al (2019) Human cardiomyocytes undergo enhanced maturation in embryonic stem cell-derived organoid transplants. Biomaterials 192:537–550. https://doi.org/10.1016/j.biomaterials.2018.11.033

Vasyutin I, Zerihun L, Ivan C, Atala A (2019) Bladder organoids and spheroids: potential tools for normal and diseased tissue modelling. Anticancer Res 39:1105–1118. https://doi.org/10.21873/anticanres.13219

Volkner M, Zschatzsch M, Rostovskaya M, Overall RW, Busskamp V, Anastassiadis K et al (2016) Retinal organoids from pluripotent stem cells efficiently recapitulate Retinogenesis. Stem Cell Reports 6:525–538. https://doi.org/10.1016/j.stemcr.2016.03.001

Walsh AJ, Castellanos JA, Nagathihalli NS, Merchant NB, Skala MC (2016) Optical imaging of drug-induced metabolism changes in murine and human pancreatic Cancer organoids reveals heterogeneous drug response. Pancreas 45:863–869. https://doi.org/10.1097/MPA.0000000000000543

Wang S, Wang X, Tan Z, Su Y, Liu J, Chang M et al (2019) Human ESC-derived expandable hepatic organoids enable therapeutic liver repopulation and pathophysiological modeling of alcoholic liver injury. Cell Res. https://doi.org/10.1038/s41422-019-0242-8

Watson CL, Mahe MM, Múnera J, Howell JC, Sundaram N, Poling HM et al (2014) An in vivo model of human small intestine using pluripotent stem cells. Nat Med 20:1310–1314. https://doi.org/10.1038/nm.3737

Weiswald LB, Bellet D (2015) Dangles-Marie V. spherical Cancer models in tumor biology. Neoplasia (United States) 17:1–15. https://doi.org/10.1016/j.neo.2014.12.004

Weren RDA, Ligtenberg MJL, Kets CM, De Voer RM, Verwiel ETP, Spruijt L et al (2015) A germline homozygous mutation in the base-excision repair gene NTHL1 causes adenomatous polyposis and colorectal cancer. Nat Genet 47:668–671. https://doi.org/10.1038/ng.3287

Wörsdörfer P, Rockel A, Alt Y, Kern A, Ergün S (2020) Generation of vascularized neural organoids by co-culturing with mesodermal progenitor cells. STAR Protoc 1:100041. https://doi.org/10.1016/j.xpro.2020.100041

Wroblewski LE, Shen L, Ogden S, Romero-Gallo J, Lapierre LA, Israel DA et al (2009) Helicobacter pylori dysregulation of gastric epithelial tight junctions by urease-mediated myosin II activation. Gastroenterology 136:236–246. https://doi.org/10.1053/j.gastro.2008.10.011

Xian B, Luo Z, Li K, Li K, Tang M, Yang R et al (2019) Dexamethasone provides effective immunosuppression for improved survival of retinal organoids after Epiretinal transplantation. Stem Cells Int 2019. https://doi.org/10.1155/2019/7148032

Xie Y, Park ES, Xiang D, Li Z (2018) Long-term organoid culture reveals enrichment of organoid-forming epithelial cells in the fimbrial portion of mouse fallopian tube. Stem Cell Res 32:51–60. https://doi.org/10.1016/j.scr.2018.08.021

Xinaris C, Benedetti V, Rizzo P, Abbate M, Corna D, Azzollini N et al (2012) In vivo maturation of functional renal organoids formed from embryonic cell suspensions. J Am Soc Nephrol 23:1857–1868. https://doi.org/10.1681/ASN.2012050505

Yan HHN, Siu HC, Law S, Ho SL, Yue SSK, Tsui WY et al (2018) A comprehensive human gastric Cancer organoid biobank captures tumor subtype heterogeneity and enables therapeutic screening. Cell Stem Cell 23:882–897.e11. https://doi.org/10.1016/j.stem.2018.09.016

Yang Y, Opara EC, Liu Y, Atala A, Zhao W (2017) Microencapsulation of porcine thyroid cell organoids within a polymer microcapsule construct. Exp Biol Med 242:286–296. https://doi.org/10.1177/1535370216673746

Yin X, Mead BE, Safaee H, Langer R, Karp JM, Levy O (2016) Engineering stem cell organoids. Cell Stem Cell 18:25–38. https://doi.org/10.1016/j.stem.2015.12.005

Yu J, Parasuraman S, Shah A, Weisdorf D (2019) Mortality, length of stay and costs associated with acute graft-versus-host disease during hospitalization for allogeneic hematopoietic stem cell transplantation. Curr Med Res Opin 35:983–988. https://doi.org/10.1080/03007995.2018.1551193

Yui S, Nakamura T, Sato T, Nemoto Y, Mizutani T, Zheng X et al (2012) Functional engraftment of colon epithelium expanded in vitro from a single adult Lgr5 + stem cell. Nat Med 18:618–623. https://doi.org/10.1038/nm.2695

Zhou T, Tan L, Cederquist GY, Fan Y, Hartley BJ, Mukherjee S et al (2017) High-content screening in hPSC-neural progenitors identifies drug candidates that inhibit Zika virus infection in fetal-like organoids and adult brain. Cell Stem Cell 21:274–283.e5. https://doi.org/10.1016/j.stem.2017.06.017

Adv Exp Med Biol - Cell Biology and Translational Medicine (2021) 14: 65–82
https://doi.org/10.1007/5584_2021_634

Published online: 5 May 2021

Aldo Keto Reductases AKR1B1 and AKR1B10 in Cancer: Molecular Mechanisms and Signaling Networks

Sreeparna Banerjee

Abstract

Deregulation of metabolic pathways has increasingly been appreciated as a major driver of cancer in recent years. The principal cancer-associated alterations in metabolism include abnormal uptake of glucose and amino acids and the preferential use of metabolic pathways for the production of biomass and nicotinamide adenine dinucleotide phosphate (NADPH). Aldo-keto reductases (AKRs) are NADPH dependent cytosolic enzymes that can catalyze the reduction of carbonyl groups to primary and secondary alcohols using electrons from NADPH. Aldose reductase, also known as AKR1B1, catalyzes the conversion of excess glucose to sorbitol and has been studied extensively for its role in a number of diabetic pathologies. In recent years, however, high expression of the AKR1B and AKR1C family of enzymes has been strongly associated with worse outcomes in different cancer types. This review provides an overview of the catalysis-dependent and independent data emerging on the molecular mechanisms of the functions of AKRBs in different tumor models with an emphasis of the role of these enzymes in chemoresistance, inflammation, oxidative stress and epithelial-to-mesenchymal transition.

S. Banerjee (✉)
Department of Biological Sciences, Orta Dogu Teknik Universitesi (ODTU/METU), Ankara, Turkey
e-mail: banerjee@metu.edu.tr

Keywords

Aldo keto reductases · Cancer · Chemoresistance · Epithelial-to-mesenchymal transition · Inflammation · Oxidative stress

Abbreviations

4HNE	4-hydroxy-trans-2-nonenal
AKR	Aldo-keto reductase
BLBC	Basal-like breast cancer
CRC	Colorectal carcinoma
EMT	epithelial-to-mesenchymal transition
HCC	Hepatocellular carcinoma
NADPH	nicotinamide adenine dinucleotide phosphate
NSCLC	Non-small cell lung carcinoma
PAAD	Pancreatic adenocarcinoma
ROS	Reactive oxygen species
SORD	sorbitol dehydrogenase

1 Introduction to AKRs, Substrate Classification and Catalytic Activity

The aldo-keto reductase (AKR) enzymes are 34–37 kDa monomeric proteins that are localized in the cytosol. Although there is high functional diversity among AKRs, the protein structure consists of a conserved $(\beta/\alpha)_8$ barrel and a pyridine nucleotide binding site with high sequence identity, which provides them with high structural similarity (Mindnich and Penning 2009). The common $(\beta/\alpha)_8$ barrel motif is involved in the oligomerization of tertiary and quaternary structures, and binding of cofactors and metals, so that they can generate an active site geometry (Wierenga 2001). Although AKRs are known as oxidoreductases, they are likely to function primarily as reductases *in vivo* utilizing reducing electrons from NADPH. Many substrates have been described for AKRs including glucose, lipid aldehydes, keto forms of steroids and prostaglandins, various carcinogens and carcinogen derivatives (Penning 2015). In the nomenclature of AKR proteins, the first number is representative of the family (e.g. AKR1), the next letter represents the subfamily (e.g. AKR1B), while the second number represents the unique protein (e.g. AKR1B1). The AKR superfamily consists of 16 members identified on the basis of sequence alignment (AKR1 – AKR16) with over 190 proteins (Penning 2015). Of the 15 AKRs expressed in humans (Penning 2015), the AKR1B and AKR1C family of proteins have been strongly implicated in cancer. The AKR1B family of proteins consist of three well defined members: AKR1B1, AKR1B10 and AKR1B15. AKRs mainly prefer to use reducing electrons from NADPH rather than NADH. Most metabolically active cells have a high NADPH/NADP+ ratio, which is reflective of biomass production (Pollak et al. 2007).

AKR1B1 is a part of the polyol pathway that is best characterized in the context of diabetic pathophysiology (Saraswathy et al. 2014). Under normal glycemic conditions, this pathway uses ~3% of total glucose flux for reduction to sorbitol (Fig. 1a). Under hyperglycemic conditions, as much as 30% of the total glucose can be reduced to sorbitol with the use of reducing electrons from NADPH via the polyol pathway (Srivastava et al. 2005). NADPH is used for reduction reactions in many metabolic pathways. One such reaction is the detoxification of ROS with reduced glutathione (GSH), converting it to its oxidized form glutathione disulfide (GSSG) in the presence of the enzyme glutathione peroxidase (GPx) (Fig. 1b). Restoration of reduced glutathione from GSSG is carried out with the enzyme glutathione reductase (GR) with the use of electrons from NADPH (Fig. 1b). Therefore, a large drain on the NADPH pool could compromise the ability of the cell to protect itself from oxidative stress. The sorbitol generated in the polyol pathway can be oxidized to fructose via the enzyme sorbitol dehydrogenase (SORD) with an accompanying reduction of NAD+ to NADH. Fructose can be converted to Fructose-1-phosphate via the enzyme Ketohexokinase (KHK) and subsequently to dihydroxyacetone phosphate (DHAP) and glyceraldehyde (GA); the latter metabolites can then enter glycolysis. Alternately, fructose can be phosphorylated to fructose-6-phosphate via the enzyme Hexokinase (HK), and then to fructose-1,6-bisphosphate via phosphofructokinase (PFK). PFK is a rate limiting enzyme of glycolysis and is inhibited with high ATP/ADP ratio, low pH, hypoxia and citrate levels. These conditions are particularly common in cancer cells. This suggests that the conversion of glucose to fructose via the polyol pathway and the preferential metabolism of fructose via KHK rather than PFK can keep glycolysis going in cancer cells (Krause and Wegner 2020), although how KHK is preferred over PFK and the role of AKR1B1 (if any) in this choice is currently unclear.

AKR1B1 inhibitors were reported to successfully ameliorate secondary diabetic complications in preclinical studies. However, none of these inhibitors have passed a Phase III clinical trial for the prevention of diabetic complications due to poor potency, except Epalrestat, which is licensed for use in diabetic neuropathy in Japan and India (Suzen and Buyukbingol 2005). Of note, a recent study suggested that the transcriptome of AKR1B1 knock-out HeLa cells was significantly divergent from Epalrestat treated HeLa cells (Ji et al. 2020), suggesting that Epalrestat may affect gene expression in cells well beyond the inhibition of AKR1B1.

AKR1B10 exhibits a more restricted substrate specificity compared to AKR1B1, showing high specificity for farnesal, geranylgeranyl, retinal and carbonyls (Gallego et al. 2007; Ma et al.

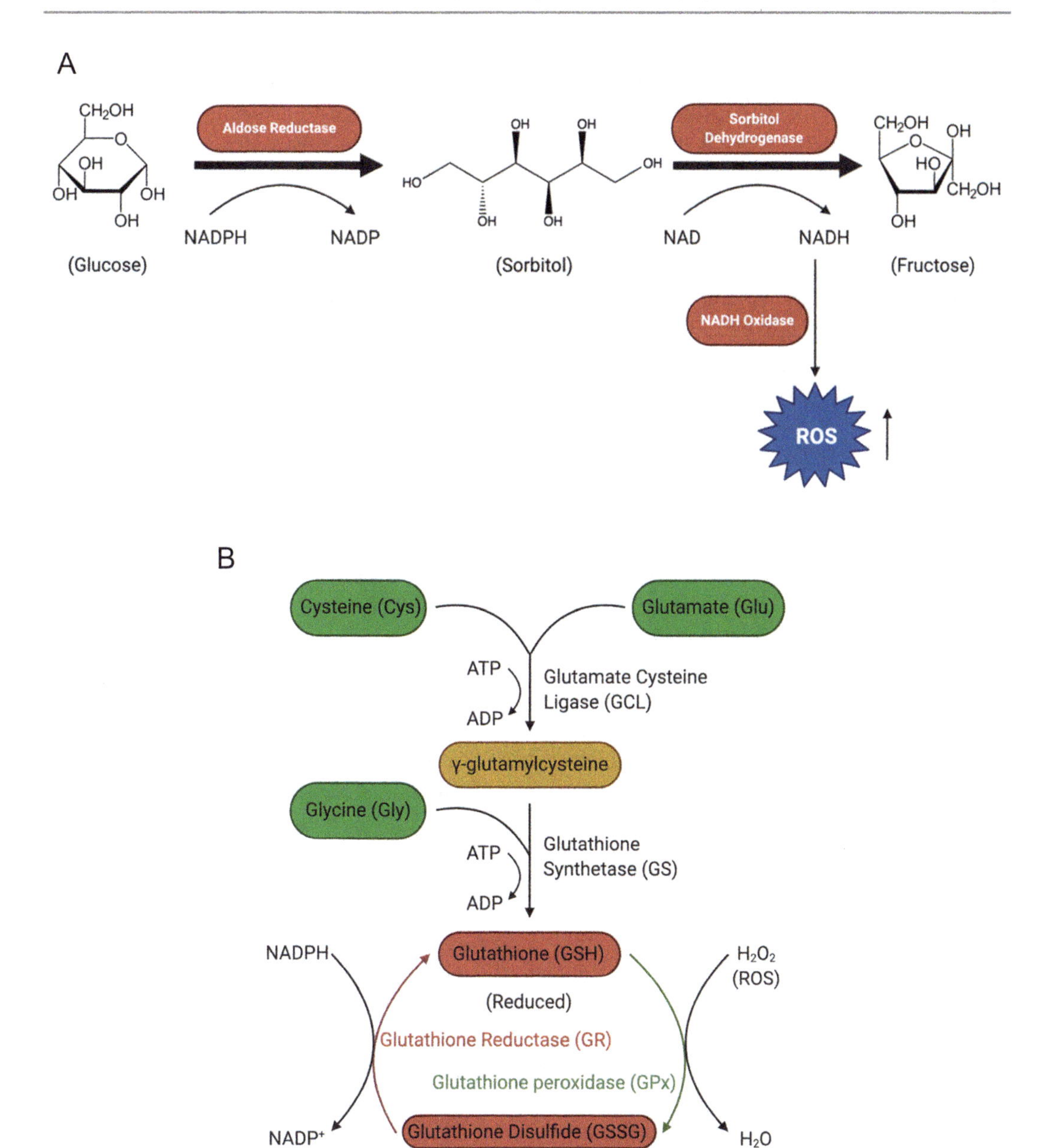

Fig. 1 High AKR activity can impair the antioxidant balance in cells. (**a**). Enzymatic reduction of glucose with AKR1B1 enzyme results in the production of sorbitol using electrons from NADPH, thereby decreasing the NADPH/NADP ratio. Sorbitol can be further metabolized to fructose with sorbitol dehydrogenase (SORD) using NAD as the electron acceptor. The NADH generated in then oxidized by NADH oxidase to generate reactive oxygen species (ROS). (**b**) In the presence of ROS, the reduced form of the glutathione antioxidant system (GSH) donates electrons and gets oxidized to glutathione disulfide (GSSG). The conversion of GSSG to GSH is catalyzed with the enzyme glutathione reductase (GR) with the use of electrons from NADPH

2008; Quinn et al. 2008). AKR1B10 was shown to play an important role in the reduction of electrophilic carbonyl compounds from the diet or the gut microbiome into more harmless alcohols (Zhong et al. 2009). The enzyme can also reduce the isoprenyl aldehydes farnesal and geranylgeranial into their respective alcohol forms thereby affecting the activity of prenylated mitogenic proteins such as k-Ras (Endo et al. 2011). Additionally, the AKR1B10 catalyzed reduction of retinal to retinol was shown to interfere with retinoic acid signaling and suppress cellular differentiation (Ruiz et al. 2009). The expression of AKR1B10 has been reported to be regulated by p53 and by Nuclear factor erythroid-derived 2-like 2 - Kelch-like ECH associated protein 1 (NRF2-KEAP1) system (Ohashi et al. 2013; Penning 2017) (See Sect. 3.1 for more details). In hepatocellular carcinoma (HCC) cells, induction of oxidative stress was reported to upregulate AKR1B10 via the activation of NRF2 (Liu et al. 2019b). In hereditary and sporadic Type 2 papillary renal cell carcinoma, mutations in the enzyme fumarate hydratase led to the accumulation of fumarate, which could post-translationally inhibit KEAP1 and thereby activate the expression of genes with Antioxidant-Response Elements (ARE) in their promoter (Ooi et al. 2011). One of the genes that was significantly upregulated in the presence of fumarate was AKR1B10 (Ooi et al. 2011). Mitigation of oxidative stress is likely one of the mechanisms for oncogenic properties of AKR1B10 observed in different tumor types, although there continue to be conflicting reports in the literature.

2 AKRs in Cancer

According to The Human Protein Atlas, AKR1B1 is expressed widely is most tissues, and is highly expressed in the adrenal gland and other endocrine tissues such as seminal vesicles and placenta. AKR1B10 is highly expressed in cells of the gastrointestinal tract. We carried out a pan-cancer bioinformatics analysis of mRNA expression of AKR1B1 and AKR1B10 in 20 different cancer types from RNA sequencing data of cancer and matched normal tissue available at The Cancer Genome Atlas (TCGA, Fig. 2).

The expression of AKR1B1 did not change dramatically between cancer and normal tissues in some tumor types. An upregulation in cancer compared to normal was seen in bladder urothelial carcinoma (BLCA), cholangiocarcinoma (CHOL), and kidney renal papillary cell carcinoma (KIRP). Downregulation in cancer compared to normal was seen in breast invasive carcinoma (BRCA), cervical squamous cell carcinoma (CESC), colon adenocarcinoma (COAD), along with a dramatic downregulation in rectal adenocarcinoma (READ), pancreatic adenocarcinoma (PAAD) and pheochromocytoma and paraganglioma (PCPG) and prostate adenocarcinoma (PRAD). Contrary to the mRNA expression; however, the protein expression of AKR1B1 via immunohistochemistry was reported to be increased in many different tumor tissues including basal-like breast cancer (Wu et al. 2017), ovarian (Saraswat et al. 2006), pancreatic (Saraswat et al. 2006) and cervical cancer (Saraswat et al. 2006). This suggests either the presence of extensive post-transcriptional regulation in the expression of AKR1B1, or differences arising from distinct patient cohorts and the number of patients being evaluated.

The expression of AKR1B10 was increased in CESC, CHOL, lung adenocarcinoma (LUAD), lung squamous cell carcinoma (LUSC), PAAD and uterine corpus endometrial carcinoma (UCEC) compared to the corresponding normals (Fig. 2). Very high variation in expression was observed in KIRP and liver hepatocellular carcinoma (LIHC) tumor samples, along with an overall increase in tumor compared to normal samples. Robust reduction in mRNA expression of AKR1B10 was seen in gastrointestinal tumors such as COAD, READ and stomach adenocarcinoma (STAD), along with decreased expression in KIRC, PRAD and PCPG. Of note, published studies on the protein expression of AKR1B10 by immunohistochemistry show remarkable concordance with the mRNA data. Thus, the expression of AKR1B10 was reported to be lost in colorectal cancer and colitis associated cancer (Shen et al. 2015) while hepatocellular (Jin et al. 2016), breast (Ma et al. 2012; Reddy et al. 2017; van Weverwijk et al. 2019), pancreatic (Chung et al. 2012), type 2 papillary renal cell carcinoma (Ooi et al. 2011), and lung cancer (Hung et al. 2018) were reported to have high expression of AKR1B10.

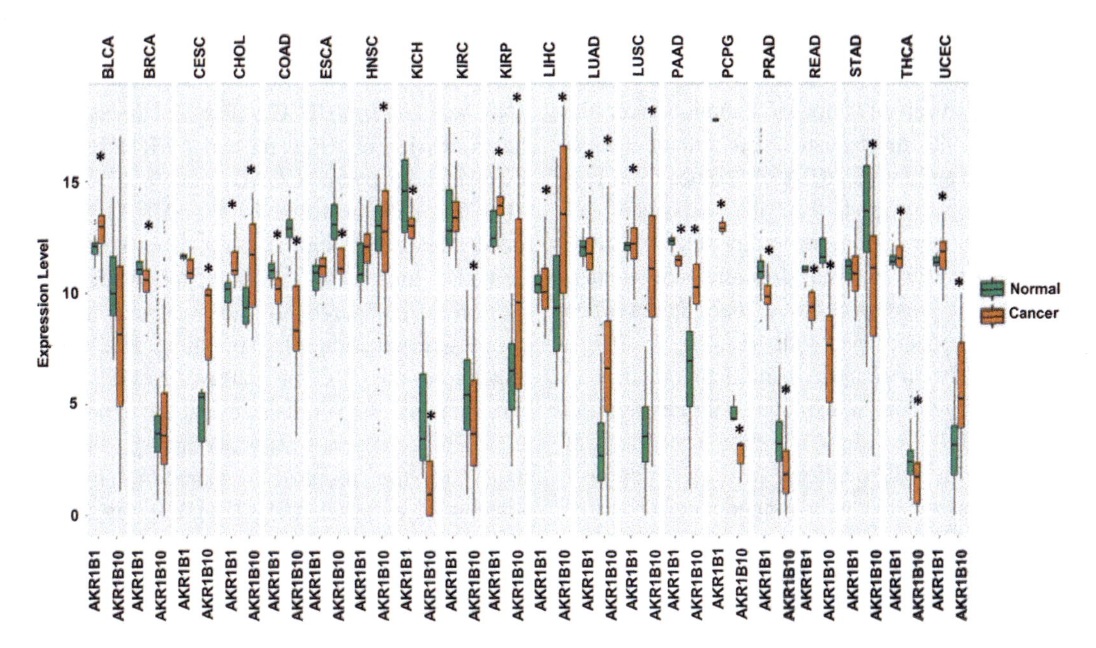

Fig. 2 A pan cancer analysis of AKR1B1 and AKR1B10 expression in matched normal (teal bars) and cancer (orange bars) samples from The Cancer Genome Atlas (TCGA). Raw HTSeq read counts were downloaded from TCGA repository using TCGABiolink package (Colaprico et al. 2016; Mounir et al. 2019) and normalized using DESeq2 (Love et al. 2014) using a general linear model according to tumor cohort and status (tumor vs normal) and statistical significance was determined by Wald's test (*$p < 0.05$). Normalized read counts for AKR1B1 and AKR1B10 were used to construct the boxplots in R

In the following sections I will summarize current findings on AKR1B1 and AKR1B10 in different cancer types. These cancer types were selected on the basis of the high difference in expression of AKR1B1 or AKR1B10 between normal and tumor samples, as well as data available in the literature.

2.1 Hepatocellular Carcinoma (HCC)

Analysis of matched normal and LIHC samples from TCGA (Fig. 2) suggests that the expression of AKR1B1, although statistically significant ($p = 6.22E\text{-}03$), was not dramatically different between normal and tumor tissues, while the expression of AKR1B10 was considerably higher in the tumor tissues compared to normal ($p = 2.86E\text{-}14$). Increased expression of AKR1B10 in HCC tissue or serum samples of HCC patients has been confirmed in several different studies (Distefano and Davis 2019; Ha et al. 2014; Han et al. 2018). However, the expression of AKR1B1 and AKR1B10 was reported to be comparable in a panel of HCC cell lines and the normal like human fetal hepatocyte cell line L-02 (Yang et al. 2013). The same study also showed a significantly higher expression of AKR1B10 in HCC tissue samples from 55 patients compared to matched non-HCC (adjacent) tissues (Yang et al. 2013). Treatment of HepG2 cells, which express AKR1B10, with a combination of the kinase inhibitor sorafenib and the AKR inhibitor epalrestat was shown to enhance the cytotoxicity of sorafenib (Geng et al. 2020). However, epalrestat inhibits both AKR1B10 and AKR1B1 (IC_{50} values of 0.33 and 0.21 μM for AKR1B10 and AKR1B1, respectively, Zhang et al. 2013) and AKR1B1 has been reported to be expressed in HepG2 cells (Zhao et al. 2017). Therefore, whether the enhanced efficacy of sorafenib and epalrestat is unequivocally related to the expression and/or activity of AKRs remains unclear.

The diagnostic biomarker potential and prognostic significance of AKR1B10 expression in HCC has received a lot of attention over the years, but the mechanisms still remain unclear (Distefano and Davis 2019). Outcomes from several studies suggest that the expression of AKR1B10 increases at the earlier stages of HCC tumors, mostly in well differentiated tissues, while the expression is lost at later stages when the cells become metastatic (Liu et al. 2015); however, the mechanism behind this change in expression remains unexplored. High expression of AKR1B10 in tumor resection specimens of patients with HCC predicted favorable prognosis and was associated with longer recurrence-free survival (RFS) and disease-specific survival (DSS) compared to specimens with low AKR1B10 expression (Ha et al. 2014). On the other hand, several studies have shown that in patients with hepatitis C virus (HCV) infection, high expression of AKR1B10 could independently predict the risk of development of HCC (Sato et al. 2012; Semmo et al. 2015). The same risk has also been reported for patients with hepatitis B virus (HBV) infection (Mori et al. 2017).

Mechanistic studies on the role of AKR1B10 in HCC are few; nonetheless, the data so far indicates an oncogenic role of AKR1B10 in HCC. One study reported that AKR1B10 mediated cell cycle progression and proliferation in Hep3B cells while another suggested that miR-383-5p, a tumor suppressive microRNA, could target the 3'UTR of AKR1B10 and lead to its downregulation (Wang et al. 2018). Additionally, a role of interleukin-1 receptor-associated kinase 1 (IRAK1) has been suggested in the ability of AKR1B10 expressing cells to form spheres in 3D culture and develop resistance to sorafenib (Cheng et al. 2018). A role of AKR1B10 in lipid metabolism has also been suggested whereby sphingosine-1 phosphate in conditioned medium of HepG2 cells expressing AKR1B10 was implicated in the higher proliferation of normal hepatocytes cultured in this conditioned medium (Jin et al. 2016).

2.2 Colorectal Carcinoma (CRC)

RNA seq data from TCGA (Fig. 2) suggests that the expression of AKR1B1 and AKR1B10 were lower in both colon (COAD, $p = 9.15E\text{-}05$ for AKR1B1 and $p = 1.29E\text{-}34$ for AKR1B10) and rectal adenocarcinoma (READ, $p = 0.042$ for AKR1B1 and $p = 1.61E\text{-}06$ for AKR1B10) specimens compared to the matched normal samples. Evaluation of expression data across stages from the CRC microarray data GSE39582 (Marisa et al. 2013) indicated that the expression of AKR1B1 did not vary across the different stages of the disease compared to normal, whereas the expression of AKR1B10 was significantly lower in tumor versus normal across all stages (Taskoparan et al. 2017). Supporting the *in silico* data, RT-qPCR analysis also showed no change in AKR1B1 expression and a significant decrease in AKR1B10 expression in CRC versus normal samples (Taskoparan et al. 2017). This was also corroborated by a report of no significant difference in AKR1B1 expression in CRC, normal colon and ulcerative colitis tissue samples (Nakarai et al. 2014). A number of *in vitro* and *in vivo* studies, however, have shown that aldose reductase (AR) inhibitors (ARi) such as Fidarestat or Sorbinil that primarily target AKR1B1, could reduce cell proliferation or colonic polyp formation via inhibition of mitogenic signaling pathways (Ramana et al. 2010; Saxena et al. 2013; Tammali et al. 2011). While a lot of these data were confirmed in animal models, the *in vitro* data involved the use of cell lines such as HT-29 and SW480 that are known to not express any AKR1B1 within the detection limits of a western blot (Taskoparan et al. 2017). Additionally, most ARi do not discriminate between AKR1B1 and AKR1B10 with very high specificity (Zhang et al. 2013). Therefore, the contribution of the specific AKRs on the inhibitory effects of ARi on mitogenic signaling is unclear.

The expression of AKR1B1 and AKR1B10 has prognostic significance in CRC. Previously, low expression of AKR1B10 from four different

CRC datasets was shown to significantly predict poor disease free survival (DFS) and overall survival (OS) (Ohashi et al. 2013). Data from our lab has shown that an AKR signature of high AKR1B1 and low AKR1B10 expression led to a highly significant prognostic stratification (log rank p value <0.001) and was associated with poor DFS independent of age, gender, TNM stage and presence of *KRAS* or *BRAF* mutations (Taskoparan et al. 2017). Validation experiments carried out in a cohort of 51 patients (Demirkol Canli et al. 2020) indicated that high AKR1B1 expression was associated with a shorter OS with borderline significance in colon cancer patients (log rank p = 0.0503). However, the same comparisons carried out with AKR1B10 did not show any statistical difference suggesting that the prognostic power of AKR1B10 was less when compared to AKR1B1 (Taskoparan et al. 2017); therefore, a good stratification was not possible in small cohorts. Moreover, the AKR signature was shown to predict RFS and DFS rather than OS in the microarray data GSE39582 and GSE17536, respectively, suggesting that this signature was more relevant to disease relapse or aggressiveness rather than time to death.

2.3 Breast Carcinoma

TCGA data indicate that the expression of AKR1B1 (p = 2.98E-05) but not AKR1B10 (p = 0.920) was significantly different between matched cancer and normal samples. The expression of 13 different members of the AKR family (including AKR1B1 and AKR1B10) that play a role in drug metabolism were evaluated in breast carcinoma patients treated by neoadjuvant chemotherapy and compared to non-neoplastic tissues (Hlaváč et al. 2014). The expression of AKR1B10, but not AKR1B1, was found to be significantly higher in tumor versus normal samples, particularly in post-menopausal women. The expression of aldose reductases however, did not contribute towards DFS in post-treatment patients in that cohort. In breast tumor samples (n = 100) that were classified into Grades I–III compared to normal tissues (n = 60), the expression of both AKR1B1 and AKR1B10 was reported to be higher across all grades; additionally, the expression of both proteins remained high in the post-chemotherapy specimens (Reddy et al. 2017). The expression of AKR1B1 was shown to be the highest in basal-like breast cancer (BLBC) patients (from transcriptome data) as well as cell lines compared to Luminal A, B, or HER-2 positive cell lines (Wu et al. 2017). BLBC is generally an aggressive cancer that is prone to metastasis and recurrence. Analysis of the TCGA cohort indicated that the expression of AKR1B10 was significantly higher in HER-2 positive and BLBC specimens compared to Luminal A or B patients (van Weverwijk et al. 2019).

A tissue microarray (TMA) study revealed that protein expression of AKR1B10 was seen in 184 of 220 (84%) breast tumors, but not in the normal breast tissue (Ma et al. 2012). The TMA data was supported by western blot and RT-qPCR analyses in an independent cohort. Kaplan Meier plots showed that AKR1B10 expression in the TMA cohort was significantly negatively correlated with OS and disease-related survival. Additionally, early stage breast tumor patients (TNM stage I, tumor <2 cm3, no lymph node or distant metastases) with low AKR1B10 expression showed greater than 90% probability of 25-year survival compared to patients with high expression of AKR1B10 (Ma et al. 2012). De novo fatty acid synthesis is known to be higher in many tumor types for both structural support and signaling functions (Röhrig and Schulze 2016). AKR1B10 was shown to be physically associated with the rate-limiting enzyme of de novo fatty acid synthesis, acetyl-CoA carboxylase-alpha (ACCA), and enhance fatty acid synthesis in breast cancer cells (Ma et al. 2008).

2.4 Lung Cancer

Analysis of expression data from TCGA (Fig. 2) indicates that the expression of AKR1B10 was remarkably higher in the lung adenocarcinoma (LUAD, p = 3.43E-37) and lung squamous carcinoma (LUSC, p = 4.42E-58) samples compared to matched normal samples. The difference in expression of AKR1B1 between cancer and matched normal was statistically significant, but not as robust as AKR1B1. However, a dramatic increase in AKR1B1 expression was reported in smokers compared to non-smokers suggesting that AKR1B1 may play a role in the development of smoking related lung cancer (Schwab et al. 2018). Two microarray datasets (GSE43580 and GSE40588) that included 77 lung adenocarcinoma tissues, 73 lung squamous cell carcinoma (SCC) tissues as well as 60 normal lung tissues adjacent to lung SCC specimens suggested that the expression of AKR1B10 was significantly higher in lung cancer compared to the normal tissues (Zhou et al. 2018). The same study indicated that silencing of AKR1B10 in A549 cells resulted in a decrease in proliferation via reduced mitogenic signaling through the MAP Kinase pathway, cell cycle arrest at G0/G1 and increased apoptosis.

A systems biology approach showed that AKR1B10 was one of three genes (the other two were CYP1A1 and HS3ST3A1) that was expressed significantly more in samples from small airways and large airways of smokers compared to non-smokers (Cubillos-Angulo et al. 2020). The same study also showed that the expression of AKR1B10 was higher in non-small cell lung cancer (NSCLC) specimens from non-smokers compared to normal matched healthy lung tissue, although the sensitivity was low indicating that AKR1B10 may not be a suitable biomarker in a clinical setting (Cubillos-Angulo et al. 2020). Microarray analysis indicated that SCC showed significantly higher expression of AKR1B10 compared to normal tissues from different organs (high expression of AKR1B10 was observed in normal intestinal and stomach tissues, as expected) (Fukumoto et al. 2005). The higher expression of AKR1B10 in SCC specimens could be confirmed at both mRNA and protein levels and the specificity of this expression was comparable to two commonly used diagnostic markers for SCC. Additionally, high AKR1B10 expression was observed in 40 out of 61 (65.6%) NSCLC specimens from smokers (Fukumoto et al. 2005). Of note, studies have also shown a concomitant upregulation of AKR1B10, AKR1C1/C2 and AKR1C3, suggesting that several AKRs may be upregulated in response to the chemicals in cigarette smoke (Macleod et al. 2016).

Exactly why AKR1B10 is upregulated in smokers and smokers who have NSCLC has not been unequivocally demonstrated, although several reasons can be postulated (Penning 2005). Metabolism of polycyclic aromatic hydrocarbons found in cigarette smoke into o-quinones that are both electrophilic and redox-active has been reported in the presence of AKR1C enzymes in A549 cells (Palackal et al. 2002). AKR1C enzymes are co-activated with AKR1B10; therefore, it is possible that the AKR1C family of enzymes, together with, or independent from AKR1B10, can convert the cigarette smoke carcinogens into a more reactive form. A potential role of the master antioxidant response transcription factor NRF2 can also be relevant here (See Sect. 3.1 for more details). Smoking leads to the exposure of lung epithelial or squamous cells to free radicals, ROS, reactive nitrogen species and xenobiotics (Müller and Hengstermann 2012). As a response to this chemical threat, the NRF2 is one of the many pathways that is activated; an upregulation of NRF2 is likely to lead to the upregulation of several different AKRs (Macleod et al. 2016).

AKR1B10 activity can lead to the reduction of all-*trans*-retinal, 9-*cis*-retinal, and 13-*cis*-retinal to the corresponding retinols rather than their oxidation via aldehyde dehydrogenase (ALDH) to retinoic acid. This can deprive retinoic acid receptors RXR and RAR of their ligand and inhibit signaling pathways related to cellular differentiation and apoptosis via RXR-RAR (Penning 2005). The cigarette smoke carcinogen nitrosamine 4-(methylnitrosamino)-1-(3-pyridyl)-1-butanone (NNK, nicotine-derived nitrosamine ketone) was reported to be detoxified by carbonyl reduction to the corresponding aldehyde NNAL in the presence of AKR1B10 (Stapelfeld et al. 2017). It is still unclear, however, whether AKR1B10 can directly metabolize chemicals in cigarette smoke condensates into a more carcinogenic form.

2.5 Pancreatic Adenocarcinoma (PAAD)

PAAD data from the TCGA suggest that the expression of AKR1B1 was significantly decreased ($p = 0.081$) while the expression AKR1B10 was significantly increased ($p = 0.023$) in tumor samples compared to the matched normal samples (Fig. 2). Several studies in the literature have suggested a role of AKR1B10 in pancreatic cancer. In an IHC based evaluation of AKR1B10 expression, morphologically normal pancreatic parenchyma was observed to be negative for AKR1B10 staining. Invasive pancreatic ductal adenocarcinomas were found to stain positively (35 out of 50 specimens, 70%) for AKR1B10 while the adjacent morphologically normal ductal epithelial cells were negative. Most of the tumors that stained positively for AKR1B10 were well- and moderately differentiated adenocarcinomas; only 1 out of 3 poorly differentiated carcinomas exhibited positive staining for AKR1B10. Moreover, 60% of PAAD patients with tumors showing overexpression of AKR1B10 were smokers with a smoking history of longer than 10 years (Chung et al. 2012). IHC evaluation of cell blocks from fine needle aspiration of cystic pancreatic lesions showed high AKR1B10 expression in epithelial tissues of 45 out of 46 (98%) mucinous lesions. None of the non-mucinous samples ($n = 59$) showed any expression of AKR1B10, giving a remarkable specificity of 100%, making a strong case for AKR1B10 as a biomarker for mucinous tumors that can be tested non-invasively (Connor et al. 2017).

PDAC is known to be one of the most Ras addicted cancers and nearly 95% of pancreatic tumors show point mutations in G12, G13, or Q61 residues of the k-Ras protein, making it constitutively active (Bryant et al. 2014). shRNA mediated knockdown of AKR1B10 in the CD18 pancreatic cancer cell line was shown to reduce the levels of GTP bound active Ras in the membrane, along with reduced signaling via the Ras-MAP Kinase pathway (Zhang et al. 2014). It is very likely that with the knockdown of AKR1B10, reduced prenylation of k-Ras protein resulted in decreased signaling via the MAPK pathway. AKR1B10 is known to catalytically reduce farnesyl and geranylgeranyl to farnesol and geranylgeraniol; these intermediates can be phosphorylated to farnesyl and geranylgeranyl pyrophosphates that are, in turn, highly involved in protein prenylation (Li et al. 2013).

3 Molecular Mechanisms and Dysregulated AKR Signaling in Cancer

3.1 Oxidative Stress, Inflammation and Chemoresistance

Reactive oxygen species (ROS) are generated both intracellularly from several endogenous pathways and from various exogenous sources. ROS levels are generally under very tight cellular control as they can regulate diverse functions including transcriptional regulation and signal transduction (Corcoran and Cotter 2013). Uncontrolled production of ROS, either through exposure to exogenous chemicals or through deregulated cellular pathways, leads to oxidative stress, which has been implicated in many human diseases including cancer (Kakehashi et al. 2013). Cancer cells generally show elevated levels of ROS as well as activation of ROS-sensitive signaling pathways; these pathways are generally pro-survival and enable metabolism as well as inflammation (Reczek and Chandel 2015). ROS molecules such as hydrogen peroxide, can act as second messengers in cellular signaling via reversible oxidation of signaling proteins such as kinases and phosphatases, including proteins in the mitogen-activated protein (MAP) kinase/ERK, phosphoinositide-3-kinase (PI3K)/Akt, and the nuclear factor κ-B (NF-κB) cascades (Liou and Storz 2010). Very high levels of ROS can induce cell death via various mechanisms such as necrosis, apoptosis or ferroptosis and is often utilized in radiotherapy and chemotherapy, which kill cancer cells by inducing high levels of ROS (Rojo de la Vega et al. 2018).

NRF2 is a cap'n'collar leucine zipper transcription factor that also acts as a ROS sensor and its protein level is very tightly regulated through proteasome-dependent proteolysis. KEAP1 is a negative regulator of NRF2 that binds to and enhances ubiquitin-dependent proteasomal degradation of NRF2 under normal reducing conditions. Reactive cysteine residues in KEAP1 get modified upon exposure to ROS or electrophilic toxins and act as sensors of oxidative stress. This modification also leads to an escape of NRF2 from ubiquitylation and proteosomal degradation, thereby allowing it to transcribe numerous genes that encode detoxifying and antioxidant enzymes (Taguchi and Yamamoto 2021).

Somatic mutations in NRF2 and KEAP1 have been identified in several cancer types that result in constitutive activation of NRF2 through disruptions in protein-protein interactions; accumulation of NRF2 enable cancer cells to survive despite the presence of high levels of ROS. Such mutations are very common in NSCLC; these tumors are therefore considered to be NRF2 addicted (Singh et al. 2006). An antioxidant response element (ARE) has been identified in the promoter region of AKR1B10 and other proteins of the AKRC family (Nishinaka et al. 2011), suggesting that high NRF2 accumulation in NSCLC can be implicated, at least in part, to the high expression of AKR1B10 in this tumor type. In fact, the AKR gene response is considered to be one of the most robust signatures of NRF2 activation in humans. The high oxidative stress and NRF2 mediated expression of AKR1B10 in HCC was associated, via oxidoreduction and detoxification, with an amelioration of ROS mediated damage in these cells (Liu et al. 2019b), most likely leading to cell survival.

Glutathione (GSH) is one of the most important endogenous cellular antioxidants (Fig. 1b). NRF2 is one of the key regulators of GSH metabolism as the enzymes glutamate-cysteine ligase and glutathione synthetase are target genes of NRF2 (Finkel 2011). SLC7A11, the light chain subunit of the Xc- antiporter system (xCT) that helps to import cysteine into the cell in exchange of glutamate was also reported to be an NRF2 target (Habib et al. 2015). The imported cysteine can then be utilized in the synthesis of glutathione (Fig. 1b). Maintenance of the glutathione antioxidant system in a reduced state [high reduced glutathione/oxidized glutathione (GSH/GSSG) ratio] requires the presence of NADPH. AKRs utilize NADPH as donors of electrons; increased activity of the polyol pathway (Fig. 1a) was shown to decrease the NADPH/NADP ratio and attenuate the glutathione reductase (GR) and glutathione peroxidase (GPx) system. This, in turn, led to a decrease in the GSH/GSSG ratio, leading to oxidative stress and complications in diabetes models (Srivastava et al. 2005). Whether the altered GSH/GSSG ratio in cancer cells expressing high AKRs contributes to carcinogenesis has not been unequivocally shown.

Oxidative stress can lead to lipid peroxidation via an attack of ROS on polyunsaturated fatty acids (PUFA) in the plasma membrane to form various reactive and cytotoxic aldehydes such as 4-hydroxy-trans-2-nonenal (HNE). 4HNE is a highly toxic aldehyde with a long half-life, can bind to many macromolecules and is associated with oxidative stress related disorders such as cancer and neurodegenerative diseases (Uchida 2003). In fact, the levels of 4HNE were reported to be high in primary CRC, making this tumor highly reliant on oxidative stress for progression (Skrzydlewska et al. 2005). Treatment of lung cancer cells with low doses of 9,10-phenanthrenequinone (9,10-PQ), a major quinone in diesel exhaust, was shown to increase cell proliferation and survival, which could be reversed upon co-treatment with an AKR1B10 inhibitor as well as the antioxidant N-acetyl cysteine. Mechanistically, it was suggested that 9,10-PQ led to an upregulation in ROS levels and the expression of AKR1B10, which in turn activated MAP Kinase signaling and activation of several metalloproteases (Matsunaga et al. 2014a).

Carbonyl compounds such as 4HNE and its glutathione conjugate (GS-HNE; a phase II reaction that can occur either spontaneously or enzymatically via glutathione-S-transferase), can be reduced by both AKR1B1 and AKR1B10 to their corresponding alcohols (Martin and Maser 2009; Srivastava et al. 1995). However, AKR1B1 was shown to reduce GSH-conjugated carbonyls efficiently while AKR1B10 was more efficient in reducing free carbonyls (Shen et al. 2011). Such a detoxification reaction was reported in HCT-8 CRC cells; moreover, individuals with low intestinal expression of AKR1B10 were suggested to be at a greater risk of development of cancer in the presence of dietary and microbial derived carbonyl carcinogens (Zu et al. 2016). Reduced carbonyl compounds, however, can serve as inflammatory signals (Srivastava et al. 2011). Thus, it was shown that the AKR1B1 inhibitor fidarestat could ameliorate inflammatory signals induced by tumor necrosis factor alpha, lipopolysaccharide and environmental allergens in different disease models such as diabetes, cardiovascular disease, uveitis, asthma, and carcinomas (Srivastava et al. 2011). In a seminal manuscript, treatment of murine peritoneal macrophages with three different ARis, sorbinil, tolrestat or zopolrestat, was shown to suppress LPS-induced production of pro-inflammatory molecules such as tumor necrosis factor (TNF)-α, interleukin (IL)-6, IL-1β, IFN-γ and MCP-1 in along with the production of nitric oxide, and prostaglandin E2 as well as the expression of iNOS and COX-2 proteins (Ramana and Srivastava 2006). Subsequently, it was shown that incubation of the CRC cell line Caco-2 with the ARi sorbinil could inhibit growth factor induced activation of NF-κB and protein kinase C (PKC), as well as the expression and activity of the inflammatory enzyme COX-2 (Tammali et al. 2006). The same study showed that treatment of Caco-2 cells with 4HNE, GS-HNE or the AKR1B1 catalyzed reduction product of GS-HNE, glutathionyl-1,4- dihydroxynonane (GS-DHN), resulted in increased COX-2 expression and prostaglandin E2 production (Tammali et al. 2006). We have shown that shRNA mediated stable silencing of AKR1B1 in HCT-116 cells resulted in reduced nuclear localization and transcriptional activity of NF-κB. Additionally, re-expression of AKR1B10 in two different cell lines (LoVo and HCT-116) that do not endogenously express any AKR1B10 was shown to reduce nuclear translocation and transcriptional activity of NF-κB (Taskoparan et al. 2017). This was also supported by a significant positive correlation between the expression of AKR1B1 and a number of key inflammatory genes, while a significant negative correlation was observed between the same genes and the expression of AKR1B10 in the TCGA colon adenocarcinoma - rectal adenocarcinoma (COAD-READ) samples (Taskoparan et al. 2017). These data also support our finding that a gene signature of high AKR1B1 and low AKR1B10 expression in CRC was significantly associated with poor disease free survival in *in silico* and *ex-vivo* samples (Demirkol Canli et al. 2020).

Doxorubicin (Dox)-resistant MKN45 gastric cancer cells showed a remarkable increase in the expression of AKR1B10; treatment of these cells with the specific AKR1B10 inhibitor oleanolic acid resulted in re-sensitization of the cells to Dox. Mechanistically, an upregulation of NRF2 was considered to lead to the upregulation of AKR1B10; NRF2 activation, in turn, could ameliorate ROS induced cell death via Dox (Morikawa et al. 2015). In another study, resistance of A549 lung cancer cells to cisplatin, which also kills cancer cells via the formation of DNA adducts and ROS, was attributed to increased synthesis of nitric oxide (NO) via inducible NO synthase (iNOS). NO mediated upregulation of AKR1B10 was implicated in the reduction of cytotoxic aldehydes such as 4HNE and 4ONE (generated with cisplatin treatment) to their less toxic forms, thereby increasing resistance to death (Matsunaga et al. 2014b). On the other hand, AKR1B1 inhibition with ARi in CRC cell lines was shown to confer sensitivity to tumor necrosis factor (TNF)-related apoptosis-inducing ligand (TRAIL)- induced apoptosis by increasing intrinsic apoptosis pathways and death receptor signaling (Shoeb et al. 2013). AKR expression in the tumor microenvironment is also of high relevance. AKR activity in the adipose tissue was implicated in the metabolism of daunorubicin to daunorubicinol, reducing its antileukemia effect (Sheng et al. 2017).

Overall, the current studies suggest that the oncogenic role of AKR1B10 in many different tumor types may be related to the direct enzymatic conversion of drugs to a less effective form or products of drug activity to a form that contributes towards cell survival. An exceptional role of AKR1B10 as a tumor suppressor in the gastrointestinal (GI) tract warrants further attention. It has been suggested that the GI tract is exposed to a number of carbonyl compounds either from food, or from the vast microbiome and that AKR1B10 expression in the intestine reduces and detoxifies these carbonyls to their alcohol forms, thereby protecting the tissues from damage (Shen et al. 2015). However, exactly how high expression of AKR1B10 in colorectal tumors leads to better prognosis is still unclear.

3.2 Epithelial to Mesenchymal Transition (EMT)

Several recent studies have implicated AKRs in EMT, cellular motility and metastasis, suggesting an important role of cellular metabolism in metastasis and drug resistance. AKR1B1 expression was shown to be a direct transcriptional target of the mesenchymal transcription factor Twist2; moreover, the expression of AKR1B1 and was shown to be correlated with several other EMT proteins, leading to enhanced motility and enrichment of stem cell markers in BLBC cells (Wu et al. 2017). In the same study, it was shown that the conversion of prostaglandin (PG) H_2 to $PGF_{2\alpha}$ in the presence of AKR1B1 could enhance the invasion and migration of BLBC cells (Wu et al. 2017). Analysis of data from the NCI-60 panel indicated the presence of a strong positive correlation between a ratio of the expression of Vimentin/E-cadherin (VIM/CDH1) and AKR1B1, along with several other mesenchymal genes (Schwab et al. 2018). The same study showed that the expression of AKR1B1 was higher in mesenchymal cell lines. Moreover, shRNA mediated knockdown of AKR1B1 was shown to reduce motility and migration of different cancer cell lines along with a reduction in the expression of stem cell markers when grown as 3D spheroids (Schwab et al. 2018). Additionally, knockdown of SORD, the enzyme that converts sorbitol to fructose (Fig. 1a), could phenocopy the effects of AKR1B1 knockdown strongly implicating a role of metabolism of glucose via the polyol pathway in enhancing the metastatic behavior of cells (Schwab et al. 2018). Using expression data from COAD and READ samples in TCGA, we have shown a strong positive correlation between the mRNA expression of AKR1B1 and several mesenchymal markers (VIM, TWIST1, TWIST2, SNAI1, SNAI2, ZEB1, ZEB2) and a negative correlation between AKR1B1 and CDH1. A significant negative correlation was observed between AKR1B10 and the mesenchymal markers (Demirkol Canli et al. 2020). Additionally, the positive correlation between the expression of AKR1B1 and VIM and the negative correlation in expression between AKR1B1 and CDH1 could be confirmed by RT-qPCR using RNA from fresh frozen CRC tumor samples (Demirkol Canli et al. 2020).

Administration of nanoparticles containing the phostosensitizer Chlorin e6 (Ce6) for photodynamic therapy and the ARi epalrestat in BLBC tumor-bearing SCID mice was shown to dramatically reduce expression of mesenchymal and stem cell markers in the primary tumor and also significantly reduce lung metastasis. An increase in ROS generation with Ce6 and inhibition of EMT with epalrestat were considered as the mechanism for lower metastasis (Zhang et al. 2021). Similarly, treatment of the colon cancer cell line HT-29 with the ARi sorbinil or fidarestat resulted in reduced invasion and migration of these cells, which was also reflected by reduced metastasis to the liver in rodents when the ARi treated cells were injected to the spleen (Saxena et al. 2012). Our group developed an EMT score based on the expression of CDH1 and VIM whereby a higher score indicated mesenchymal phenotype while a lower score indicated an epithelial phenotype (Demirkol 2018). We observed strong statistically significant correlation between the EMT score and AKR1B1 expression and a weak but statistically significant negative correlation between EMT score and AKR1B10 expression in CRC microarray datasets and *ex vivo* tumor specimens (Demirkol Canli et al. 2020).

An shRNA screen of putative enhancers of breast cancer metastasis in a mouse model showed strong enrichment of *Akr1b8* (mouse homolog of AKR1B10) along with other well defined regulators of breast cancer progression and metastasis (van Weverwijk et al. 2019). Silencing of *Akr1b8* in the mouse mammary epithelial cells 4 T1 cell line did not show any difference in proliferation, but showed significantly lower lung colonization upon intravenous administration. Furthermore, orthotopic inoculation of the *Akr1b8* silenced cells into the fourth mammary fat pad of BALB/c mice did not affect tumor growth at the primary site but reduced lung metastasis significantly (van Weverwijk et al. 2019). Similarly, ectopic expression of AKR1B10 in the MDA-MB-231 human mammary cancer cell line resulted in increased tumor burden in the lung compared to control cells (van Weverwijk et al. 2019). *In silico* analysis of an unclassified breast cancer dataset indicated reduced distant metastasis-free survival in patients with low expression of AKR1B10 (van Weverwijk et al. 2019). AKR1B10 was also shown to be upregulated in brain metastasis samples of NSCLC; silencing of AKR1B10 in the PC9-BrM3 NSCLC cancer cells reduced their extravasation through the blood brain barrier via a downregulation of matrix metalloproteinase (MMP)-2 and MMP-9 through reduced activation of the MEK/ERK signaling pathway (Liu et al. 2019a).

3.3 Nutrient Sensing Pathways

Most cancer cells undergo metabolic reprogramming whereby cells preferentially use glycolysis for the generation of energy even in the presence of oxygen and synthesize lactate via the so called Warburg effect (Heiden et al. 2009). The lactate is then released from the cell into the circulation, preventing its conversion to pyruvate and oxidative phosphorylation in cancer cells. Interestingly recent studies suggest that lactate from the circulation can be taken up again by tumor cells, converted to pyruvate, which can then enter the TCA cycle (Faubert et al. 2017). Aberrant activation of the polyol pathway in the presence of excess glucose can enhance oxidative stress and inflammation, thereby increasing the risk of cancer (Tammali et al. 2011). Fructose is known to promote tumorigenesis, in part by inducing metabolic reprogramming and enhanced production of lactate (Liu and Heaney 2011) and is a product of hyperactivation of the polyol pathway (Fig. 1b). SORD, the enzyme of the polyol pathway that catalyzes the synthesis of fructose, was implicated in enhanced metastatic behavior of CRC cells (Schwab et al. 2018).

AKR1B1 was shown to physically interact with AKT1 and overexpression of AKR1B1 in HepG2 cells resulted in increased phosphorylation of AKT1, mTORC1 and pyruvate kinase M2 (PKM2) as well as increased synthesis of lactate and activity of lactate dehydrogenase (LDH) (Zhao et al. 2017). Supporting these data, inhibition of AKR1B1 with sorbinil prevented the growth factor mediated activation of phosphoinositide 3-kinase/AKT and ROS generation in colon cancer cells (Ramana et al. 2010). Treatment with ARi also led to the inhibition of miR-21 leading to the reactivation of the putative miR-21 target PTEN (Saxena et al. 2012) and PDCD4 (Saxena et al. 2013), along with an inhibition of several proteins of the mTOR pathway (Saxena et al. 2013). Drugs such as gedunin (Kishore et al. 2016) and (−)-Kusunokinin (Tanawattanasuntorn et al. 2021) that are known to inhibit signaling via the AKT pathway were shown to mediate this inhibition via the direct binding and strong inhibition of AKR1B1. Reverse phase protein array (RPPA) data from our lab indicated that *ex vivo* human colon tumor specimens expressing high levels of AKR1B1 were significantly associated with hyper-activation of several members of the AKT and GSK-3β family (Demirkol Canli et al. 2020).

AKR1B10 is known to be a poor reductant of glucose and does not participate in the polyol pathway. AKR1B10 over-expressing breast cancer cells were shown to rely on fatty acid oxidation for energy rather than glycolysis and were able to survive even when grown in the presence of low glucose (van Weverwijk et al. 2019). AKR1B10 was also shown to enhance fatty acid synthesis in breast cancer cells via physical association with acetyl-CoA carboxylase-alpha (ACCA), the rate-limiting enzyme of de novo fatty acid synthesis (Ma et al. 2008). It is highly unlikely for both fatty acid synthesis and fatty acid oxidation to occur concurrently in cells, suggesting the presence of exquisite regulation that is yet to be identified.

Enhanced fatty acid oxidation was also implicated in increased metastatic colonization of breast cancer tumor cells expressing high levels of AKR1B10 to the lungs (van Weverwijk et al. 2019). These data are suggestive of a highly divergent role of AKR1B10 on energy sensing pathways when compared to AKR1B1. This was reflected in RPPA data on colon tumor specimens whereby tumors with high expression of AKR1B10 but not AKR1B1 was associated with reduced phosphorylation of p70S6 kinase (indicative of reduced mTORC1 activation) and increased expression of PDCD4 (Demirkol Canli et al. 2020).

4 Conclusions and Future Perspectives

Metabolic perturbations in cancer cells have been increasingly implicated in many oncogenic processes suggesting a close association between the production and/or regulation of the levels of metabolites and mitogenic signaling. Although the AKRs have been studied extensively for diabetic complications, studies in the last decade, mostly through the use of existing AKR inhibitors, indicate the presence of extensive crosstalk between AKRs and inflammation, oxidative stress, drug metabolism, chemoresistance and EMT in many different tumor types. Nonetheless, several outstanding questions have remained in the field. One aspect of AKR1B1 signaling that has not been addressed to date is the synthesis of fructose via the polyol pathway. SORD was reported to phenocopy AKR1B1 in the induction of EMT (Schwab et al. 2018), providing tantalizing clues on a potential role of fructose in AKR1B1 mediated oncogenic signaling. Moreover, through reversible enzymatic reactions, a conversion of fructose to glucose in conditions of low glucose and low NADPH can be envisaged, suggesting the activation of glycolysis in hypoxic and acidic microenvironments (Krause and Wegner 2020). However, whether this pathway is active in tumor cells is currently unclear. Recent studies have also indicated the activation of nutrient sensing pathways via AKRs, although it is currently unknown whether this aids in tumor development or whether it can be targeted for therapeutic purposes. Yet another unsolved question in the AKR literature is the mechanism behind highly disparate outcomes of AKR1B10 expression in CRC versus HCC or NSCLC. It is highly likely that ROS levels and NRF2 mediated signaling play important roles in both HCC and NSCLC; nonetheless, it is unclear whether such signaling processes exist in CRC and if so, why the outcomes of such a signaling pathway are so different between the different tumor types. It is clear that much still needs to be learnt about this highly versatile group of enzymes and their role in cancer. The recent resurgence of interest in metabolic signaling in cancer as well as excellent manuscripts that have utilized unbiased screening approaches to highlight the importance of AKRs in many signaling pathways relevant to cancer may encourage scientists to tackle the long existing problem of development of highly specific clinically relevant inhibitors of AKR1B1 and AKR1B10.

Acknowledgements Ilir Sheraj is gratefully acknowledged for analyzing TCGA data and generating Fig. 2. Cagdas Ermis is gratefully acknowledged for generating the images used in Fig. 1. Dr. Seçil Demirkol Canli, Esin Gulce Seza, Ilir Sheraj, Ismail Guderer and Cagdas Ermis are gratefully acknowledged for critically reading the manuscript. This work was supported by TUBITAK project 118Z688 and the COST action CA17118.

References

Bryant KL, Mancias JD, Kimmelman AC, Der CJ (2014) KRAS: feeding pancreatic cancer proliferation. Trends Biochem Sci 39:91–100. https://doi.org/10.1016/j.tibs.2013.12.004

Cheng BY et al (2018) Irak1 augments cancer stemness and drug resistance via the ap-1/akr1b10 signaling cascade in hepatocellular carcinoma. Cancer Res 78:2332–2342. https://doi.org/10.1158/0008-5472.CAN-17-2445

Chung YT et al (2012) Overexpression and oncogenic function of aldo-keto reductase family 1B10 (AKR1B10) in pancreatic carcinoma. Mod Pathol 25:758–766. https://doi.org/10.1038/modpathol.2011.191

Colaprico A et al (2016) TCGAbiolinks: an R/Bioconductor package for integrative analysis of TCGA data. Nucleic Acids Res 44(8):e71. https://doi.org/10.1093/nar/gkv1507

Connor JP, Esbona K, Matkowskyj KA (2017) AKR1B10 expression by immunohistochemistry in surgical resections and fine needle aspiration cytology material in patients with cystic pancreatic lesions; potential for improved nonoperative diagnosis. Hum Pathol 70:77–83. https://doi.org/10.1016/j.humpath.2017.10.006

Corcoran A, Cotter TG (2013) Redox regulation of protein kinases. FEBS J 280:1944–1965. https://doi.org/10.1111/febs.12224

Cubillos-Angulo JM et al (2020) Systems biology analysis of publicly available transcriptomic data reveals a critical link between AKR1B10 gene expression, smoking and occurrence of lung cancer. PLoS One 15(2): e0222552. https://doi.org/10.1371/journal.pone.0222552

Demirkol S (2018) Prediction of prognosis and chemosensitivity in gastrointestinal cancers. Bilkent University. Retrieved from http://repository.bilkent.edu.tr/handle/11693/35716

Demirkol Canli S et al (2020) Evaluation of an aldo-keto reductase gene signature with prognostic significance in colon cancer via activation of epithelial to mesenchymal transition and the p70S6K pathway. Carcinogenesis 41:1219–1228. https://doi.org/10.1093/carcin/bgaa072

Distefano JK, Davis B (2019) Diagnostic and prognostic potential of akr1b10 in human hepatocellular carcinoma. Cancers 11:1–13. https://doi.org/10.3390/cancers11040486

Endo S et al (2011) Roles of rat and human aldo-keto reductases in metabolism of farnesol and geranylgeraniol. Chem Biol Interact 191:261–268. https://doi.org/10.1016/j.cbi.2010.12.017

Faubert B et al (2017) Lactate metabolism in human lung tumors. Cell 171:358–371.e9. https://doi.org/10.1016/j.cell.2017.09.019

Finkel T (2011) Signal transduction by reactive oxygen species. J Cell Biol 194:7–15. https://doi.org/10.1083/jcb.201102095

Fukumoto SI et al (2005) Overexpression of the aldo-keto reductase family protein AKR1B10 is highly correlated with smokers' non-small cell lung carcinomas. Clin Cancer Res 11:1776–1785. https://doi.org/10.1158/1078-0432.CCR-04-1238

Gallego O et al (2007) Structural basis for the high all-trans-retinaldehyde reductase activity of the tumor marker AKR1B10. Proc Natl Acad Sci U S A 104:20764–20769. https://doi.org/10.1073/pnas.0705659105

Geng N, Jin Y, Li Y, Zhu S, Bai H (2020) AKR1B10 inhibitor epalrestat facilitates sorafenib-induced apoptosis and autophagy via targeting the mtor pathway in hepatocellular carcinoma. Int J Med Sci 17:1246–1256. https://doi.org/10.7150/ijms.42956

Ha SY, Song DH, Lee JJ, Lee HW, Cho SY, Park CK (2014) High expression of aldo-keto reductase 1B10 is an independent predictor of favorable prognosis in patients with hepatocellular carcinoma. Gut Liver 8:648–654. https://doi.org/10.5009/gnl13406

Habib E, Linher-Melville K, Lin HX, Singh G (2015) Expression of xCT and activity of system xc- are regulated by NRF2 in human breast cancer cells in response to oxidative stress. Redox Biol 5:33–42. https://doi.org/10.1016/j.redox.2015.03.003

Han C et al (2018) Immunohistochemistry detects increased expression of Aldo-Keto reductase family 1 member B10 (AKR1B10) in early-stage hepatocellular carcinoma. Med Sci Monit 24:7414–7423. https://doi.org/10.12659/MSM.910738

Heiden MGV, Cantley LC, Thompson CB (2009) Understanding the Warburg effect: the metabolic requirements of cell proliferation. Science 324:1029–1033. https://doi.org/10.1126/science.1160809

Hlaváč V et al (2014) The role of cytochromes P450 and aldo-keto reductases in prognosis of breast carcinoma patients. Medicine (Baltimore) 93:e255. https://doi.org/10.1097/MD.0000000000000255

Hung JJ, Yeh YC, Hsu WH (2018) Prognostic significance of AKR1B10 in patients with resected lung adenocarcinoma. Thorac Cancer 9:1492–1499. https://doi.org/10.1111/1759-7714.12863

Ji J et al (2020) The AKR1B1 inhibitor epalrestat suppresses the progression of cervical cancer. Mol Biol Rep 47:6091–6103. https://doi.org/10.1007/s11033-020-05685-z

Jin J, Liao W, Yao W, Zhu R, Li Y, He S (2016) Aldo-keto reductase family 1 member B 10 mediates liver cancer cell proliferation through sphingosine-1-phosphate. Sci Rep 6:22746. https://doi.org/10.1038/srep22746

Kakehashi A, Wei M, Fukushima S, Wanibuchi H (2013) Oxidative stress in the carcinogenicity of chemical carcinogens. Cancers (Basel) 5:1332–1354. https://doi.org/10.3390/cancers5041332

Kishore TKK, Ganugula R, Gade DR, Reddy GB, Nagini S (2016) Gedunin abrogates aldose reductase, PI3K/Akt/mToR, and NF-κB signaling pathways to inhibit angiogenesis in a hamster model of oral carcinogenesis. Tumour Biol 37:2083–2093. https://doi.org/10.1007/s13277-015-4003-0

Krause N, Wegner A (2020) Fructose metabolism in cancer. Cell 9:2635. https://doi.org/10.3390/cells9122635

Li H, Yang AL, Chung YT, Zhang W, Liao J, Yang GY (2013) Sulindac inhibits pancreatic carcinogenesis in LSL-krasg12d-LSL-Trp53R172H-Pdx-1-Cre mice via suppressing aldo-keto reductase family 1B10 (AKR1B10). Carcinogenesis 34:2090–2098. https://doi.org/10.1093/carcin/bgt170

Liou GY, Storz P (2010) Reactive oxygen species in cancer. Free Radic Res 44:479–496. https://doi.org/10.3109/10715761003667554

Liu H, Heaney AP (2011) Refined fructose and cancer. Expert Opin Ther Targets 15:1049–1059. https://doi.org/10.1517/14728222.2011.588208

Liu TA et al (2015) Regulation of Aldo-keto-reductase family 1 B10 by 14-3-3 epsilon and their prognostic impact of hepatocellular carcinoma. Oncotarget 6:38967–38982. https://doi.org/10.18632/oncotarget.5734

Liu W et al (2019a) AKR1B10 (Aldo-keto reductase family 1 B10) promotes brain metastasis of lung cancer cells in a multi-organ microfluidic chip model. Acta Biomater 91:195–208. https://doi.org/10.1016/j.actbio.2019.04.053

Liu Y et al (2019b) Compensatory upregulation of aldo-keto reductase 1B10 to protect hepatocytes against oxidative stress during hepatocarcinogenesis. Am J Cancer Res 9:2730–2748

Love MI, Huber W, Anders S (2014) Moderated estimation of fold change and dispersion for RNA-seq data with DESeq2. Genome Biol 15:550. https://doi.org/10.1186/s13059-014-0550-8

Ma J, Yan R, Zu X, Cheng JM, Rao K, Liao DF, Cao D (2008) Aldo-keto reductase family 1 B10 affects fatty acid synthesis by regulating the stability of acetyl-CoA carboxylase-α in breast cancer cells. J Biol Chem 283:3418–3423. https://doi.org/10.1074/jbc.M707650200

Ma J et al (2012) AKR1B10 overexpression in breast cancer: association with tumor size, lymph node metastasis and patient survival and its potential as a novel serum marker. Int J Cancer 131:E862–E871. https://doi.org/10.1002/ijc.27618

Macleod AK et al (2016) Aldo-keto reductases are biomarkers of NRF2 activity and are co-ordinately overexpressed in non-small cell lung cancer. Br J Cancer 115:1530–1539. https://doi.org/10.1038/bjc.2016.363

Marisa L et al (2013) Gene expression classification of colon cancer into molecular subtypes: characterization, validation, and prognostic value. PLoS Med 10:e1001453. https://doi.org/10.1371/journal.pmed.1001453

Martin HJ, Maser E (2009) Role of human aldo-keto-reductase AKR1B10 in the protection against toxic aldehydes. Chem Biol Interact 178:145–150. https://doi.org/10.1016/j.cbi.2008.10.021

Matsunaga T et al (2014a) Exposure to 9,10-phenanthrenequinone accelerates malignant progression of lung cancer cells through up-regulation of aldo-keto reductase 1B10. Toxicol Appl Pharmacol 278:180–189. https://doi.org/10.1016/j.taap.2014.04.024

Matsunaga T et al (2014b) Nitric oxide confers cisplatin resistance in human lung cancer cells through upregulation of aldo-keto reductase 1B10 and proteasome. Free Radic Res 48:1371–1385. https://doi.org/10.3109/10715762.2014.957694

Mindnich RD, Penning TM (2009) Aldo-keto reductase (AKR) superfamily: genomics and annotation. Hum Genomics 3:362–370. https://doi.org/10.1186/1479-7364-3-4-362

Mori M et al (2017) Aldo-keto reductase family 1 member B10 is associated with hepatitis B virus-related hepatocellular carcinoma risk. Hepatol Res 47:E85–E93. https://doi.org/10.1111/hepr.12725

Morikawa Y et al (2015) Acquisition of doxorubicin resistance facilitates migrating and invasive potentials of gastric cancer MKN45 cells through up-regulating Aldo-keto reductase 1B10. Chem Biol Interact 230:30–39. https://doi.org/10.1016/j.cbi.2015.02.005

Mounir M et al (2019) New functionalities in the TCGAbiolinks package for the study and integration of cancer data from GDC and GTEx. PLoS Comput Biol 15(3):e1006701. https://doi.org/10.1371/journal.pcbi.1006701

Müller T, Hengstermann A (2012) Nrf2: friend and foe in preventing cigarette smoking-dependent lung disease. Chem Res Toxicol 25:1805–1824. https://doi.org/10.1021/tx300145n

Nakarai C et al (2014) Expression of AKR1C3 and CNN3 as markers for detection of lymph node metastases in colorectal cancer. Clin Exp Med 15:333–341. https://doi.org/10.1007/s10238-014-0298-1

Nishinaka T, Miura T, Okumura M, Nakao F, Nakamura H, Terada T (2011) Regulation of aldo-keto reductase AKR1B10 gene expression: involvement of transcription factor Nrf2. Chem Biol Interact 191:185–191. https://doi.org/10.1016/j.cbi.2011.01.026

Ohashi T, Idogawa M, Sasaki Y, Suzuki H, Tokino T (2013) AKR1B10, a transcriptional target of p53, is downregulated in colorectal cancers associated with poor prognosis. Mol Cancer Res 11:1554–1563. https://doi.org/10.1158/1541-7786.MCR-13-0330-T

Ooi A et al (2011) An antioxidant response phenotype shared between hereditary and sporadic type 2 papillary renal cell carcinoma. Cancer Cell 20:511–523. https://doi.org/10.1016/j.ccr.2011.08.024

Palackal NT, Lee SH, Harvey RG, Blair IA, Penning TM (2002) Activation of polycyclic aromatic hydrocarbon trans-dihydrodiol proximate carcinogens by human aldo-keto reductase (AKR1C) enzymes and their functional overexpression in human lung carcinoma (A549) cells. J Biol Chem 277:24799–24808. https://doi.org/10.1074/jbc.M112424200

Penning TM (2005) AKR1B10: a new diagnostic marker of non-small cell lung carcinoma in smokers. Clin Cancer Res 11:1687–1690. https://doi.org/10.1158/1078-0432.CCR-05-0071

Penning TM (2015) The aldo-keto reductases (AKRs): overview. Chem Biol Interact 234:236–246. https://doi.org/10.1016/j.cbi.2014.09.024

Penning TM (2017) Aldo-Keto reductase regulation by the Nrf2 system: implications for stress response, chemotherapy drug resistance, and carcinogenesis. Chem Res Toxicol 30:162–176. https://doi.org/10.1021/acs.chemrestox.6b00319

Pollak N, Dölle C, Ziegler M (2007) The power to reduce: pyridine nucleotides – small molecules with a multitude of functions. Biochem J 402:205–218. https://doi.org/10.1042/BJ20061638

Quinn AM, Harvey RG, Penning TM (2008) Oxidation of PAH trans-dihydrodiols by human aldo-keto reductase AKR1B10. Chem Res Toxicol 21:2207–2215. https://doi.org/10.1021/tx8002005

Ramana KV, Srivastava SK (2006) Mediation of aldose reductase in lipopolysaccharide-induced inflammatory signals in mouse peritoneal macrophages. Cytokine 36:115–122. https://doi.org/10.1016/j.cyto.2006.11.003

Ramana KV, Tammali R, Srivastava SK (2010) Inhibition of aldose reductase prevents growth factor-induced g 1-s phase transition through the AKT/phosphoinositide 3-kinase/E2F-1 pathway in human colon cancer cells. Mol Cancer Ther 9:813–824. https://doi.org/10.1158/1535-7163.MCT-09-0795

Reczek CR, Chandel NS (2015) ROS-dependent signal transduction. Curr Opin Cell Biol 33:8–13. https://doi.org/10.1016/j.ceb.2014.09.010

Reddy AK, Uday Kumar P, Srinivasulu M, Triveni B, Sharada K, Ismail A, Bhanuprakash Reddy G (2017) Overexpression and enhanced specific activity of aldoketo reductases (AKR1B1 and AKR1B10) in human breast cancers. Breast 31:137–143. https://doi.org/10.1016/j.breast.2016.11.003

Röhrig F, Schulze A (2016) The multifaceted roles of fatty acid synthesis in cancer. Nat Rev Cancer 16:732–749. https://doi.org/10.1038/nrc.2016.89

Rojo de la Vega M, Chapman E, Zhang DD (2018) NRF2 and the hallmarks of cancer. Cancer Cell 34:21–43. https://doi.org/10.1016/j.ccell.2018.03.022

Ruiz FX et al (2009) Aldo-keto reductases from the AKR1B subfamily: retinoid specificity and control of cellular retinoic acid levels. Chem Biol Interact 178:171–177. https://doi.org/10.1016/j.cbi.2008.10.027

Saraswat M, Mrudula T, Kumar PU, Suneetha A, Rao TS, Srinivasulu M, Reddy GB (2006) Overexpression of aldose reductase in human cancer tissues. Med Sci Monit 12:CR525–CR529

Saraswathy R, Anand S, Kunnumpurath SK, Kurian RJ, Kaye AD, Vadivelu N (2014) Chromosomal aberrations and exon 1 mutation in the AKR1B1 gene in patients with diabetic neuropathy. Ochsner J 14:339–342

Sato S et al (2012) Up-regulated aldo-keto reductase family 1 member B10 in chronic hepatitis C: association with serum alpha-fetoprotein and hepatocellular carcinoma. Liver Int 32:1382–1390. https://doi.org/10.1111/j.1478-3231.2012.02827.x

Saxena A, Tammali R, Ramana KV, Srivastava SK (2012) Aldose reductase inhibition prevents colon cancer growth by restoring phosphatase and tensin homolog through modulation of miR-21 and FOXO3a. Antiox Redox Signal 18:1249–1262. https://doi.org/10.1089/ars.2012.4643

Saxena A, Shoeb M, Ramana KV, Srivastava SK (2013) Aldose reductase inhibition suppresses colon cancer cell viability by modulating microRNA-21 mediated programmed cell death 4 (PDCD4) expression. Eur J Cancer 49:3311–3319. https://doi.org/10.1016/j.ejca.2013.05.031

Schwab A et al (2018) Polyol pathway links glucose metabolism to the aggressiveness of cancer cells. Cancer Res 78:1604–1618. https://doi.org/10.1158/0008-5472.CAN-17-2834

Semmo N, Weber T, Idle JR, Beyoglu D (2015) Metabolomics reveals that aldose reductase activity due to AKR1B10 is upregulated in hepatitis C virus infection. J Viral Hepat 22:617–624. https://doi.org/10.1111/jvh.12376

Shen Y, Zhong L, Johnson S, Cao D (2011) Human aldo-keto reductases 1B1 and 1B10: a comparative study on their enzyme activity toward electrophilic carbonyl compounds. Chem Biol Interact 191:192–198. https://doi.org/10.1016/j.cbi.2011.02.004

Shen Y et al (2015) Impaired self-renewal and increased colitis and dysplastic lesions in colonic mucosa of AKR1B8-deficient mice. Clin Cancer Res 21:1466–1476. https://doi.org/10.1158/1078-0432.CCR-14-2072

Sheng X et al (2017) Adipocytes sequester and metabolize the chemotherapeutic daunorubicin. Mol Cancer Res 15:1704–1713. https://doi.org/10.1158/1541-7786.MCR-17-0338

Shoeb M, Ramana KV, Srivastavan SK (2013) Aldose reductase inhibition enhances TRAIL-induced human colon cancer cell apoptosis through AKT/FOXO3a-dependent upregulation of death receptors. Free Radic Biol Med 63:280–290. https://doi.org/10.1016/j.freeradbiomed.2013.05.039

Singh A et al (2006) Dysfunctional KEAP1-NRF2 interaction in non-small-cell lung cancer. PLoS Med 3: e420. https://doi.org/10.1371/journal.pmed.0030420

Skrzydlewska E, Sulkowski S, Koda M, Zalewski B, Kanczuga-Koda L, Sulkowska M (2005) Lipid peroxidation and antioxidant status in colorectal cancer. World J Gastroenterol 11:403–406. https://doi.org/10.3748/wjg.v11.i3.403

Srivastava S, Chandra A, Bhatnagar A, Srivastava SK, Ansari NH (1995) Lipid peroxidation product, 4-Hydroxynonenal and its conjugate with GSH are excellent substrates of bovine lens aldose reductase. Biochem Biophys Res Commun 217:741–746. https://doi.org/10.1006/bbrc.1995.2835

Srivastava SK, Ramana KV, Bhatnagar A (2005) Role of aldose reductase and oxidative damage in diabetes and the consequent potential for therapeutic options. Endocr Rev 26(3):380–392. https://doi.org/10.1210/er.2004-0028

Srivastava SK et al (2011) Aldose reductase inhibition suppresses oxidative stress-induced inflammatory disorders. Chem Biol Interact 191:330–338. https://doi.org/10.1016/j.cbi.2011.02.023

Stapelfeld C, Neumann KT, Maser E (2017) Different inhibitory potential of sex hormones on NNK detoxification in vitro: a possible explanation for gender-specific lung cancer risk. Cancer Lett 405:120–126. https://doi.org/10.1016/j.canlet.2017.07.016

Suzen S, Buyukbingol E (2005) Recent studies of aldose reductase enzyme inhibition for diabetic complications. Curr Med Chem 10:1329–1352. https://doi.org/10.2174/0929867033457377

Taguchi K, Yamamoto M (2021) The Keap1–Nrf2 system as a molecular target of cancer treatment. Cancers 13:1–21. https://doi.org/10.3390/cancers13010046

Tammali R, Ramana KV, Singhal SS, Awasthi S, Srivastava SK (2006) Aldose reductase regulates growth factor-induced cyclooxygenase-2 expression and prostaglandin E2 production in human colon cancer cells. Cancer Res 66:9705–9713. https://doi.org/10.1158/0008-5472.CAN-06-2105

Tammali R, Srivastava SK, Ramana KV (2011) Targeting aldose reductase for the treatment of cancer. Curr Cancer Drug Targets 11:560–571. https://doi.org/10.2174/156800911795655958

Tanawattanasuntorn T, Thongpanchang T, Rungrotmongkol T, Hanpaibool C, Graidist P, Tipmanee V (2021) (-)-Kusunokinin as a potential aldose reductase inhibitor: equivalency observed via AKR1B1 dynamics simulation. ACS Omega 6(1):606–614. https://doi.org/10.1021/acsomega.0c05102

Taskoparan B, Seza EG, Demirkol S, Tuncer S, Stefek M, Gure AO, Banerjee S (2017) Opposing roles of the aldo-keto reductases AKR1B1 and AKR1B10 in colorectal cancer. Cell Oncol (Dordr) 40:563–578. https://doi.org/10.1007/s13402-017-0351-7

Uchida K (2003) 4-Hydroxy-2-nonenal: a product and mediator of oxidative stress. Prog Lipid Res 42:318–343. https://doi.org/10.1016/S0163-7827(03)00014-6

van Weverwijk A et al (2019) Metabolic adaptability in metastatic breast cancer by AKR1B10-dependent balancing of glycolysis and fatty acid oxidation. Nat Commun 10(1):2698. https://doi.org/10.1038/s41467-019-10592-4

Wang J, Zhou Y, Fei X, Chen X, Chen Y (2018) Biostatistics mining associated method identifies AKR1B10 enhancing hepatocellular carcinoma cell growth and degenerated by miR-383-5p. Sci Rep 8:11094. https://doi.org/10.1038/s41598-018-29271-3

Wierenga RK (2001) The TIM-barrel fold: a versatile framework for efficient enzymes. FEBS Lett 492:193–198. https://doi.org/10.1016/S0014-5793(01)02236-0

Wu X et al (2017) AKR1B1 promotes basal-like breast cancer progression by a positive feedback loop that activates the EMT program. J Exp Med 214:1065–1079. https://doi.org/10.1084/jem.20160903

Yang L, Zhang J, Zhang S, Dong W, Lou X, Liu S (2013) Quantitative evaluation of aldo-keto reductase expression in hepatocellular carcinoma (HCC) cell lines. Genomics Proteomics Bioinformatics 11:230–240. https://doi.org/10.1016/j.gpb.2013.04.001

Zhang L et al (2013) Inhibitor selectivity between aldo-keto reductase superfamily members AKR1B10 and AKR1B1: role of Trp112 (Trp111). FEBS Lett 587:3681–3686. https://doi.org/10.1016/j.febslet.2013.09.031

Zhang W, Li H, Yang Y, Liao J, Yang GY (2014) Knock-down or inhibition of aldo-keto reductase 1B10 inhibits pancreatic carcinoma growth via modulating Kras-E-cadherin pathway. Cancer Lett 355:273–280. https://doi.org/10.1016/j.canlet.2014.09.031

Zhang J, Wang N, Li Q, Zhou Y, Luan Y (2021) A two-pronged photodynamic nanodrug to prevent metastasis of basal-like breast cancer. Chem Commun (Camb) Feb 3. https://doi.org/10.1039/d0cc08162k

Zhao JX et al (2017) Aldose reductase interacts with AKT1 to augment hepatic AKT/mTOR signaling and promote hepatocarcinogenesis. Oncotarget 8:66987–67000. https://doi.org/10.18632/oncotarget.17791

Zhong L, Liu Z, Yan R, Johnson S, Zhao Y, Fang X, Cao D (2009) Aldo-keto reductase family 1 B10 protein detoxifies dietary and lipid-derived alpha, beta-unsaturated carbonyls at physiological levels. Biochem Biophys Res Commun 387(2):245–250. https://doi.org/10.1016/j.bbrc.2009.06.123

Zhou Z, Zhao Y, Gu L, Niu X, Lu S (2018) Inhibiting proliferation and migration of lung cancer using small interfering RNA targeting on Aldo-keto reductase family 1 member B10. Mol Med Rep 17:2153–2160. https://doi.org/10.3892/mmr.2017.8173

Zu X et al (2016) Aldo-keto reductase 1B10 protects human colon cells from DNA damage induced by electrophilic carbonyl compounds. Mol Carcinog 56:118–129. https://doi.org/10.1002/mc.22477

Adv Exp Med Biol - Cell Biology and Translational Medicine (2021) 14: 83–113
https://doi.org/10.1007/5584_2021_637

Published online: 2 May 2021

Bilayer Scaffolds for Interface Tissue Engineering and Regenerative Medicine: A Systematic Reviews

Sheida Hashemi, Leila Mohammadi Amirabad, Fatemeh Dehghani Nazhvani, Payam Zarrintaj, Hamid Namazi, Abdollah Saadatfar, and Ali Golchin

Abstract

Purpose This systematic review focus on the application of bilayer scaffolds as an engaging structure for the engineering of multilayered tissues, including vascular and osteochondral tissues, skin, nerve, and urinary bladder. This article provides a concise literature review of different types of bilayer scaffolds to understand their efficacy in targeted tissue engineering.

Methods To this aim, electronic search in the English language was performed in PMC, NBCI, and PubMed from April 2008 to December 2019 based on the PRISMA guidelines. Animal studies, including the "bilayer scaffold" and at least one of the following items were examined: osteochondral tissue, bone, skin, neural tissue, urinary bladder, vascular system. The articles which didn't include "tissue engineering" and just in vitro studies were excluded.

Results Totally, 600 articles were evaluated; related articles were 145, and 35 full-text English articles met all the criteria. Fifteen articles in soft tissue engineering and twenty items in hard tissue engineering were the results of this exploration. Based on selected papers, it was revealed that the bilayer scaffolds were used in the regeneration of the multilayered tissues. The highest multilayered tissue regeneration has been achieved when bilayer scaffolds were used with mesenchymal stem cells and differentiation medium before implanting. Among the studies being reported in this review, bone marrow mesenchymal stem cells are the most studied mesenchymal

S. Hashemi
School of Advanced Technologies in Medicine, Shahid Beheshti University of Medical Sciences, Tehran, Iran

L. M. Amirabad (✉)
Marquette University School of Dentistry, Milwaukee, WI, USA

Cultured Decadence, Inc., Madison, WI, USA
e-mail: leila.mohammadi.am@gmail.com

F. D. Nazhvani and H. Namazi
Bone and Joint Diseases Research Center, Shiraz University of Medical Sciences, Shiraz, Iran

P. Zarrintaj
Polymer Engineering Department, Faculty of Engineering, Urmia University, Urmia, Iran

Advanced Materials Group, Iranian Color Society (ICS), Tehran, Iran

A. Saadatfar
Department of urology, Kermanshah University of Medical Science, Kermanshah, Iran

A. Golchin (✉)
Department of Clinical Biochemistry and Applied Cell Science, School of Medicine, Urmia University of Medical Sciences, Urmia, Iran
e-mail: agolchin.vet10@yahoo.com; Golchin.a@umsu.ac.ir

stem cells. Among different kinds of multilayer tissue, the bilayer scaffold has been most used in osteochondral tissue engineering in which collagen and PLGA have been the most frequently used biomaterials. After osteochondral tissue engineering, bilayer scaffolds were widely used in skin tissue engineering.

Conclusion The current review aimed to manifest the researcher and surgeons to use a more sophisticated bilayer scaffold in combinations of appropriate stem cells, and different can improve multilayer tissue regeneration. This systematic review can pave a way to design a suitable bilayer scaffold for a specific target tissue and conjunction with proper stem cells.

Keywords

Interface · Tissue engineeringBilayer scaffolds · Multilayered tissues · Osteochondral tissue · Skin · Tissue engineering

1 Introduction

Todays, recapitulating the tissues' exact structures is a challenging issue in tissue engineering and regenerative medicine. The scaffolds should be similar to chemically and structurally with targeted tissue to achieve the best regeneration. However, the conventional scaffolds cannot fulfill such aims, and more complicated scaffolds should be designed to mimic the biological and mechanical properties of tissues by inspiring from native tissue. Polymeric scaffolds are designed to conduct the reconstruction pathway, and research in this field resulted in the development of different scaffold topographic designing to lead a particular tissue repair (Langer and Vacanti 1993; Mohammadi Amirabad et al. 2017). Recently, extended studies in tissue structures have suggested new promising approaches in tissue defect reconstruction. Bilayer scaffolds can recapitulate the appropriate biological milieu similar to the extracellular matrix (ECM) in layered tissues, such as osteochondral, cutaneous, osseous, nervous, vascular tissues and urinary bladder to achieve functional engineered constructs (Atoufi et al. 2017; Raghunath et al. 2007). Bones are the layered tissues in which reconstruction with the equivalent structures of inner high porous cortical and outer dense trabecular layers is one of the significant challenges in their tissue engineering (Cai et al. 2010).

The osteochondral tissue is layered tissues consist of zone cartilage and subchondral bone, with different biophysical properties. The preparation of biomimetic scaffolds for this tissue should follow the natural tissue structure to optimize the recapitulation of the osteochondral interface's continuous gradients in the early phase (Martin et al. 2007). Commonly used monolayer scaffolds with isotropic structural and functional properties are inadequate to restore the defects in the cartilage-bone interface. Hence, various bilayer scaffolds have been introduced to repair these layered tissues, providing optimum environments for the cell/cell and cell/matrix interaction.

Skin is another multilayered tissue whose defects cannot repair spontaneously due to dermis and epidermis loss in third-degree burns or other conditions. During wound healing, the epidermis can be formed throughout the cellular chemotaxis in the presents of growth factors secreted from the dermis. Moreover, a high density of fibroblast cells in the dermis is a thicker layer with higher flexibility and porosity. Simultaneously, the epidermis is the thinner layer composed of squamous cells of keratinocytes and makes a protective barrier against pathogenic microorganisms with its lower porosity characteristic (Guo et al. 2010). Designing an equivalent layered construct by the art of tissue engineering, which simulates epidermis and dermis, is drastically required to make wound healing/reconstruction pattern. Bilayer scaffolds derived from biopolymers with desired physical, chemical, and biological properties should be engineered in a manner that can reflect the anatomical and physiological properties of skin tissue niches.

Recapitulation of longitudinal contact guidance in one layer for attachment and guiding the glial cells or axons of neurons with isotropic mechanical support and isolation properties in another layer is a critical determinant in tissue engineering of nerve and spinal cord. Synthetic nerve guidance conduits (NGCs) with the properties mentioned above can simulate the niche structure of neurons (Zarrintaj et al. 2020). Here, the bilayer structure of the scaffolds restrains the growth of soft tissue into the lumen due to small porosity or insulating properties of outer layer (Kozlovsky et al. 2009).

A similar structure of scaffold is also useful for urinary bladder reconstruction where the inner porous layer of the scaffold makes a niche for cell attachment and proliferation, and the outer hydrophobic layer act as a liquid isolator or barrier (Yudintceva et al. 2016; Zhao et al. 2015).

In tissue engineering of vessels, it is so difficult to resemble the different properties of vessel together, such as high mechanical strength, low liquid leakage, and thrombogenicity through a unique homogenous structure (Liu et al. 2013). The studies showed that bilayer scaffolds could simulate the vessel structure with higher cell infiltration during *in vivo* implantation. In these cases, the scaffold's outer layer should provide a suitable environment for the growth of smooth muscle cells and fibroblasts with appropriate mechanical strength to resist aneurysm formation and blood leakage (McClure et al. 2012). The internal layer should also provide a proper condition for proliferation and ingrowth of endothelial cells with the lowest thrombogenicity (Liu et al. 2013). Studies showed that some synthetic biomaterials, such as polyurethane, can provide suitable mechanical strength for the scaffold's outer layer. Natural biomaterials have the desired compatibility and low thrombogenicity, which is appropriate for the inner layer.

There are numerous bilayer scaffolds that are useful for tissue engineering of multilayer tissues, which make it challenging to select the best scaffold among them fabricated from the best biomaterials with appropriate structure and mechanical properties. To this aim, the current systematic review provides the critical determinants in designing the bilayer scaffold for each kind of multilayer tissue by investigating the results of some commonly used bilayer scaffolds in animal studies of multilayer tissue engineering. The following multilayer tissues were selected in this study are bone, skin, and urinary bladder, osteochondral, vascular, and neural tissues.

This systematic review can be beneficial for researchers to determine and design a suitable bilayer scaffold for a specific target tissue and conjunction with proper stem cells. The most detailed studies in this systematic review had focused on the used bilayer scaffolds, which suitable for different types of tissue reconstruction.

2 Methods

2.1 Eligibility Criteria

Studies in the field of "tissue engineering" and "bilayer scaffolds" were considered. Two reviewers screened the articles based on inclusion and exclusion criteria.

Inclusion criteria:

1. Studies in the field of "bilayer or biphasic scaffolds" designed by the art of "tissue engineering."
2. Animal studies of applying scaffolds.

Exclusion criteria:

1. Studies that do not include "tissue engineering" and/or "reconstruction medicine".
2. Studies with no bilayer scaffold (in means of no inclusion criteria).
3. Just *in vitro* studies.
4. Studies not published in English.
5. Unpublished studies.

It must be mentioned that any dissension about article accommodation to the inclusion criteria such as study type, scaffold function, and safety and treatment efficacy were consulted.

2.2 Source of Information and Search

Full text published and English edited articles from April 2008 to December 2019 were searched via PMC, NCBI, and PubMed database. The articles were found using the following keywords and search items: (Bilayer or biphasic) and scaffold and ("tissue engineer*" or regenerat* or bioengin*) and (bone or osseous or osteo* or skin or osteochond* or cartilage or chondro* or nerv* or neuron or neural or bladder or vascular or vessel* or vein or arter* or capillar*) (Pérez-Silos et al. 2019).

: ((Bilayer or biphasic) and (scaffold or membrane or composite) and ("tissue engineer*" or regenerat* or bioengin*)

2.3 Study Selection

The articles were selected primarily using their titles and abstracts according to the inclusion criteria. Then, the eligibility of the articles was determined by reviewing the full text. The latest and more relevant information of different reports of each experiment is investigated. Detailed studies are demonstrated briefly in Fig. 1, according to PRISMA guidelines [14].

2.4 Population Selection

Animal studies in which tissue regeneration was examined in specific multilayer tissue defects using different bilayer scaffolds.

3 Results

3.1 Study Selection

After screening steps, 1487 articles, 600 articles were assessed, and 145 were related to this review's aim. Finally, 35 full-text English articles could meet all the criteria. Fifteen articles in soft tissue engineering, twenty articles in hard tissue engineering were the results of this exploration. In those studies of hard tissue engineering, 18 cases reported *in vivo* evaluation of cartilage tissue engineering for focal articular defects. Three articles evaluated osseous tissue engineering for the reconstruction of various bone defects. Whereas in studies of soft tissue engineering, seven cases explained the application of bilayer

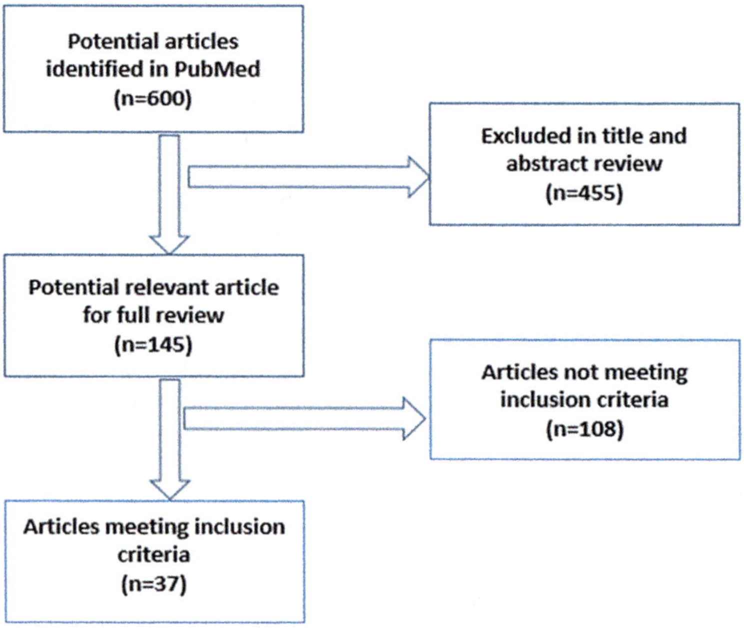

Fig. 1 The flow chart of literature selection for this systematic review

Potential articles identified in PubMed (n=600)

Excluded in title and abstract review (n=455)

Potential relevant article for full review (n=145)

Articles not meeting inclusion criteria (n=108)

Articles meeting inclusion criteria (n=37)

scaffolds in cutaneous tissue engineering, and eight cases pointed reconstruction of the urinary bladder, vascular, and nervous tissues. The articles are described in detail in the tables, and also different aspects were assessed, which are summarized and discussed later. Here, the qualifying data were investigated as much as possible since the studies were designed differently with wide variations.

4 Osteochondral Tissue

Among the 17 articles on bilayer scaffold investigation for synovial cartilage reconstruction (Table 1), one case had used bilayer bacterial nano-cellulose scaffold; and others had evaluated natural and synthetic polymers alone or in combination. Widely used natural and synthetic polymers for cartilage regeneration were collagen, gelatin, cellulose, polycaprolactone, and poly LCG. Animal studies in the field included mice, rats, rabbits, pigs, goats, sheep, and horses. Four studies used MSCs alone, two cases used the combination of MSCs and chondrocytes, and one used osteoblasts. These assays included micro-CT imaging, Histological evaluation (HIE), IHC, RT-PCR analysis, microscopic and macroscopic observation, biochemistry, and biomechanical tests to achieve precise results and follow-ups varied from 4 to 24 weeks. Osteogenic differentiation assay includes the below criteria and tests:

- Bone formation: q-PCR, μCT, ALP
- Histological analysis: Safranin-O/Fast Green staining, Hematoxylin and Eosin, Immunohistochemistry
- Mechanical strength: Compressive modulus, tensile modulus

4.1 Mouse

There were two *in vivo* studies that were conducted in a mouse model. A study was conducted using human nasoseptal chondrocytes (h-NC) to evaluate the osteochondral regeneration potential of the scaffold fabricated from Bacterial Nano Cellulose (BNC) (Ávila et al. 2015). The bilayer scaffold was composed of a dense nanocellulose layer (1%) joined with a macroporous composite layer of nanocellulose (17%) and alginate. For *in vitro* study, nasoseptal chondrocytes (NCs) were seeded on the scaffolds for 6 weeks. The chondrogenic marker genes' expression, including *aggrecan* and collagen type IIA1 (*COLIIA1*), and dedifferentiation markers, was assessed. *Aggrecan* and *COLIIA1* were upregulated by factors 3.4- and 4.9 after 6 weeks, respectively, compared with their expression after 2 weeks. However, the dedifferentiation marker did not express even after 6 weeks. The results confirmed the chondrogenic capacity of the NCs in the bilayer BNC scaffolds. The cell viability on the bilayer scaffold was $97.8 \pm 4.7\%$. For *in vivo* study, the human NCs and bone marrow-derived human mononuclear cells (BM-MNC) cultured on bilayer scaffolds were implanted subcutaneously in nude mice for 8 weeks. The results showed that glycosaminoglycan content was significantly higher (12 fold) in cell-seeded bilayer BNC scaffolds than cell-free ones (Ávila et al. 2015). P Giannoni et al. developed a bilayer monolithic osteochondral poly(ε-caprolactone) (PCL) –based by combining solvent casting/particulate leaching (Giannoni et al. 2015). The bone-facing layer of the scaffolds composed of hydroxyapatite granules dispersed in a highly porous PCL mesh. They used bovine trabecular bone-derived mesenchymal stem cells (BTB-MSCs) and bovine articular chondrocytes (B-ACs) techniques. The results showed that the tensile strength of the bilayer scaffold was 0.34 ± 0.05 Mpa. For *in vivo* study, BTB-MSCs and B-ACs were seeded in the bony layer and the chondral layer, respectively, and then the cell-seeded scaffolds were implanted subcutaneously in nude mice. After 9 weeks, the bony and compact structures were generated in BTB-MSCs seeded region of scaffold and chondrocyte lacunae with a cartilaginous alcianophilic matrix were observed in the chondral layer. Moreover, in the bony layer, the number of blood vessels (average lumen size~10 mm) was significantly higher than those

Table 1 Different bilayer scaffolds in osteochondral tissue engineering

Scaffold	Study model	Defect size and location	Sample size	Stem cell	Follow-up	Characterization tests	Results	References
Bacterial nano cellulose	Mouse	10 mm, subcutaneous region	8 × 3 mm	h-NC BM-MNC	2,4,6 w CG: Cell-free scaffolds	RT-qPCR; Histological analysis; Mechanical test	BF:- HA: 12 fold> GC MS: Bony layer: - Chondral layer: -whole ≈0.16 Mpa	Ávila et al. (2015)
PCL	**Mouse**	**Subcutaneous region**	**3 × 3 × 5 mm**	**BTB-MSCs** B-ACs	**9 w** CG: cell-**free** scaffolds	**Mechanical test**	**BF:** - HA: - **MS:** Bony layer≈ 0.46 Mpa Chondral layer≈ 0.21 Mpa Whole≈0.34 Mpa	**Giannoni** et al. (2015)
BC-HA/BC-GAG	Rat	1 × 3 mm, patellar groove of the knee joints	2 × 2 mm	–	4,12 w CG: Untreated	μCT; Histological analysis;	BF: ~3 and ~1.5 folds better than CG after 4 and 12 weeks, respectively HA: ~2 and ~3 folds better than CG after 4 and 12 weeks, respectively MS: -	Kumbhar et al. (2017)
GO-SA Gel/ aptamer-GBF	Rat	2 mm, knee joint	8 × 10 mm	–	8,12 w CG: Aptamer gel	μCT; Histological analysis; Mechanical test	BF: >four fold more than GC HA: 1.5 fold>GC MS: Bony layer≈ 0.135 Mpa Chondral layer≈ 0.02 Mpa Whole: -	Hu et al. (2017)
Collagen	Rabbit	4 × 3 mm, deep-patellar groove	4 × 3 mm	–	6,12 weeks CG: Untreated	Histological analysis	BF: > > GC HA: 1.5 fold>GC MS:-	Jiang et al. (2013)
PLGA	Rabbit	4 × 4 mm, in both knee joints	4 × 4 mm	–	4, 12 w CG: Untreated	μCT; Histological analysis	BF: Bilayer PLGA scaffold>CG HA: 1.2 fold more than CG MS: -	Zhang et al. (2017)

Collagen	Rabbit	4 × 3 mm, in both knee joints	4 × 3 mm	rBM-MSCs	6,12 w CG: Untreated	Histological analysis; Mechanical test	BF:- HA: ~1.5 and ~2 better than CG after 6 and 12 weeks, respectively MS (compressive): Bony layer: - Chondral layer: - Whole≈ 0.18	Qi et al. (2012)
PLGA	Rabbit	4 × 3 mm, in depth- patellar groove	4× 3 mm	rBM-MSCs	6,12 w CG: PLGA	Histological analysis; Mechanical test	BF: - HA: ~2.3 and ~1.8 better than CG after 6 and 12 weeks, respectively. MS:-	Qi et al. (2014)
SPU/PLGA	Rabbit	4.5 × 4 mm parapatellar incision was made in the knee	4.5 × 4 mm	–	6,12 w CG: Untreated	Histological analysis	BF: - HA: Bilayer PU/PLGA/BMP2 group>four fold more than CG SPU/PLGA ~1.5 and ~1.3 better than CG after 6 and 12 weeks, respectively. MS:-	Reyes et al. (2014)
Yarn Collagen-I/ Hhyaluronate / TCP	Rabbit	5 × 6 mm patellar groove of the distal femur	5 × 5 mm	rBMSCs	12 w CG: Untreated	Histological analysis; Mechanical test	BF: - HA:~1.5 fold better than CG MS(compressive strength): Bony layer≈ 2.6 Mpa Chondral layer≈0.1 Mpa Whole≈ 0.25 Mpa MPa-Young's modulus~0.2 Mpa	Liu et al. (2015)
PLGA/TCP/ collagenI-bCECM	Rabbit	Patella femoral groove with 5 × 6 mm	5 × 6 mm	rBMSCs	12, 24 w CG: Untreated	μCT; Histological analysis; Mechanical test	BF: >CG HA~14 and ~8 better than CG after 12 and 24 weeks, respectively. MS: Bony layer: - Chondral layer: - Whole≈ Tensile strength ~0.1 Mpa Shear strength ~0.13 Mpa	Da et al. (2013)

(continued)

Table 1 (continued)

Scaffold	Study model	Defect size and location	Sample size	Stem cell	Follow-up	Characterization tests	Results	References
CAN/PAC hydrogel	Rabbit	4 × 5 mm, chondral and subchondral Bone layers of the patellar groove	10 mm	–	6,12,18 w CG: Untreated	μCT; Histological analysis; Mechanical test	BF: ~1.6 fold >GC HA: Two fold better than CG MS: Bony layer≈ 0.261 Mpa Chondral layer≈0.065 Mpa Whole≈ 0.154 Mpa MPa	Liao et al. (2017)
PLGA/β-TCP	Pig	8 × 8 femoral condyles	8 × 8 mm	pC	6 m CG: Untreated	Histological analysis; Mechanical test	HA: More than two folds >CG MS: Bony layer: - Chondral layer: - Whole≈(tensile modulus) 93 Mpa	Jiang et al. (2007)
Aragonite-hyaluronate	Goat	6 × 10 mm femoral condyle goats	6 × 10 mm	–	24, 48 weeks CG: Filled by blood clots	μCT; Histological analysis	BF: ~1.6 and ~1.3 folds better than CG after 24 and 48 weeks, respectively. HA: ~4 and ~2 folds better than CG after 24 and 48 weeks, respectively MS: -	Kon et al. (2015)
Collagen–GAG/ calcium phosphate	Goat	5.8 × 6.0 mm, medial femoral condyle and lateral trochlear sulcus	6 × 6 mm	–	26 w CG: Untreated	Histological analysis	BF: - HA: > 1.3 fold more than CG MS:-	Getgood et al. (2012)
Collagen	Sheep	8 × 10 mm medial femoral condyle	8 × 10 mm	–	4 w CG: Untreated	Histological analysis	BF: - HA: > four fold more than CG MS:-	Howard et al. (2015)
TMPG	Sheep	20 mm medial femoral condyles	20 mm	–	16 w CG: Untreated	Histological analysis	BF: - HA: ≈CG MS:-	Mrosek et al. (2016)

Gelatin, PRP/β-TCP, BMP2	Horse	4.5 × 10 mm lateral trochlear ridges of the talus	5 × 5 mm	h –MSCs h-chondrocytes	16 w CG: BGTS	μCT; Histological analysis; Mechanical test	BF: ~1.3 fold >GC HA: ~1.3 fold better than CG MS: Bony layer: - Chondral layer: - Whole≈(compression modulus) ~1.13 Mpa	Seo et al. (2013)

Abbreviations: Gelatin/b-Tricalcium Phosphate (GT), Bacterial Nano Cellulose(BNS), Poly(e-Caprolactone) (PCL), GT Sponge loaded with MSCs and BMP-2 (MSC/BMP2/GT), Bilayer GT Sponges (BGTS), Bacterial Cellulose (BC), Polyurethane (PU), Poly-(Lactic-co-Glycolic Acid) (PLGA), Trabecularmetal/Periosteal Graft (TMPG), Beta-Tricalcium Phosphate (TCP), Radiography Evaluation (RE), Quantitative Computed Tomography (QCT), Scanning Electron Microscopy (SEM), Toluidine Blue Staining (TBS), Biochemical Analysis(BA), sGAG Content (sGAG- C), dsDNA Content (dsDNA –C), Micro-Computed Tomography (μCT), Attenuated Total Reflectance Fourier Transform Infra-Red (ATR-FTIR) Spectroscopy, Alcian Blue Staining (ABS), Bovine Trabecular Bone-derived MSCs (BTB-MSCs), Bovine ACs (B-ACs), Real-Time quantitative PCR (RT-qPCR), Autologous Platelet Rich Plasma (PRP), Bone Marrow-derived MSCs (BMSCs), Growth Factor Release Assays (GFRA), X-Ray Diffraction (XRD), Fourier Transform Infrared (FTIR) spectra, mice Chondrocytes (mC), mice Osteoblasts (mO), pig Chondrocytes (pC), horse MSCs (hoMSCs), rabbit MSCs (rMSCs), 4′,6-diamidino-2-phenylindole, dihydrochloride (DAPI), Alkaline Phosphatase Staining (APS), human Adipose tissue-derived MSCs (hATMSCs), Type II Collagen Content (col II –C), Masson's Trichrome (MTS), Macroscopic Evaluation (ME), Safranin-O/Fast Green staining (SOFG), Confocal Microscopy (CM), Flow Cytometry Analysis (FCA), Mechanical Test (MT), Hematoxylin and Eosin (H&E), Immunohistochemistry (IHC), Bone Morphogenetic Protein (BMP-2), Transforming Growth Factor-b1 (TGF-b1), and Platelet Rich Plasma (PRP); bovine cartilage extracellular matrix (bCECM) in acellular controls of osteochondral scaffolds. However, blood vessels in cavities were hardly detectable in the chondral layer (Giannoni et al. 2015).

4.2 Rat

There were two *in vivo* studies that were performed in rat model. X Hu et al. in 2017 designed and fabricated an aptamer-bilayer scaffold which recruited autologous mesenchymal stem cell (MSC) from a marrow clot in the region of osteochondral defect and induce chondrogenic and osteogenic simultaneous differentiation of MSCs in different layers in the osteochondral defect (Hu et al. 2017). The chondral layer was fabricated by crosslinking graphene oxide with a sodium alginate containing kartogenin. The bony layer was also fabricated using a 3D graphene oxide-based biomineral framework. After assembling the aforementioned two layers, MSC specific aptamers (Apt19S) were immobilized on them. The mechanical stregth of the bony layer was ≈135 kPa, while that of the chondral layer was ≈2 kPa. For *in vivo* study, the bilayer scaffolds were implanted in full-thickness osteochondral defects in knee of Sprague Dawley rat model. After 8 weeks, the aptamer-bilayer scaffold group had significantly higher score (1.5 fold) than aptamer gel scaffold confirming the positive effect of bilayer structure in osteochondral regeneration. Furthermore, the bone formation in the aptamer-bilayer scaffold group was >four fold higher than aptamer gel scaffold (Hu et al. 2017).

J V Kumbhar et al. in 2017 developed a Bacterial cellulose (BC)-bilayer scaffolds composed of BC-hydroxyapatite as bony layer and BC-glycosaminoglycans as chondral layer (Kumbhar et al. 2017). For *in vivo* study, the acellular bilayered scaffolds implanted in the osteochondral defect in Wistar rat knees and defect without any treatment was considered as control group. The bone mineral density in bilayer scaffolds was ~3 and ~1.5 times more than control group after 4 and 12 weeks, respectively. Histological score also showed that the bilayer scaffolds were ~2 and ~3 times better than control group after 4 and 12 weeks, respectively (Kumbhar et al. 2017).

4.3 Rabbit

Y Jiang et al. in 2013 used Povidone-iodine (PVP-I) in a bilayer collagen scaffolds (Jiang et al. 2013). For in vivo study, the scaffolds were implanted in knee joint osteochondral defect of male New Zealand white rabbits Histological score evaluation showed that the PVP-I-bilayer collagen scaffolds worked about 5 times better than control group (Jiang et al. 2013).

Y Zhang et al. in 2017 fabricated by room temperature compression molding and particulate leaching technique using PLGA (Zhang et al. 2017). The bony layer and chondral layer were different in pore size and thickness. For in vivo study, the bilayer PLGA scaffolds were implanted in the osteochondral defect of New Zealand white rabbit knee joint (Zhang et al. 2017). The untreated group was considered as control group. The histological score showed that the bilayer scaffolds worked about 1.2 times better than control group after 4 and 12 weeks (Zhang et al. 2017).

Y Qi in 2012 used rabbit bone marrow (rBM) MSC-seeded bilayer collagen scaffolds for cartilage repair (Qi et al. 2012). The compressive modulus of bilayer scaffolds with and without MSCs were ~0.27 Mpa and ~0.18 Mpa. For in vivo study, the MSC-seeded scaffolds were implanted into male New Zealand white rabbit knee cartilage defects and the untreated defects was considered as control group. Histological scores showed that the rBM-MSC-seeded bilayer collagen scaffold acted ~1.5 and ~2 times better than control group after 6 and 12 weeks, respectively (Qi et al. 2012). Whereas, bilayer collagen scaffolds worked ~1.3 times better than control group after both 6 and 12 weeks. In another study conducted by Y Qi et al. rabbit bone marrow MSC- seeded bilayer PLGA scaffolds fabricated by a porogen-leaching technique were investigated for osteochondral regeneration (Qi et al. 2012). A similar *in vivo* study was conducted on osteochondral defects of rabbit knee joint and the PLGA scaffolds considered as control group. The histological analysis showed that the MSC-PLGA seeded scaffolds increased chondrogenic regeneration ~2.3 and ~1.8 times compared with untreated control group after 6 and 12 weeks, respectively (Qi et al. 2012).

R Reyes et al. in 2014 investigated the different dosage of transforming growth factor-β1 (TGF-β1) and bone morphogenetic protein-2 (BMP-2) in the bilayer scaffold of segmented polyurethane/polylactic-co-glycolic (SPU/PLGA) fabricated by gas foaming (Reyes et al. 2014). For *in vivo* study, the bilayer scaffolds were implanted into the osteochondral defects in the rabbit knees. Histological scores showed that the bilayer scaffolds with 5 μg BMP-2 increased the osteochondral regeneration compared the other groups. However, the SPU/PLGA bilayer scaffold increased osteochondral regeneration ~1.5 and ~1.3 times compared with untreated control group after 6 and 12 weeks, respectively (Reyes et al. 2014).

Shen Liu in 2015 fabricated a bilayer scaffold which bony layer was composed of porous β-tricalcium phosphate (β-TCP) and chondral layer was composed of poly(L-lacticacid)-copoly(e-caprolactone) P(LLA-CL)/collagen type I(Col-I) nanofiber and Col-I/Hhyaluronate sponge(Shen Liu et al. 2015). The Yong modulus whole bilayer scaffolds was ~0.2 Mpa. The compressive strength of bony layer, chondral layer, and whole bilayer scaffolds were measured as ~2.6 Mpa, ~0.1 Mpa, and ~0.25Mpa, respectively. For *in vivo* study, rBM-MSCs seeded scaffolds were implanted into osteochondral defect in rabbit knees and untreated defect was considered as control group (Liu et al. 2015). Histological scores showed that the bilayer scaffolds increased osteochondral regeneration by factor of ~1.5 more than control group (Liu et al. 2015).

Hu Da et al. in 2013 fabricated a bilayer scaffold with bony layer composed of a PLGA/TCP skeleton wrapped in collagen I and chondral layer composed of bovine cartilage extracellular matrix-derived scaffold, fabricated by a modified temperature gradient-guided thermal-induced phase separation (TIPS) method (Da et al. 2013). Then the aforementioned layers were attached to each other through a compact layer by dissolving conglutination method. The tensile and shear strength of the whole scaffolds was ~0.1 Mpa and ~0.13 Mpa, respectively. For *in vivo* study, after seeding autogeneic osteoblast- or chondrocyte-induced BM-MSCs on the bony and chondral layers, respectively, the scaffolds were implanted into rabbit knee osteochondral defects. The untreated defects were considered as control group (Da et al. 2013). The bone mineral density determined that bilayered scaffold increased effectively the bone formation. The histological scores showed that the bilayer scaffolds increased the osteochondral regeneration ~14 and ~8 times more than control group after 12 and 24 weeks, respectively (Da et al. 2013).

In 2017 Liao et al. fabricated a biphasic CAN-PAC hydrogel different densities for the two layers using thermally reactive and rapid cross-linking technique (Liao et al. 2017). The chondral layer was cross-linked by methacrylated chondroitin sulfate and N-Isopropylacrylamide, and the bony layer was fabricated using acryloyl chloride-poly (ε-caprolactone)-poly (ethylene glycol)-poly(ε-caprolactone)-acryloyl chloride (PECDA), acrylamide and Polyethylene glycol diacrylate (PEGDA) (Liao et al. 2017). The two layers were first attached physically by the calcium gluconate and alginate and then chemically by carbon-carbon double bonds (Liao et al. 2017). The results showed that the tensile strength of bony and chondral layers were ~0.261 MPa and ~0.065, respectively. For *in vivo* study, the bilayer hydrogels were implanted in rabbit model defects in knee. During 18 weeks, the defect gradually restored with new subchondral tissue which only a small pinhole in the central region was observed at the end of 18 weeks. The bone volume in the hydrogel group was reached to 91.47% whereas the bone volume in control group showed 56.23% at 18 weeks. Histological scores of in hydrogel group was significantly higher compared with control group by factor 2.5 at 18 weeks, which confirmed a promising capability of the fabricated hydrogel for osteochondral regeneration (Liao et al. 2017).

4.4 Pig

C Jiang et al. in 2007 fabricated a bilayer scaffold with bony layer composed of β-TCP and chondral layer composed of PLGA (C. C. Jiang et al. 2007). The tensile modulus of the scaffolds was 0.93 Mpa similar to native cartilage (0.88 Mpa). For *in vivo* study, after seeding the Autologous chondrocytes in chondral layer, the scaffolds were implanted into load-bearing osteochondral defects in the Lee-Sung mini-pig femoral condyles. Treatment of the defect with cell-free scaffolds was considered as control (C. C. Jiang et al. 2007). The histological scores showed that the fabricated bilayer scaffolds enhanced the osteochondral regeneration more than 2 times compared with control group (C. C. Jiang et al. 2007).

4.5 Goat

E Kon et al. in 2015 fabricated an aragonite-hyaluronate bilayer scaffold for purpose of osteochondral regeneration (Kon et al. 2015). For *in vivo* study, the scaffolds were implanted into the full-thickness osteochondral defects in the weight-bearing of goat femoral condyle. The defects filled with blood clots were considered as control group. The histological scores showed that the bilayer scaffolds increased the osteochondrar differentiation by fold of ~4 and ~2 compared with control group after 24 and 48 weeks, respectively. The μCT results showed bone formation in the bilayer scaffolds was ~1.6 and ~1.3 times more than bone formation in control group after 24 and 48 weeks, respectively (Kon et al. 2015).

A Getgood et al. (2012) used a bilayer scaffold composed of collagen–GAG (chondral layer) and calcium phosphate (bony layer) for the osteochondral regeneration purpose (Getgood et al. 2012). For in vivo study, osteochondral defects in medial Boer Cross goat femoral condyle and lateral trochlear sulcus were filled with the bilayer scaffolds (Getgood et al. 2012). The empty defects served as control group. The histological scores showed that the scaffolds improved osteochondral regeneration by factor of ~1.3 more than control group (Getgood et al. 2012).

4.6 Sheep

D Howard et al. (2015) investigated the effect of bilayer collagen scaffolds contained 32 μg recombinant human fibroblastic growth factor 18 (rhFGF8) on the osteochondral regeneration (Howard et al. 2015). For *in vivo* study, the bilayer scaffolds were grafted into the osteochondral defect in sheep medial femoral condyles (Howard et al. 2015). The untreated defects served as control group. The histological scores showed that the osteochondral regeneration in the bilayer collagen scaffold with 32 μg rhFGF8 was significantly higher by factor ~4 compared with control group.

E Mrosek et al. (2016) fabricated a durable bilayer implant composed of trabecular metal and autologous periosteum for osteochondral regeneration purpose. For *in vivo* study, the bilayer scaffolds were grafted into the osteochondral defect in the load-bearing region of sheep medial femoral condyles. The untreated defect was considered as control group. Histological scores demonstrated the scaffolds acted approximately equal to control group (Howard et al. 2015).

4.7 Horse

J Seo in 2013 investigated autologous bilayer sponge scaffold composed of β-TCP and BMP-2 (mimic bony layer), gelatin and platelet rich plasma (PRP) (mimic chondral layer), for osteochondral regeneration. In the bony layer the autologous BM-MSCs were seeded and in the chondral layer autologous BM-MSCs and chondrocytes were seeded (Seo et al. 2013). The compression modulus of the fabricated bilayer scaffolds was measured as 1.13 ± 0.13 Mpa. For *in vivo* study, the scaffolds were implanted into critical sized osteochondral defects in the both lateral trochlear ridges of the horse talus (Seo et al. 2013). The scaffolds composed of β-TCP and gelatin were used for control group. According histological scores, the fabricated scaffolds increased significantly the ostechondral regeneration by factor ~1.3 compared with control group. The ϱCT showed that after 16 weeks the bone formation in test group was ~1.3 times more than control group. The radiography also showed that the size of defect decreases 2 times more than control group (Seo et al. 2013).

5 Bone

Three articles have evaluated bilayer scaffolds for bone tissue reconstruction (Table 2). The used biomaterials for bone tissue engineering were collagen, hydroxyapatite, and Poly-L-Lactic Acid. Two assays had gene therapy in parallel. Animal studies in the field included models of rabbits. One assay had ADMSCs, HLE, SEM, IHC, immunofluorescence staining, RT-PCR analysis, bone mineral assessment, gross morphology observation, micro-CT imaging, RGE and biochemical tests to evaluate the results and follow-ups varied from 14 to 112 days.

5.1 Rabbit

A study conducted by Y Zhi Cia et al. (2010) investigated the effect of bilayer porous collagenous membrane reinforced by PLLA nanofibrous membrane on bone (Cai et al. 2010). For *in vivo* study, critical sized defects of bone of rabbit tibia were filled with the bilayer PLLA membrane. The untreated defects served as control group. The results showed that the bone formation in the bilayer PLLA membrane ~3.5 and ~2.8 times more than control group after 3 and 6 weeks, respectively. Histological scores demonstrated the osteogenic effect of the bilayer PLLA membrane was ~6.5 and ~3 times more than control group after 3 and 6 weeks, respectively (Cai et al. 2010).

T Guda et al. (2011) fabricated a novel bilayer hydroxyapatite scaffolds with two porosities (Outer dense shell = 200 μm (73% total volume); Inner porous core = 450 μm (27% total volume)) for bone regeneration (Guda et al. 2011). For in vivo study, the bone segmental defects in rabbit radius were filled with the bilayer scaffolds. The untreated defects were considered as control groups. The flexural strength of the grafted bilayer scaffolds was about 26.59 ± 6.96 Mpa which is lower and higher than autologous bone graft (~43.84 ± 15.31) and the empty defects (16.13 ± 7.81 MPa), respectively. The bone formation in bilayer scaffold group was ~1.5 times more than control group (Guda et al. 2011).

6 Skin

Seven articles have been evaluated in the field of skin repair, in which various types of bilayer scaffolds were assessed to tissue reconstruction (Table 3). Widely used natural and synthetic polymers for cutaneous reconstruction investigations were collagen and chitosan, and PGE and polyurethane, respectively. And also that gene therapy was accessed parallel in two assays. Animal studies in the field included models of rabbits, pigs, rats and mice. One assay had ADMSCs, HLE, SEM, IHC, immunofluorescence staining, RT-PCR analysis, western blotting, microscopic and macroscopic observation, hematology, micro-CT imaging and biomechanical tests to achieve more precise results and follow-ups varied from 7 to 112 days.

6.1 Mouse

However, cutaneous wound healing is investigated in a large-scale range of animal models, mouse frequently studied as a full-thickness wound healing model. Recently, Golchin and Nourani (2020) investigated epidermal growth factor (EGF) releasing potency of bilayer PCL and PVA/Chitosan electrospun mat for the wound healing process in the mouse. They demonstrated that the designed bilayer mat was

Table 2 Different bilayer scaffolds in bone tissue engineering

Scaffold	Study model	Defect Model	Sample size	Stem cell	Follow-up	Characterization tests	Results	References
Hydroxyapatite	Rabbit	Segmental defect in the rabbit radius model	10 × 3 × 5 mm	–	8 weeks; CG: Untreated	μCT; Mechanical test	BF: ~1.5 fold >GC MS:(flexural strength) 26.59 ± 6.96 MPa)	Guda et al. (2011)
Collagen/PLLA nanofiberous membrane	Rabbit	10 × 15 mm-anteromedial cortex of the proximal tibia	10 × 5 mm	–	3, 6 weeks; CG: Untreated	Radiographic evaluation; Histological analysis	BF: ~3.5 and ~2.8 folds better than CG after 3 and 6 weeks, respectively. HA: ~6.5 and ~3 folds better than CG after 3 and 6 weeks, respectively. MS: -	Cai et al. (2010)

Abbreviations: human Cartilage-derived Progenitor Cells (hCPC), rabbit Bone Marrow-derived MSCs (rBMSCs), Mechanical Strength (MS), Micro-Computed Tomography (μCT), Poly-L-Lactic Acid (PLLA)

Table 3 Different bilayer scaffolds in skin tissue engineering

Scaffold	Study model	Defect size and location	Sample size	Cell source	Follow-up	Characterization tests	Results	References
Nanofiber- ADM	Rat	20 × 20 mm backs of rat	20 × 20 mm	h-dsASC	14,21 d	SEM,,MTT,H&E, IHC&,RT- qPCR, MTS	–	Mirzaei-parsa et al. (2018)
Gene-activated dermal equivalent TMC	Pig	30 mm-Backs of pigs	30 × 2 mm	–	7,10,14 d and21, 42, 70, 112 d CG: Blank BED	ME,SEM,IHC&IF, RT-qPCR, WB,MT	Relative mRNA expression in TMC/pDNA-VEGF Group:CD31: 2.5 and α-SMA: Two fold more than CG	Guo et al. (2010)
Plasmid DNA encoding VEGF-165 activated collagen – chitosan	Pig	30 mm -backs of pigs	30 × 2 mm	–	7,10,21 d and 28,56,105 d CG: Blank BED	ME,SEM,TEM, ELISA,MTT,H&E, IHC&IF,RT-qPCR, WB,MT	Relative mRNA expression in TMC/ pDNA-VEGFgroup:CD31: 3and α-SMA: 3 foldmore than CG	Guo et al. (2011)
Bilayer dermal equivalent collagen/chitosan/ silicone	Pig	30 mm -backs of pigs	30 × 0.4 mm	–	7,14,21 d and 28 d CG:ADM	ME,SEM,IHC&IF, IHI,RT-qPCR, WB	Relative mRNA expression in BDE group:CD31: 1.5 and α-SMA: 1.2 fold less than CG	Guo et al. (2014)
Collagen-PEG fibrin-based bilayer hydrogel	Rat	15 mm cm dorsum of the rat down to the panniculus	50 mm	h-dsASC	4,8,12,16 d CG: Untreated	Me,LMI,IHC&if, MTS,MT	–	Natesan et al. (2013)
PVP-I/ gelatin cryogel	Rabbit	20 mm	20 mm	–	7,14,21,28 d CG: Untreated	ME,SEM,FTIR, MT,μ CT,MTT,DRA,PT, HA,H&E	–	Priya et al. (2016)
PU-PLGAm/CCS	Rat	20 × 20 mm dorsum proximal to the head	20 × 20 mm	–	1,2,3,4,8,12 w CG: PELNAC	ME,SEM,IHC&IF, RT-qPCR, WB,MT	Relative mRNA expression in PU-PLGAm/CCSgroup: CD31: 1.3 and α-SMA: 1.3 fold more than CG	Wang et al. (2016)
SIS membrane	Mouse	10 × 10 mm on the dorsal surface of the mice	10 × 10 mm	–	7 d CG: Untreated	ME,SEM,H&E,RT-qPCR,MT	Relative mRNA expression in SIS membranegroup: CD31: 4.7 fold more thanCG	Wang et al. (2018)

(continued)

Table 3 (continued)

Scaffold	Study model	Defect size and location	Sample size	Cell source	Follow-up	Characterization tests	Results	References
PCL-PVA/Cs	Mouse	10 × 10 mm on the dorsal surface of the mice	10 × 10 mm	hBFP-MSCs	7d-14d CG: Vaseline	SEM Water absorption MTT DAPI Histopathology	The results recommend the potential of the designed bilayer nanofibrous scaffold (PCL-PVA/Cs) containing EGF for the use in full-thickness wound dressing and skin regeneration	Golchin and Nourani, (2020)

Abbreviations: acellular dermal matrix (ADM),Bilayer Dermal Equivalent (BDE), N,N,N-Trimethyl Chitosan Chloride (TMC), Polyethylene Glycol (PEG), Polyvinyl Pyrrolidone-iodine (PVP-I), Polyurethane (PU) membrane, Membrane/knitted Mesh-Rein forced Collagen–Chitosan Bilayer Dermal Substitute (PU-PLGAm/CCS), Acellular Dermal Matrix (ADM), Small Intestinal Submucosa (SIS),Vascular Endothelial Growth Factor (VEGF), Smooth Muscle Actin (α-SMA), Plasmid DNA encoding VEGF (pDNA-VEGF), Macroscopic Evaluation (ME), Scanning Electron Microscopy (SEM), Human Adipose-Derived Stem Cells (h-dsASC), Hematoxylin and Eosin (H&E), Immunohistochemistry and Immunofluorescence (IHC&IF), Real-Time quantitative PCR (RT-qPCR), Mechanical Test (MT), Western Blotting(WB), Degradation Rate (DR), Peel Test (PT), Hematological Analysis (HA), day (d), week (w), Chitosan (Cs), Human Buccal Fat Pad (hBFP)

capable of cell seeding and increased their growth and proliferation. In this study, bilayer PCL and PVA/Chitosan electrospun mat showed meaningful differences for wound closure and improved histological healing in an *in vivo* condition (Golchin and Nourani 2020). Wang et al. investigated the wound healing potency of a bilayer patch contains the top layer for providing humidity and antibacterial condition and the cryogel layer for promoting cell proliferation (Wang et al. 2018). This SIS bilayer wound dressing demonstrated noteworthy mechanical properties to protect the wound area and could also maintain a humid condition for cell proliferation and migration at the wound site in mouse skin wounds (Wang et al. 2018).

6.2 Rat

In this regard, Mirzaei-Parsa et al. evaluated the feasibility of PCL and fibrinogen combining nanofibers with the adipose tissue-derived stem cells (ATSC) seeded on the acellular dermal matrix (ADM) as a bilayer scaffold for the treatment of full-thickness skin wounds in a single-step system in the rat model (Mirzaei-parsa et al. 2018). They reported that ATSC seeded on ADM, PCL-fibrinogen nanofibrous patch, and bilayer scaffolds can support angiogenesis, re-epithelialization, and collagen remodeling in comparison with cell-free scaffolds (Mirzaei-parsa et al. 2018). Natesan et al. investigated a skin equivalent, made by a collagen-polyethylene glycol (PEG) fibrin-based bilayer hydrogel in an excision wound model in athymic rats (Natesan et al. 2013). They added debrided skin adipose stem cells (dsASCs) on the structure and demonstrated that within the bilayer hydrogels, cells proliferate and differentiate, maintain a spindle-shaped morphology in collagen, and also develop a tubular microvascular network in the PEGylated fibrin. Moreover, the results of *in vivo* study indicated that dsASC-bilayer hydrogels contribute significantly to wound healing and vascularization of skin equivalent (Natesan et al. 2013). Wang et al. investigated and characterized a bilayer dermal substitute that prepared by integrating a hybrid dermal scaffold with a PU membrane to fabricate a PU membrane/knitted mesh-reinforced collagen–chitosan bilayer skin equivalent (PU-PLGAm/CCS) (Wang et al. 2016). This structure was evaluated in full-thickness skin wound model in rats (Wang et al. 2016). The results showed a balance between porous structure, biocompatibility, and mechanical properties as a skin substitute by integrating the advantages of biological and synthetic polymers (Wang et al. 2016).

6.3 Rabbit

Priya et al. investigated the potential of cryogel bilayer wound dressing and skin regenerating equivalent for the healing of surgically operated full-thickness wounds in the rabbit model (Priya et al. 2016). They reported that the gelatin cryogel sheet supported the cell viability of fibroblasts and keratinocytes. They also demonstrated injuries implanted with cryogel having mannose-6-phosphate showed better and faster skin regeneration with no scar formation (Priya et al. 2016).

6.4 Pig

Guo et al. developed a gene-activated bilayer dermal substitute in a full-thickness wound model of the pig, that contained the complexes with the plasmid DNA encoding vascular endothelial growth factor-165 (VEGF-165), which incorporated into collagen–chitosan/silicone membrane scaffold (Guo et al. 2010). They demonstrated that this skin substitute was transplanted onto the dermis regenerated by the gene-activated complexes on day 10 and well survived. At the follow-up period, the healing zone of skin shown a similar structure and $\sim$80% tensile strength of the normal skin (Guo et al. 2010). Also, in another study, they reported similar results of using this construct to treat burn wounds in the pig model (Guo et al. 2011). This group also compared the effects upon skin repair between the collagen/chitosan porous scaffold and a silicone membrane and ADM in the full-thickness excisional and burned wounds of pigs (Guo et al. 2014). In this study, they reported no significant difference between ADM and collagen/chitosan/silicone bilayer dermal construct on the formation of a new formed blood vessels for the cutaneous wounds on different days (Guo et al. 2014).

7 Vessels

Three articles of bilayer scaffolds investigation for vascular tissue regeneration had used biomaterials such as polyethylene glycol, poly lactic acid and poly caprolactone (Table 4). Animal studies in this field included models of rabbits, rats and mice. Adipose-derived stem cells were used in one study. The HLE, SEM, IHC, immunofluorescence staining, RT-PCR analysis, microscopic and macroscopic observation were the used tests and follow-ups varied from 5 to 56 days. In these studies, there wasn't any quantitative parameters between the group used the bilayer scaffolds and untreated defect.

7.1 Mouse

In 2016 M Cherubino et al. to investigate human adipose-derived stem cells seeded on the new vascularization, implanted the cell-seeded Bilayer and Flowable Integra® scaffolds in the lower backside of nude mice between subcutaneous and muscle tissue. The scaffold without the cells were considered as control group (Cherubino et al. 2016).

7.2 Rat

A bilayer PLLA fibrous scaffold was designed in 2014 by J Pu et al. for vascularization purpose (Pu et al. 2015). The align-oriented thin fibers and random-oriented thick fibers were fabricated to achieve a cylindered bilayer scaffold using two parallel disks and electrospinning method. Control scaffolds were composed of just random fibers and were fabricated using same electrospinning parameters. To test the vascularization capacity of the scaffolds, two incisions were created on left (underneath cavity for implantation place of control scaffolds) and right side (underneath implantation place of bilayer scaffolds) of the abdominal wall of each rat (Pu et al. 2015). The scaffolds are placed so that the align-oriented thin fibers faced the superficial fascia and random-oriented thick fibers faced rat fat tissue and muscles (Pu et al. 2015).

7.3 Rabbit

Zhou et al. in 2016 fabricated a bilayer vascular scaffolds with inner layer composed of microRNA-126 (miR-126) and poly (ethylene glycol)-b-poly(L-lactide-co-e-caprolactone) (PELCL) and with outer layer composed of PCL and gelatin (Zhou et al. 2016). The inner layer was used to deliver miR-126 to endothelial cells of vessels thereby induced vascularization. The cell viability of the vascular endothelial cells cultured on the bilayer scaffolds was ~1.4, ~1.2, and ~1.5 times more than the viability of the cells culture on PELCL scaffolds after 3, 6, and 9 days (Zhou et al. 2016). The outer layer of the scaffolds also provided mechanical support. For *in vivo* study, after removing 10 mm rabbit left common carotid artery the scaffolds transplanted end-to-end. Native tissue was considered as control group. The proangiogenic actions of miR-126 correlated with its repression of SPRED-1, a negative regulator of MAP kinase signaling (McClure et al. 2012). In the absence of miR-126, the increased expression of SPRED-1 diminishes the transmission of intracellular angiogenic signals generated by VEGFs and FGFs the qualitative comparison between the bilayer scaffolds, PELCL scaffolds and native tissue showed that the fabricated bilayer scaffold loaded with miR-126 could enhance endothelialization *in vivo* (Zhou et al. 2016).

Table 4 Different bilayer scaffolds in vessel tissue engineering

Scaffold	Study model	Defect size and location	Sample size	Stem cell	Follow-up	Characterization tests	Results	References
PLLA	Rat	50 × 50 mm Lower abnormal cavity	50 × 50 mm	–	5, 14 d CG: Randomly distributed fibers	Hystological analysis	The results of this study illustrate the high prospect of the fabricated bilayer fibrous scaffolds in tissue engineering and regeneration	Pu et al. (2015)
Integra®	Mouse	15 mm Lower back side	10 × 10 mm	hAdMSCs	30 d CG: Cell-free scaffold	Histological analysis	Results support the therapeutic potential of human adipose-derived stem cells to induce new vascular networks of engineered organs and tissues	Cherubino et al. (2016)
PELCL/ PCL / gelatin	Rabbit	10 mm left common carotid artery	15 × 12 mm	HUVECs	4, 8 w	SEM,TEM, FCA,IF, RT-sqPCR,H&E	These results demonstrated the potential of this approach towards a new and more effective Delivering system for local delivery of miRNAs to facilitate blood vessel regeneration	Zhou et al. (2016)

Abbreviations: Poly(Ethylene glycol)-b-poly(L-lactide-co-e-Caprolactone) (PELCL), Poly(L-lactic Acid) (PLA), Aligned-Fiber Layer (AFL), Scanning electron microscopy (SEM),Macroscopic evaluation(ME), hematoxylin and eosin (H&E),Human adipose -derived mesenchymal stem cells (hAdMSCs),immunofluorescence and cytofluorimetric analysis (ICFA), transmission electron microscope (TEM), Human umbilical vein endothelial cells (HUVECs), Real-time semi quantitative PCR(RT-sqPCR), Flow cytometry analysis(FCA), Immunohistochemistry (IHC)

8 Urinary Bladder

Three studies had investigated the urinary bladder repair by bilayer scaffolds (Table 5). One had used poly-l-lactide/silk fibroin and the other had used a cell matrix graft-silk fibroin. Animal models have been rabbits and rats. Bone marrow stromal cells (BMSCs) were evaluated in one case. The HLE, SEM, IHC, local tissue response, cystography, functional and systemic evaluations of the reconstructed urinary bladder were used to assess the results and follow-ups varied from 2 to 12 weeks.

8.1 Rat

In 2015 a bilayer scaffold with waterproof property in a natural bladder acellular matrix graft layer and a second layer appropriate for cell proliferation (silk fibroin) was designed and fabricated by Y Zhao et al. (2015). The tensile strength of the bilayer scaffolds was 0.39 ± 0.09 MPa. The longitudinal cystotomy defects in male Sprague Dawley rats were created for *in vivo* study of bladder augmentation model. The untreated cystotomy defects were considered as control group (Zhao et al. 2015). Bladder capacity measurement showed that the capacity of the bladders implanted with bilayer scaffolds was ~1.8 times of the capacity of ones in the control groups. The number and diameter of micro vessels, 12 weeks after implantation of the bilayer scaffolds, was equal to control group (Zhao et al. 2015).

8.2 Rabbit

In a study conducted by N Yudintceva et al. (2016) an allogenic rBM-MSC-seeded bilayer scaffold was fabricated for bladder tissue engineering. The silk fibroin inner layer of the scaffold is porous and appropriate for cell proliferation (Yudintceva et al. 2016). The PLLA outer layer make a water proof barrier. For *in vivo* study, the scaffolds with and without rBM-MSC (control group) were transplanted to partial bladder wall cystectomy defect in rabbits (Yudintceva et al. 2016). The Electromyography results showed that the bladder volume capacity of the PL-SF /rBMSCs group is ~2.5 fold more than the control group. The number of microvessels 8 and 12 weeks after implantation of the bilayer scaffolds was ~2.6 and ~1.3 times more than the control group. The diameter of microvessels 8 and 12 weeks after implantation of the bilayer scaffolds was ~1.8 and ~1.6 times more than control group (Yudintceva et al. 2016).

8.3 Dog

X Lv et al. (2018) developed a novel bilayer scaffold composed of silk fibroin microporous network and bacterial nano cellulose for urethral tissue engineering purposes (Lv et al. 2018). The Young's modulus of the scaffolds was measured as 6.85 ± 0.26 MPa. For *in vivo* study, the scaffolds without cells (control group) and the scaffolds seeded with dog-derived keratinocytes and muscle cells were grafted into dog long-segment urethral defects (Lv et al. 2018). The cell-seeded scaffolds improved the urethra more efficiently reconstruction after 12 weeks of implantation (Lv et al. 2018).

9 Nervous System

Three studies had investigated the bilayer scaffolds of collagen and chitosan for nervous system repair (Table 6). Animal models have been rabbits and rats. Adipose-derived stem cells were used in one study. The HLE and histomorpho-assessment of functional nerve, immune-staining and F-actin staining, SEM, and electrophysiological assessment were used to assess the results. Follow-ups varied from 10 to 180 days.

Table 5 Different bilayer scaffolds in urinary bladder tissue engineering

Scaffold	Study model	Defect size and location	Sample size	Stem cell	Follow-up	Characterization tests	Results	References
Silk fibroin/ PLLA	Rabbit	10 × 10 mm^2- partial bladder wall cystectomy	18 × 3 mm	rBMSCs	8, 12 w CG: The scaffold without cells	Electromyography, H&E staining	***Bladder volume capacity:*** Bilayer scaffold group ~2.5 fold >than CG ***Microvessels***: Number: ~2.6 and ~1.3 times >CG after 8 and 12 weeks, respectively Diameter: ~1.8 and ~1.6 times >CG after 8 and 12 weeks, respectively	Yudintceva et al. (2016)
Silk fibroin / BAMG	Rat	15 mm longitudinal cystotomy incision at the dome of bladder	–	–	12 w CG: Cystotomy	Urodynamic studies; Immunofluorescence analyses; Mechanical test	***Bladder volume capacity:*** Bilayer scaffold group ~1.8 fold >than CG ***Microvessels***: Number ≈CG Diameter≈ CG ***MS***: Whole ~0.39 ± 0.09 MPa MPa	Zhao et al. (2015)
Silk fibroin/ bacterial cellulose	Dog	50 mm long segment of urethral	2 × 5 cm	Dog lingual keratinocytes and muscle cells	12 w CG: The scaffold without cells	Histological analysis; Mechanical test	Urethra reconstructed with the SF-BC scaffold seeded with keratinocytes and muscle cells displayed superior structure compared to those with only SF-BC scaffold ***MS:*** Whole ~6.85 ± 0.26 MPa	Lv et al. (2018)

Abbreviations: Control Group (CG), Bladder Acellular Matrix Graft (BAMG), rabbit Bone Marrow Stromal Cells (rBMSCs), Mechanical Strength (MS)

Table 6 Different bilayer scaffolds in nervous tissue engineering

Scaffold	Study model	Defect size and location	Sample size	Stem cell	Follow-up	Characterization tests	Results	References
ENF	Rat	16 mm-sciatic nerve defect	2 × 16 mm	DRG, Sch C	12w Positive CG: Sciatic nerve defects were repaired with 14 mm reversed nerve grafts	SEM,TBS, DAPIS,FAS, IHC, RPhS	Bilayer NGCs hold great potential in facilitating motor axon regeneration and functional motor recovery	Xie et al. (2014)
Collagen	Rat	10 mm-sciatic nerve defect	10 mm	–	20 w Negative CG: Untreated Positive CG: Sciatic nerve defects were repaired with reversed nerve grafts	SEM,H&E,LS, TEM,ME, MPhA,	The hybrid system of bilayer collagen conduit and GDNF-loaded gelatin microspheres combined with gelatin-methacrylamide hydrogels could serve as a new biodegradable artificial nerve guide for nerve tissue engineering	Zhuang et al. (2016)
BChS	Rabbit	Removed a 4 × 4 mm section of the dura	5 × 5 mm	–	10, 21, or 180 d Positive CG: Duraplasty with autologous dura mater	MT,SEM,FLPT	BChS is an ideal alternative for a watertight dural closure because it can be sutured, and it induces organized regeneration with fibroblasts without evidence of fibrosis	Sandoval-Sánchez et al. (2012)

Abbreviations: Scanning Electron Microscopy (SEM),Transmission Electron Microscopy (TEM), Dorsal Root Ganglia (DRG), Schwann Cells (SchC), Functional Assessment Station (FAS), Rhodamine Phalloidin Staining (RPhS), Electrophysiological Assessment (EPHA), Loyez Staining (LS), Toluidine Blue Staining (TBS), Bilayer Chitosan Scaffolding (BChS), Electrospun Nanofibers (ENF), Fluid Leakage Pressure Test (FLPT), Mechanical Test (MT), Immunohistochemistry (IHC)

10 Discussion

Whereas "tissue engineering" art is grafting a bio-factor (cell, genome and/or protein) inside a porous bio-matrix, which is biocompatible and biodegradable, its goal is the fabrication of a bio-contexture with the analogous complexity of human tissues in structure and function. Usually, an engineered tissue has two main components, cells and scaffolds. The cells provide biological processes, while the scaffolds play a critical role in providing 3D media for organizing the ECM and cellular interactions (Bakhshandeh et al. 2017). Approaches for scaffold designing must follow a template of hierarchical porosity so the structures could achieve the desired mechanical properties and mass transport (permeability and diffusion) and the creation of 3D anatomical shapes and complexity inside the structures. A variety of techniques have been used to make a 3D complex of biological forms. More extended and precise studies about tissue structures' details had opened a new window for researchers toward enhancing the efficiency of tissue defect repair. Bilayer scaffolds are used for assembling the biological function and structure complex of 3D media of ECM and inducing significant regeneration in different tissues (Zarrintaj et al. 2017a).

Osteochondral defects are severe articular cartilage defects that are extended to the deep subchondral bone layer. In fact, articular cartilage and the subchondral defect will be repaired utilizing regeneration and repair in both cartilaginous and osseous tissues. For osteochondral defects, the primary focusing of tissue engineering was just on upper cartilaginous layers; therefore, not considering the subchondral layer and the results were disappointing (Gomoll et al. 2010). The providing bio-mimetic properties, stratified (bilayer) scaffolds can supply adequate 3D media to include cellular diffusion, proliferation, differentiation, and enhancing the inductive interactions between bone and cartilage (Hashemi et al. 2021). The bio-mimetic designed scaffold should have the ability to make a natural structure with the primary goal of integration of osteochondral tissues (Fig. 2) (Oliveira et al. 2006). It means that bilayer scaffolds must have

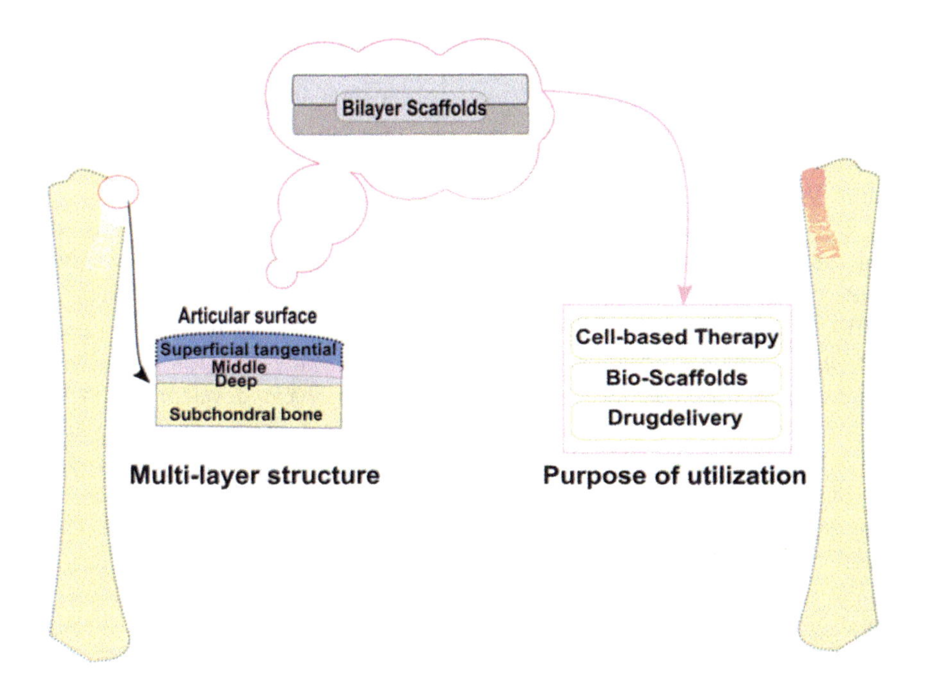

Fig. 2 Osteochondral unit structure; the multilayer structure of this target tissue makes it convenient for using bilayer structures as therapeutic composition

an appropriate capacity of hyaline-like cartilage deposition and bony mineralization simultaneously. The mechanical and biological reasons which support the development of bilayer scaffolds include gathering mechanical resistance and ameliorating natural synovial structure, monotonic marking on the osteochondral junction, and merging the bilayer implant to the host tissue for biological function maintenance. For good physical properties, an osteochondral scaffold should have a rigid osseous layer to support the overlying cartilage and the fusion of the native bone; and also have a chondral layer to seed and penetration of chondrocytes and MSCs for final cartilaginous ECM production (Dodel et al. 2017; Rajaei et al. 2017; Zamanlui et al. 2018). As mentioned in the studies, a bilayer scaffold should be made of two parts and the natural anatomical structure of osteochondral tissues. These two parts include the cartilage and the subchondral part. Various materials have been evaluated for bilayer scaffolds synthesis with different properties. Up to now, natural biomaterials such as gelatin, collagen, cellulose, polyurethane, and artificial polymers, such as PLC, PGA, and other stiff materials and their compositions are the most used materials (Mohebbi et al. 2018; Zarrintaj et al. 2018c). Natural biomaterials usually have high biocompatibility and bioactivity, but no good stiffness and biodegradation time are difficult to control. Usually, the upper cartilage layers are made of low resistance hydrogels, and subchondral underlying layers include high strength materials such as TCP compositions (Da et al. 2013; Moradi et al. 2018). The apparent osteogenic inductive ability of bilayer scaffolds was next examined by gene expression analysis. Comparing ALP gene expression between the bilayer PLLA nanofiber group and the control group showed a six fold increasing ratio. This finding suggested enhanced osteogenic differentiation of MSC stimulated by PLLA nanofibrous bilayer scaffold more than other scaffolds (Birhanu et al. 2018).

Stem cells are one the most critical part of tissue engineering (Ardeshirylajimi et al. 2018; Niknam et al. 2021), especially for osseous tissues due to their ability of proliferation and chondrogenic differentiation that is approved by many studies (Ardeshirylajimi et al. 2018). In this review, most MSCs of different sources have been evaluated for osteochondral reconstruction, and just a few articles had used differentiated chondrocyte. Mesenchymal stem cells have different advantages which we reviewed them in separate study (Golchin et al. 2018, 2020). However, suitable stem cell source finding and using the high-tech methods in manipulating them can be suitable tools in developing bone tissue engineering and its regenerative medicine (Golchin et al. 2018, 2020). The ICRS Visual Histological Assessment Scale was used to evaluate the healing of the defect site. This scoring system was primarily based on the surface, matrix, cell distribution, cell population viability, subchondral bone, and cartilage mineralization (Mainil-Varlet et al. 2003). Host tissue scores in the bilayer scaffolds and the control groups showed significant differences in the mean scores of the matrix, cell morphology, matrix staining, surface regularity of cartilage, and the thickness of cartilage and tissue integration. Among all the scaffolds, the bilayer collagen scaffold designed by Y Qi et al. has a higher histological score in comparison to the control group. The author believes this improvement may be explained by the fact that progenitor cell migration from the bone marrow could be captured and organized by the scaffolds (Qi et al. 2012).

Millions of people suffering of skin defects which need alternative treatments such as wound dressing, allograft and autograft, so traditional autografts are restricted in use due to on time availability and low frequency of donor sites. The recent development in regenerative medicine and tissue engineering has extended the conception of wound healing and causes various novel methods in skin reconstruction and healing (Diaz-Flores et al. 2012). Full-thickness skin defects formed because of different factors like burning, diabetic injuries cannot heal autologous due to loss of dermis. In these cases, an appropriate equivalent pattern of the dermis is hugely needed. Wound healing is a dynamic process based on cellular growth and reconstruction, including ECM interactions, cell growth factors, and various types of resident cells. Appropriate scaffold design can provide ECM role and thought help the skin reconstruction. At the moment, most of the skin scaffolds are monolayer, such as hydrogels and nanofibers that can promote wound healing (Farokhi et al. 2018; Nilforoushzadeh et al. 2018). Wound dressing should be biocompatible, biodegradable, antibacterial, and transfer the proper gas and nutrient with the outer environment (Farokhi et al. 2018; Nilforoushzadeh et al. 2018). Various scaffolds have been designed as a wound dressing, such as nanofibrous, hydrogel, and conductive scaffolds (Zarrintaj et al. 2017b, Zarrintaj et al. 2018a; b). These scaffolds are biodegradable in biological media so that they can be degraded and replaced by reconstructed tissues. Full-thickness skin defects cannot heal by primary closure due to the absence of dermal tissue. Therefore, to achieve optimum full-thickness wound repair, there will be a significant requirement of artificial skin equivalent to replace autogenous skin grafts. To overcome the mentioned problems, researchers focused on skin bilayer scaffolds, which are usually made of upper dense elastic layer and lower soft layer. The upper layer serves as a barrier against bacterial infection of the wound and also keeps moist in the microenvironment. In contrast, the lower layer acts as a cellular scaffold to promote cell migration and tissue reconstruction (Burke et al. 1981; Dagalakis et al. 1980). There are different methods for fabricating bilayer scaffolds to mimic the natural skin structure and provide a normal healing process of defected or injured skin (Fig. 3). Natural and artificial polymers such as fibrin, PEG, collagen, chitosan, PVP-I, gelatin, and polyurethane are used for various matrix production. Besides, active researchers in this field access the bioactive molecular role in the wound healing process and develop composite scaffolds of biopolymers and bioceramics with desired Physico-chemical and biological properties for designing the ideal scaffold systems for skin tissue reconstruction (Guo et al. 2010, 2011; Natesan et al. 2013; Priya et al. 2016; Wang et al. 2018). These bilayer scaffolds have a briliant aspect in skin tissue engineering for the better resembling the skin niches. One of the critical issues for the transplantation of the skin graft is the sufficient vascularization of the implanted artificial dermis, which is essential for delivering nutrients and metabolic wastes and thereby the survival of the graft. It is known that angiogenesis, a necessary process of vascularization, is extensively enhanced by various angiogenic factors such as bFGF, VEGF, PDGF etc. These factors have shown great promise and have been applied singly or simultaneously in matrices to improve angiogenesis (Ehrbar et al. 2008; Nillesen et al. 2007; Perets et al. 2003; Richardson et al. 2001). To compare the angiogenesis properties of the different scaffolds, immunochemical staining for CD31 (a transmembrane protein expressed early in vascular development) was performed to evaluate the newly formed vessels. α-SMA is a plasma protein secreted by the vascular smooth muscle cells enveloping mature blood vessels (Valarmathi et al. 2009). Therefore, co-staining for CD31 and α-SMA was used to identify the mature blood vessels. So, in our systematic review we focused mainly on the biological responses, including relevant factors such as CD31 and a-SMA and dermal repair in different bilayer scaffolds. Our review results showed that The SIS membrane that designed by Wang, L., et al. had a significantly higher relative mRNA expression of CD31 (4.7 fold), which was used to identify the mature blood vessels (L. Wang et al. 2018). Relative mRNA expression of CD31 and a-SMA in a plasmid DNA encoding VEGF-165 activated collagen–chitosan bilayer scaffold was threefold more than control group (Guo et al. 2011). Therefore, these two bilayer scaffolds' significant difference was shown in the higher density of the mature vessels among the other groups.

When nerve injury occurs, the proximal stump of the peripheral nerve begins with a scenario of the regeneration process, while the distal stump undergoes degeneration. After the injury, the distal nerve stump and target tissues are chronically denervated along with nerve and target tissues atrophy. Various scaffolds have been designed for connecting the distal and proximal of the injured nerve (Daly et al. 2012; Zarrintaj et al. 2018d). To regenerate the damaged nerve, axon

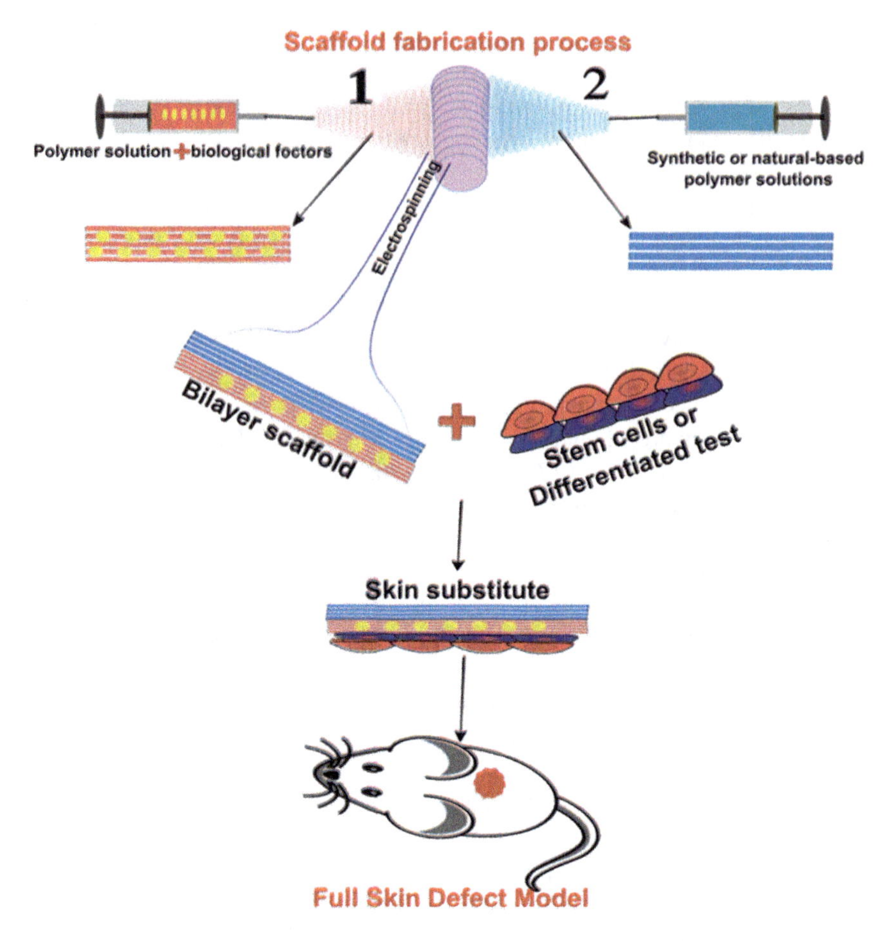

Fig. 3 Electrospinning method for fabricating skin patch; there are several methods and divers polymers for fabricating bilayer scaffolds to mimic the natural skin structure and provide a normal healing process of defected or injured skin

regeneration is initiated by the cell body from proximal toward distal through the synthesis of proteins to lead the repairing axons and connect the inter-stump gap, forming a structure known as a growth cone. During regeneration, Schwann cells at distal and proximal ends provide the Büngner bands to lead the axon. Nerve regeneration is a complex process in which cell-cell, cell-ECM, and cell-scaffold play an essential role. Nerve conduit should be biocompatible, flexible and biomimetic to regenerate damaged nerve properly and stimulate nerve growth. Different strategies have been developed to enhance regeneration such as genetic manipulation, electrical stimulation, the addition of growth factor, physical stimulation, and ECM-like polymer usage (Zarrintaj et al. 2017a; b, 2018b, d).

The bilayer scaffolds could effectively improve nerve fiber sprouting and motor recovery following implantation in a sciatic nerve injury/repair model. H Zhuang et al. (2016) first fabricated a collagen tube with a commercial bilayer collagen membrane (Bio-Gide). To improve nerve growth, they synthesized GDNF-loaded microspheres. The microspheres were coated with GelMA hydrogel onto the inner conduit. The tube with microspheres formed a novel biodegradable nerve conduit to bridge a 10 mm long sciatic nerve defect in rats. Oh et al. fabricated the asymmetrical nerve conduit based on PCL/Pluronic with nano and micro surface. Such a structure enhanced the cell growth on the nano surface and allowed the nutrient permeation through the conduit. Conduit with nerve growth factor concentration gradient exhibited the proper regeneration than the uniform concentration (Sarker et al. 2018).

Bladder augmentation or partial substitution is one of the primary surgical challenges in urology required for different clinical conditions, including contracted bladder, tumors, bladder fibrosis, and neurogenic bladder. Application of transplantable autologous urinary bladder neo-organs produced by tissue engineering demonstrated the feasibility and efficacy of this method in reconstructing the bladder (Oberpenning et al. 1999; Tariverdian et al. 2018). Subsequent clinical application of autologous engineered bladder tissues reported by Atala further proved the efficacy of engineered tissues for cystoplasty (Atala, 2011). A suitable scaffold plays an essential role during tissue engineering bladder regeneration. Recently, some new bilayered scaffold combines the advantages of each scaffold layer for bladder TE and serves as a barrier between urine and the viscera while accommodating sufficient numbers and various types of cells in the regenerated bladder wall, facilitating bladder reconstruction (Eberli et al. 2009; Horst et al. 2013) application of a bilayer PL-SF scaffold seeded with BMSCs was effective for bladder wall reconstruction *in vivo*. The functional activity of the urinary bladder was tested by filling with saline under EMG control. Bladder volume capacity following implantation of the PL-SF /rBMSCs group was 2.5 fold more than CG (Yudintceva et al. 2016).

11 Conclusion

Because of the tissue complexity, scaffold architecture should be a subtle balance between simplicity and sophistication to mimic the native tissue behavior. Biomimic materials are proper substrates for cell growth, but the structure of them should be designed based on target tissue to recapitulate the tissue complexity and tissue interface. Multilayered tissue regeneration necessitates the architecting the unique and impressive construction to reflect healthy tissue performance. This systematic review endeavors to display the studied multilayered tissue regeneration with a growing interest in the bilayer scaffolds. It is understood that there is an increasing fascination with the role of bilayer scaffolds in tissue engineering. The growing trend in bilayer scaffold designs indicates the importance of such scaffold applications in tissue engineering. However, online data reveals that most of the studies performed *in vitro* characterizations, *in vivo* investigations were neglected in the studies for many of the scaffolds. Such a gap between *in vitro* and *in vivo* research necessitates focusing on animal studies to illuminate the bilayer scaffold performances *in vivo* and clinical situations.

Conflict of Interest The authors hold no conflicts of interest.

Funding No funding was received for this article.

References

Ardeshirylajimi A, Golchin A, Khojasteh A, Bandehpour M (2018) Increased osteogenic differentiation potential of MSCs cultured on nanofibrous structure through activation of Wnt/β-catenin signalling by inorganic polyphosphate. Artif Cells Nanomed Biotechnol 46 (sup3):S943–S949. https://doi.org/10.1080/21691401.2018.1521816

Atala A (2011) Tissue engineering of human bladder. Br Med Bull 97(1):81–104

Atoufi Z, Zarrintaj P, Motlagh GH, Amiri A, Bagher Z, Kamrava SK (2017) A novel bio electro active alginate-aniline tetramer/agarose scaffold for tissue engineering: synthesis, characterization, drug release and cell culture study. J Biomater Sci Polym Ed 28 (15):1617–1638. https://doi.org/10.1080/09205063.2017.1340044

Ávila HM, Feldmann E-M, Pleumeekers MM, Nimeskern L, Kuo W, de Jong WC, Schwarz S, Müller R, Hendriks J, Rotter N (2015) Novel bilayer bacterial nanocellulose scaffold supports neocartilage formation in vitro and in vivo. Biomaterials 44:122–133

Bakhshandeh B, Zarrintaj P, Oftadeh MO, Keramati F, Fouladiha H, Sohrabi-jahromi S, Ziraksaz Z (2017) Tissue engineering; strategies, tissues, and biomaterials. Biotechnol Genet Eng Rev 33 (2):144–172. https://doi.org/10.1080/02648725.2018.1430464

Birhanu G, Akbari Javar H, Seyedjafari E, Zandi-Karimi-A, Dusti Telgerd M (2018) An improved surface for enhanced stem cell proliferation and osteogenic

differentiation using electrospun composite PLLA/P123 scaffold. Artif Cells Nanomed Biotechnol 46(6):1274–1281. https://doi.org/10.1080/21691401.2017.1367928
Burke JF, Yannas IV, Quinby WC Jr, Bondoc CC, Jung WK (1981) Successful use of a physiologically acceptable artificial skin in the treatment of extensive burn injury. Ann Surg 194(4):413
Cai YZ, Wang LL, Cai HX, Qi YY, Zou XH, Ouyang HW (2010) Electrospun nanofibrous matrix improves the regeneration of dense cortical bone. J Biomed Mater Res A 95(1):49–57
Cherubino M, Valdatta L, Balzaretti R, Pellegatta I, Rossi F, Protasoni M, Tedeschi A, Accolla RS, Bernardini G, Gornati R (2016) Human adipose-derived stem cells promote vascularization of collagen-based scaffolds transplanted into nude mice. Regen Med 11(3):261–271
Da H, Jia S-J, Meng G-L, Cheng J-H, Zhou W, Xiong Z, Mu Y-J, Liu J (2013) The impact of compact layer in biphasic scaffold on osteochondral tissue engineering. PLoS One 8(1):e54838. https://doi.org/10.1371/journal.pone.0054838
Dagalakis N, Flink J, Stasikelis P, Burke JF, Yannas IV (1980) Design of an artificial skin. Part III. Control of pore structure. J Biomed Mater Res 14(4):511–528
Daly W, Yao L, Zeugolis D, Windebank A, Pandit A (2012) A biomaterials approach to peripheral nerve regeneration: bridging the peripheral nerve gap and enhancing functional recovery. J R Soc Interface 9(67):202–221. Royal Society. https://doi.org/10.1098/rsif.2011.0438
Diaz-Flores L, Gutierrez R, Madrid JF, Acosta E, Avila J, Diaz-Flores L, Martin-Vasallo P (2012) Cell sources for cartilage repair; contribution of the mesenchymal perivascular niche. Front Biosci Scholar S4(4):1275–1294. https://doi.org/10.2741/s331
Dodel M, Hemmati Nejad N, Bahrami SH, Soleimani M, Mohammadi Amirabad L, Hanaee-Ahvaz H, Atashi A (2017) Electrical stimulation of somatic human stem cells mediated by composite containing conductive nanofibers for ligament regeneration. Biologicals 46:99–107. https://doi.org/10.1016/j.biologicals.2017.01.007
Eberli D, Freitas Filho L, Atala A, Yoo JJ (2009) Composite scaffolds for the engineering of hollow organs and tissues. Methods 47(2):109–115
Ehrbar M, Zeisberger SM, Raeber GP, Hubbell JA, Schnell C, Zisch AH (2008) The role of actively released fibrin-conjugated VEGF for VEGF receptor 2 gene activation and the enhancement of angiogenesis. Biomaterials 29(11):1720–1729
Farokhi M, Mottaghitalab F, Fatahi Y, Khademhosseini A, Kaplan DL (2018) Overview of silk fibroin use in wound dressings. Trends Biotechnol 36(9):907–922. Elsevier Ltd. https://doi.org/10.1016/j.tibtech.2018.04.004
Getgood AMJ, Kew SJ, Brooks R, Aberman H, Simon T, Lynn AK, Rushton N (2012) Evaluation of early-stage osteochondral defect repair using a biphasic scaffold based on a collagen–glycosaminoglycan biopolymer in a caprine model. Knee 19(4):422–430
Giannoni P, Lazzarini E, Ceseracciu L, Barone AC, Quarto R, Scaglione S (2015) Design and characterization of a tissue-engineered bilayer scaffold for osteochondral tissue repair. J Tissue Eng Regen Med 9(10):1182–1192
Golchin A, Nourani MR (2020) Effects of bilayer nanofibrillar scaffolds containing epidermal growth factor on full-thickness wound healing. Polym Adv Technol 31(11):2443–2452. https://doi.org/10.1002/pat.4960
Golchin A, Rekabgardan M, Taheri RA, Nourani MR (2018) Promotion of cell-based therapy: special focus on the cooperation of mesenchymal stem cell therapy and gene therapy for clinical trial studies. In: Advances in experimental medicine and biology. Springer, Cham, pp 103–118. https://doi.org/10.1007/5584_2018_256
Golchin A, Shams F, Karami F (2020) Advancing mesenchymal stem cell therapy with CRISPR/Cas9 for clinical trial studies. Adv Exp Med Biol 1247:89–100. https://doi.org/10.1007/5584_2019_459
Gomoll AH, Madry H, Knutsen G, van Dijk N, Seil R, Brittberg M, Kon E (2010) The subchondral bone in articular cartilage repair: current problems in the surgical management. Knee Surg Sports Traumatol Arthrosc 18(4):434–447
Guda T, Walker JA, Pollot BE, Appleford MR, Oh S, Ong JL, Wenke JC (2011) In vivo performance of bilayer hydroxyapatite scaffolds for bone tissue regeneration in the rabbit radius. J Mater Sci Mater Med 22(3):647–656
Guo R, Xu S, Ma L, Huang A, Gao C (2010) Enhanced angiogenesis of gene-activated dermal equivalent for treatment of full thickness incisional wounds in a porcine model. Biomaterials 31(28):7308–7320
Guo R, Xu S, Ma L, Huang A, Gao C (2011) The healing of full-thickness burns treated by using plasmid DNA encoding VEGF-165 activated collagen–chitosan dermal equivalents. Biomaterials 32(4):1019–1031
Guo R, Teng J, Xu S, Ma L, Huang A, Gao C (2014) Comparison studies of the in vivo treatment of full-thickness excisional wounds and burns by an artificial bilayer dermal equivalent and J-1 acellular dermal matrix. Wound Repair Regen 22(3):390–398
Hashemi S, Mohammadi Amirabad L, Farzad-Mohajeri S, Rezai Rad M, Fahimipour F, Ardeshirylajimi A, Dashtimoghadam E, Salehi M, Soleimani M, Dehghan MM, Tayebi L, Khojasteh A (2021) Comparison of osteogenic differentiation potential of induced pluripotent stem cells and buccal fat pad stem cells on 3D-printed HA/β-TCP collagen-coated scaffolds. Cell Tissue Res. https://doi.org/10.1007/s00441-020-03374-8
Horst M, Madduri S, Milleret V, Sulser T, Gobet R, Eberli D (2013) A bilayered hybrid microfibrous PLGA–

acellular matrix scaffold for hollow organ tissue engineering. Biomaterials 34(5):1537–1545

Howard D, Wardale J, Guehring H, Henson F (2015) Delivering rhFGF-18 via a bilayer collagen membrane to enhance microfracture treatment of chondral defects in a large animal model. J Orthop Res 33 (8):1120–1127

Hu X, Wang YY, Tan Y, Wang J, Liu H, Wang YY, Yang S, Shi M, Zhao S, Zhang Y, Yuan Q (2017) A Difunctional regeneration scaffold for knee repair based on aptamer-directed cell recruitment. Adv Mater 29(15):1605235. https://doi.org/10.1002/adma.201605235

Jiang CC, Chiang H, Liao CJ, Lin YJ, Kuo TF, Shieh CS, Huang YY, Tuan RS (2007) Repair of porcine articular cartilage defect with a biphasic osteochondral composite. J Orthop Res 25(10):1277–1290. https://doi.org/10.1002/jor.20442

Jiang Y, Chen L, Zhang S, Tong T, Zhang W, Liu W, Xu G, Tuan RS, Heng BC, Crawford R (2013) Incorporation of bioactive polyvinylpyrrolidone–iodine within bilayered collagen scaffolds enhances the differentiation and subchondral osteogenesis of mesenchymal stem cells. Acta Biomater 9 (9):8089–8098

Kon E, Filardo G, Shani J, Altschuler N, Levy A, Zaslav K, Eisman JE, Robinson D (2015) Osteochondral regeneration with a novel aragonite-hyaluronate biphasic scaffold: up to 12-month follow-up study in a goat model. J Orthop Surg Res 10(1):81

Kozlovsky A, Aboodi G, Moses O, Tal H, Artzi Z, Weinreb M, Nemcovsky CE (2009) Bio-degradation of a resorbable collagen membrane (Bio-Gide®) applied in a double-layer technique in rats. Clin Oral Implants Res 20(10):1116–1123

Kumbhar JV, Jadhav SH, Bodas DS, Barhanpurkar-Naik A, Wani MR, Paknikar KM, Rajwade JM (2017) In vitro and in vivo studies of a novel bacterial cellulose-based acellular bilayer nanocomposite scaffold for the repair of osteochondral defects. Int J Nanomedicine 12:6437

Langer, R., & Vacanti, J. P. (1993). Tissue engineering. Science 260: 920–926. Tissue Engineering: The Union of Biology And Engineering, 98. https://doi.org/https://scholar.google.com/scholar?hl=en&as_sdt=0%2C5&q=Langer%2C+R.%2C+%26+Vacanti%2C+J.+%281993%29.+Tissue+engineering.+Science+260%3A+920-926.+TISSUE+ENGINEERING%3A+THE+UNION+OF+BIOLOGY+AND+ENGINEERING%2C+98.+&btnG

Liao J, Tian T, Shi S, Xie X, Ma Q, Li G, Lin Y (2017) The fabrication of biomimetic biphasic CAN-PAC hydrogel with a seamless interfacial layer applied in osteochondral defect repair. Bone Res 5:17018

Liu S, Dong C, Lu G, Lu Q, Li Z, Kaplan DL, Zhu H (2013) Bilayered vascular grafts based on silk proteins. Acta Biomater 9(11):8991–9003

Liu S, Wu J, Liu X, Chen D, Bowlin GL, Cao L, Lu J, Li F, Mo X, Fan C (2015) Osteochondral regeneration using an oriented nanofiber yarn-collagen type I/hyaluronate hybrid/TCP biphasic scaffold. J Biomed Mater Res A 103(2):581–592

Lv X, Feng C, Liu Y, Peng X, Chen S, Xiao D, Wang H, Li Z, Xu Y, Lu M (2018) A smart bilayered scaffold supporting keratinocytes and muscle cells in micro/nano-scale for urethral reconstruction. Theranostics 8 (11):3153

Mainil-Varlet P, Aigner T, Brittberg M, Bullough P, Hollander A, Hunziker E, Kandel R, Nehrer S, Pritzker K, Roberts S (2003) Histological assessment of cartilage repair: a report by the Histology Endpoint Committee of the International Cartilage Repair Society (ICRS). JBJS 85:45–57

Martin I, Miot S, Barbero A, Jakob M, Wendt D (2007) Osteochondral tissue engineering. J Biomech 40 (4):750–765

McClure MJ, Simpson DG, Bowlin GL (2012) Tri-layered vascular grafts composed of polycaprolactone, elastin, collagen, and silk: optimization of graft properties. J Mech Behav Biomed Mater 10:48–61

Mirzaei-parsa MJ, Ghanbari H, Alipoor B, Tavakoli A, Najafabadi MRH, Faridi-Majidi R (2018) Nanofiber acellular dermal matrix as a bilayer scaffold containing mesenchymal stem cell for healing of full-thickness skin wounds. Cell Tissue Res 75(3):709–721

Mohammadi Amirabad L, Massumi M, Shamsara M, Shabani I, Amari A, Mossahebi Mohammadi M, Hosseinzadeh S, Vakilian S, Steinbach SK, Khorramizadeh MR, Soleimani M, Barzin J (2017) Enhanced cardiac differentiation of human cardiovascular disease patient-specific induced pluripotent stem cells by applying unidirectional electrical pulses using aligned electroactive nanofibrous scaffolds. ACS Appl Mater Interfaces 9(8):6849–6864. https://doi.org/10.1021/acsami.6b15271

Mohebbi S, Nezhad MN, Zarrintaj P, Jafari SH, Gholizadeh SS, Saeb MR, Mozafari M (2018) Chitosan in biomedical engineering: a critical review. Curr Stem Cell Res Ther 14(2):93–116. https://doi.org/10.2174/1574888x13666180912142028

Moradi SLSL, Golchin A, Hajishafieeha Z, Khani M-M, Ardeshirylajimi A (2018) Bone tissue engineering: adult stem cells in combination with electrospun nanofibrous scaffolds. J Cell Physiol 233 (10):6509–6522. https://doi.org/10.1002/jcp.26606

Mrosek EH, Chung HW, Fitzsimmons JS, O'Driscoll SW, Reinholz GG, Schagemann JC (2016) Porous tantalum biocomposites for osteochondral defect repair: a follow-up study in a sheep model. Bone Joint Res 5 (9):403–411

Natesan S, Zamora DO, Wrice NL, Baer DG, Christy RJ (2013) Bilayer hydrogel with autologous stem cells derived from debrided human burn skin for improved skin regeneration. J Burn Care Res 34(1):18–30

Niknam Z, Golchin A, Rezaei–Tavirani M, Ranjbarvan P, Zali H, Omidi MMV (2021) Osteogenic differentiation potential of adipose-derived mesenchymal stem cells cultured on magnesium oxide/polycaprolactone nanofibrous scaffolds for improving bone tissue reconstruction. Adv Pharm Bull, In press

Nilforoushzadeh MA, Amirkhani MA, Zarrintaj P, Salehi Moghaddam A, Mehrabi T, Alavi S, Mollapour Sisakht M (2018) Skin care and rejuvenation by cosmeceutical facial mask. J Cosmet Dermatol 17 (5):693–702. Blackwell Publishing Ltd. https://doi.org/10.1111/jocd.12730

Nillesen STM, Geutjes PJ, Wismans R, Schalkwijk J, Daamen WF, van Kuppevelt TH (2007) Increased angiogenesis and blood vessel maturation in acellular collagen–heparin scaffolds containing both FGF2 and VEGF. Biomaterials 28(6):1123–1131

Oberpenning F, Meng J, Yoo JJ, Atala A (1999) De novo reconstitution of a functional mammalian urinary bladder by tissue engineering. Nat Biotechnol 17(2):149. https://doi.org/10.1038/6146

Oliveira JM, Rodrigues MT, Silva SS, Malafaya PB, Gomes ME, Viegas CA, Dias IR, Azevedo JT, Mano JF, Reis RL (2006) Novel hydroxyapatite/chitosan bilayered scaffold for osteochondral tissue-engineering applications: scaffold design and its performance when seeded with goat bone marrow stromal cells. Biomaterials 27(36):6123–6137

Perets A, Baruch Y, Weisbuch F, Shoshany G, Neufeld G, Cohen S (2003) Enhancing the vascularization of three-dimensional porous alginate scaffolds by incorporating controlled release basic fibroblast growth factor microspheres. J Biomed Mater Res A 65 (4):489–497

Pérez-Silos V, Moncada-Saucedo NK, Peña-Martínez V, Lara-Arias J, Marino-Martínez IA, Camacho A, Romero-Díaz VJ, Banda ML, García-Ruiz A, Soto-Dominguez A, Rodriguez-Rocha H, López-Serna N, Tuan RS, Lin H, Fuentes-Mera L (2019) A cellularized biphasic implant based on a bioactive silk fibroin promotes integration and tissue organization during osteochondral defect repair in a porcine model. Int J Mol Sci 20(20):5145. https://doi.org/10.3390/ijms20205145

Priya SG, Gupta A, Jain E, Sarkar J, Damania A, Jagdale PR, Chaudhari BP, Gupta KC, Kumar A (2016) Bilayer cryogel wound dressing and skin regeneration grafts for the treatment of acute skin wounds. ACS Appl Mater Interfaces 8(24):15145–15159

Pu J, Yuan F, Li S, Komvopoulos K (2015) Electrospun bilayer fibrous scaffolds for enhanced cell infiltration and vascularization in vivo. Acta Biomater 13:131–141

Qi Y, Zhao T, Xu K, Dai T, Yan W (2012) The restoration of full-thickness cartilage defects with mesenchymal stem cells (MSCs) loaded and cross-linked bilayer collagen scaffolds on rabbit model. Mol Biol Rep 39 (2):1231–1237

Qi Y, Du Y, Li W, Dai X, Zhao T, Yan W (2014) Cartilage repair using mesenchymal stem cell (MSC) sheet and MSCs-loaded bilayer PLGA scaffold in a rabbit model. Knee Surg Sports Traumatol Arthrosc 22 (6):1424–1433

Raghunath J, Rollo J, Sales KM, Butler PE, Seifalian AM (2007) Biomaterials and scaffold design: key to tissue-engineering cartilage. Biotechnol Appl Biochem 46 (2):73–84

Rajaei B, Shamsara M, Amirabad LM, Massumi M, Sanati MH (2017) Pancreatic endoderm-derived from diabetic patient-specific induced pluripotent stem cell generates glucose-responsive insulin-secreting cells. J Cell Physiol 232(10):2616–2625. https://doi.org/10.1002/jcp.25459

Reyes R, Delgado A, Solis R, Sanchez E, Hernandez A, Roman JS, Evora C (2014) Cartilage repair by local delivery of transforming growth factor-β1 or bone morphogenetic protein-2 from a novel, segmented polyurethane/polylactic-co-glycolic bilayered scaffold. J Biomed Mater Res A 102(4):1110–1120

Richardson TP, Peters MC, Ennett AB, Mooney DJ (2001) Polymeric system for dual growth factor delivery. Nat Biotechnol 19(11):1029

Sandoval-Sánchez JH, Ramos-Zúñiga R, de Anda SL, López-Dellamary F, Gonzalez-Castañeda R, De la Cruz Ramírez-Jaimes J, Jorge-Espinoza G (2012) A new bilayer chitosan scaffolding as a dural substitute: experimental evaluation. World Neurosurg 77 (3–4):577–582

Sarker MD, Naghieh S, McInnes AD, Schreyer DJ, Chen X (2018) Regeneration of peripheral nerves by nerve guidance conduits: influence of design, biopolymers, cells, growth factors, and physical stimuli. Prog Neurobiol 171:125–150. Elsevier Ltd. https://doi.org/10.1016/j.pneurobio.2018.07.002

Seo J, Tanabe T, Tsuzuki N, Haneda S, Yamada K, Furuoka H, Tabata Y, Sasaki N (2013) Effects of bilayer gelatin/β-tricalcium phosphate sponges loaded with mesenchymal stem cells, chondrocytes, bone morphogenetic protein-2, and platelet rich plasma on osteochondral defects of the talus in horses. Res Vet Sci 95(3):1210–1216

Tariverdian T, Zarintaj P, Milan PB, Saeb MR, Kargozar S, Sefat F, Samadikuchaksaraei A, Mozafari M (2018) Nanoengineered biomaterials for kidney regeneration. In: Nanoengineered biomaterials for regenerative medicine. Elsevier, pp 325–344. https://doi.org/10.1016/B978-0-12-813355-2.00014-4

Valarmathi MT, Davis JM, Yost MJ, Goodwin RL, Potts JD (2009) A three-dimensional model of vasculogenesis. Biomaterials 30(6):1098–1112

Wang X, Wu P, Hu X, You C, Guo R, Shi H, Guo S, Zhou H, Chaoheng Y, Zhang Y (2016) Polyurethane membrane/knitted mesh-reinforced collagen–chitosan bilayer dermal substitute for the repair of full-thickness skin defects via a two-step procedure. J Mech Behav Biomed Mater 56:120–133

Wang L, Wang W, Liao J, Wang F, Jiang J, Cao C, Li S (2018) Novel bilayer wound dressing composed of SIS membrane with SIS cryogel enhanced wound healing process. Mater Sci Eng C 85:162–169

Xie J, MacEwan MR, Liu W, Jesuraj N, Li X, Hunter D, Xia Y (2014) Nerve guidance conduits based on double-layered scaffolds of electrospun nanofibers for repairing the peripheral nervous system. ACS Appl Mater Interfaces 6(12):9472–9480

Yudintceva NM, Nashchekina YA, Blinova MI, Orlova NV, Muraviov AN, Vinogradova TI, Sheykhov MG, Shapkova EY, Emeljannikov DV, Yablonskii PK (2016) Experimental bladder regeneration using a poly-l-lactide/silk fibroin scaffold seeded with nanoparticle-labeled allogenic bone marrow stromal cells. Int J Nanomedicine 11:4521

Zamanlui S, Amirabad LM, Soleimani M, Faghihi S (2018) Influence of hydrodynamic pressure on chondrogenic differentiation of human bone marrow mesenchymal stem cells cultured in perfusion system. Biologicals 56:1–8. https://doi.org/10.1016/j.biologicals.2018.04.004

Zarrintaj P, Moghaddam AS, Manouchehri S, Atoufi Z, Amiri A, Amirkhani MA, Nilforoushzadeh MA, Saeb MR, Hamblin MR, Mozafari M (2017a) Can regenerative medicine and nanotechnology combine to heal wounds? The search for the ideal wound dressing. Nanomedicine 12(19):2403–2422. Future Medicine Ltd. https://doi.org/10.2217/nnm-2017-0173

Zarrintaj P, Rezaeian I, Bakhshandeh B, Heshmatian B, Ganjali MR (2017b) Bio - conductive scaffold based on agarose - polyaniline for tissue engineering. J Skin Stem Cell, (In Press). https://doi.org/10.5812/jssc.67394

Zarrintaj P, Ahmadi Z, Reza Saeb M, Mozafari M (2018a) Poloxamer-based stimuli-responsive biomaterials. Mater Today Proc 5(7):15516–15523. https://doi.org/10.1016/j.matpr.2018.04.158

Zarrintaj P, Manouchehri S, Ahmadi Z, Saeb MR, Urbanska AM, Kaplan DL, Mozafari M (2018b) Agarose-based biomaterials for tissue engineering. Carbohydr Polym 187:66–84. Elsevier Ltd. https://doi.org/10.1016/j.carbpol.2018.01.060

Zarrintaj P, Saeb MR, Ramakrishna S, Mozafari M (2018c) Biomaterials selection for neuroprosthetics. Curr Opin Biomed Eng 6:99–109. Elsevier B.V. https://doi.org/10.1016/j.cobme.2018.05.003

Zarrintaj P, Urbanska AM, Gholizadeh SS, Goodarzi V, Saeb MR, Mozafari M (2018d) A facile route to the synthesis of anilinic electroactive colloidal hydrogels for neural tissue engineering applications. J Colloid Interface Sci 516:57–66. https://doi.org/10.1016/j.jcis.2018.01.044

Zarrintaj P, Zangene E, Manouchehri S, Amirabad LM, Baheiraei N, Hadjighasem MR, Farokhi M, Ganjali MR, Walker BW, Saeb MR, Mozafari M, Thomas S, Annabi N (2020) Conductive biomaterials as nerve conduits: recent advances and future challenges. Appl Mater Today 20:100784. Elsevier Ltd. https://doi.org/10.1016/j.apmt.2020.100784

Zhang YT, Niu J, Wang Z, Liu S, Wu J, Yu B (2017) Repair of osteochondral defects in a rabbit model using bilayer poly (Lactide-co-Glycolide) scaffolds loaded with autologous platelet-rich plasma. Med Sci Monit Int Med J Exp Clin Res 23:5189. https://doi.org/10.12659/MSM.904082

Zhao Y, He Y, Guo J, Wu J, Zhou Z, Zhang M, Li W, Zhou J, Xiao D, Wang Z (2015) Time-dependent bladder tissue regeneration using bilayer bladder acellular matrix graft-silk fibroin scaffolds in a rat bladder augmentation model. Acta Biomater 23:91–102

Zhou F, Jia X, Yang Y, Yang Q, Gao C, Hu S, Zhao Y, Fan Y, Yuan X (2016) Nanofiber-mediated microRNA-126 delivery to vascular endothelial cells for blood vessel regeneration. Acta Biomater 43:303–313

Zhuang H, Bu S, Hua L, Darabi MA, Cao X, Xing M (2016) Gelatin-methacrylamide gel loaded with microspheres to deliver GDNF in bilayer collagen conduit promoting sciatic nerve growth. Int J Nanomedicine 11:1383

Adv Exp Med Biol - Cell Biology and Translational Medicine (2021) 14: 115–133
https://doi.org/10.1007/5584_2021_643

Published online: 15 May 2021

Pluripotent Stem Cell Derived Neurons as In Vitro Models for Studying Autosomal Recessive Parkinson's Disease (ARPD): *PLA2G6* and Other Gene Loci

Renjitha Gopurappilly

Abstract

Parkinson's disease (PD) is a neurodegenerative motor disorder which is largely sporadic; however, some familial forms have been identified. Genetic PD can be inherited by autosomal, dominant or recessive mutations. While the dominant mutations mirror the prototype of PD with adult-onset and L-dopa-responsive cases, autosomal recessive PD (ARPD) exhibit atypical phenotypes with additional clinical manifestations. Young-onset PD is also very common with mutations in recessive gene loci. The main genes associated with ARPD are *Parkin*, *PINK1*, *DJ-1*, *ATP13A2*, *FBXO7* and *PLA2G6*. Calcium dyshomeostasis is a mainstay in all types of PD, be it genetic or sporadic. Intriguingly, calcium imbalances manifesting as altered Store-Operated Calcium Entry (SOCE) is suggested in *PLA2G6*-linked PARK 14 PD. The common pathways underlying ARPD pathology, including mitochondrial abnormalities and autophagic dysfunction, can be investigated ex vivo using induced pluripotent stem cell (iPSC) technology and are discussed here. PD pathophysiology is not faithfully replicated by animal models, and, therefore, nigral dopaminergic neurons generated from iPSC serve as improved human cellular models. With no cure to date and treatments aiming at symptomatic relief, these in vitro models derived through midbrain floor-plate induction provide a platform to understand the molecular and biochemical pathways underlying PD etiology in a patient-specific manner.

Keywords

Autophagic–lysosomal pathway · Calcium · Cellular reprogramming · Dopaminergic neurons · Lewy bodies · Mitophagy · PARK-14 · Phospholipase A2 · SOCE

Abbreviations

ARPD	autosomal recessive Parkinson's disease
DA	dopaminergic
ER	endoplasmic reticulum
ESC	embryonic stem cell
iPSC	induced pluripotent stem cell
LB	Lewy bodies
NSC	neural stem cell

R. Gopurappilly (✉)
National Centre for Biological Sciences, Tata Institute of Fundamental Research, Bengaluru, India
e-mail: renjithap@ncbs.res.in

PD Parkinson's disease
PM plasma membrane
ROS reactive oxygen species
SNpc substantia nigra pars compacta
SOCE store-operated calcium entry
TH tyrosine hydroxylase

1 Introduction

Parkinson's disease (PD) is the second most common neurodegenerative disorder characterized by motor symptoms such as resting tremor, bradykinesia, rigidity, postural instability, stooped posture and freezing, as well as non-motor symptoms including cognitive and behavioural symptoms, sleep disorders, autonomic dysfunction, sensory symptoms and fatigue (Tolosa et al. 2006; de Lau et al. 2006; Jankovic 2008; O'Sullivan et al. 2008; Kalinderi et al. 2016). The pathological hallmark of PD is the progressive loss of dopaminergic (DA) neurons in the substantia nigra pars compacta (SNpc) and the subsequent loss of dopamine inputs to forebrain striatal structures along with the appearance of protein inclusions called Lewy bodies (LB) composed of α-synuclein fibrils (de Lau et al. 2006; Jankovic 2008). PD is classified into two genetic subtypes including monogenic familial forms with Mendelian inheritance and sporadic forms with no underlying genetic factors (Karimi-Moghadam et al. 2018). The sporadic forms of PD are highly prevalent whereas familial PD accounts for only 5–10% of the reported cases. Monogenic familial forms of PD are rare, caused by highly penetrant disease-causing mutations. Parkinsonism caused by dominant mutations including *alpha-synuclein (SNCA)*, *leucine-rich repeat kinase 2 (LRRK2)*, *vacuolar protein sorting 35 (VPS35)* and the like are largely similar to the common, late-onset sporadic PD (Bonifati 2012, 2014). Autosomal recessive PD (ARPD) results from mutations in different loci which have clinical signs typical of PD or can exhibit a wide range of other complex symptoms. The common pathways underlying ARPD are mitochondrial quality control, protein degradation processes and oxidative stress responses, among others (Scott et al. 2017; van der Merwe et al. 2015).

The current knowledge of PD is mostly from postmortem studies and animal models. While the former represent only the end-stage of the disease, the latter fail to reflect human disease pathology due to interspecies differences. In this context, human pluripotent stem cells (both embryonic stem cells, ESC; and induced pluripotent stem cells, iPSC) are an excellent source of cells for differentiation to DA neurons in vitro. 'Disease modelling in a dish' by recapitulating the disease phenotypes in defined cell populations would make it possible to understand the cellular and molecular mechanisms of PD, along with providing a high-throughput drug screening platform (Marchetto et al. 2011; Badger et al. 2014; Martínez-Morales and Liste 2012).

2 Autosomal Recessive Parkinsonism

The hereditary forms of parkinsonism which are transmitted in an autosomal recessive fashion are given in Table 1. Mutations have been identified most commonly in three genes in several ethnic groups spanning different geographical locations: *parkin* (*PRKN*, PARK2), *PTEN- induced putative kinase 1* (*PINK1*, PARK6), and *Parkinson protein 7* (*DJ-1*, PARK7). Point mutations, large genomic rearrangements, leading to deletions or multiplications presenting as homozygous or compound heterozygous variations are reported, particularly for *parkin* (Lucking et al. 2000). Recessive mutations in several genes, including *ATPase type 13A2* (*ATP13A2*, PARK9), *phospholipase A2, group VI* (*PLA2G6*, PARK14), *F-box only protein 7* (*FBXO7*, PARK15), *spatacsin* (*SPG11*), and *DNA polymerase gamma* (*POLG*), cause young- or juvenile-onset PD. These present with other clinical manifestations like dystonia, dementia and other disturbances (Bonifati 2012). *DNAJ subfamily C member 6* (*DNAJC6*, PARK19), *synaptojanin-1* (*SYNJ1*, PARK 20) and *vacuolar protein sorting 13C* (*VPS13C*, PARK23) are also reported to

Table 1 Genes associated with autosomal recessive Parkinson's disease (ARPD)

Locus	Gene	Protein
PARK2	*Parkin*	E3 ubiquitin-ligase
PARK6	*Pink1*	Phosphatase and tensin homolog-induced putative kinase1
PARK7	*DJ-1*	Parkinson protein 7, oncogene DJ-1
PARK9	*ATP13A2*	Lysosomal P5-type ATPase
PARK14	*PLA2G6*	Phospholipase A2, group VI
PARK15	*FBXO7*	F-box only protein 7
PARK19	*DNAJC6*	Putative tyrosine-protein phosphatase auxilin
PARK20	*SYNJ1*	Synaptojanin-1
PARK23	*VPS13C*	Vacuolar protein sorting 13C
	SPG11	Spatacsin
	POLG	DNA polymerase gamma

Genes mapped to different PARK loci and associated with ARPD are listed together with the involved protein. Rarely, mutations in *spatacsin (SPG11)* and *DNA polymerase gamma (POLG)*, cause autosomal recessive parkinsonism with juvenile onset, mostly with atypical features

have mutations causing autosomal recessive PD (Karimi-Moghadam et al. 2018).

2.1 *PLA2G6* (PARK14)

Phospholipase A2 group 6 (*PLA2G6*, *iPLA2β*) gene encodes a calcium-independent group 6 phospholipase A2 enzyme, which hydrolyzes the sn-2 ester bond of the membrane glycerophospholipids to produce free fatty acids and lysophospholipids (Pérez et al. 2004). Various mutations in this gene have been discovered in patients with neurodegenerative disorders such as infantile neuroaxonal dystrophy (INAD), atypical neuroaxonal dystrophy (ANAD), adult-onset dystonia-parkinsonism (DP) and autosomal recessive early-onset parkinsonism (AREP), together known as *PLA2G6*-asscociated neurodegeneration, PLAN (Gregory et al. 1993). PLAN can be classified as neurodegeneration with brain iron accumulation II (NBIA II); however, a wide range of clinical variability is exhibited in these phenotypes with most PD cases devoid of brain iron deposition or cortical atrophy (Guo et al. 2018). iPLA2β protein contains an N-terminal domain, Ankyrin repeats and catalytic domains (Fig. 1). iPLA2β is predominantly localized in the cytosol but can translocate to the Golgi, ER, mitochondria and nucleus under stimulation (Turk and Ramanadham 2004; Kinghorn et al. 2015; Balsinde and Balboa 2005; Ramanadham et al. 2015). Two distinct 85 kDa (VIA-1) and 88 kDa (VIA-2) human iPLA2β isoforms have been discovered along with many N-terminal truncated forms due to proteolytic cleavage and alternate splicing (Ramanadham et al. 2015). It is highly expressed in the human brain including SNpc (http://www.proteinatlas.org). The *PLA2G6* gene mutations was associated with parkinsonism almost a decade ago, with R741Q and R747W being the first to be reported in adult-onset levodopa-responsive dystonia-parkinsonism (Paisan-Ruiz et al. 2009, 2010; Sina et al. 2009). p.R741Q has also been indicated in early-onset PD without dystonia (Bohlega et al. 2016). Though PD-associated mutations in this gene are mostly homozygous, some of them are rare and specific to geographic areas (Gui et al. 2013; Shen et al. 2018; Kapoor et al. 2016; Lu et al. 2012) while others are compound heterozygous (Shen et al. 2019; Chu et al. 2020; Ferese et al. 2018). The mutations that are pathogenic and causal for PD in *PLA2G6* are detailed in Table 2.

Widespread LB pathology is seen with *PLA2G6*-linked PD (Paisán-Ruiz et al. 2012; Miki et al. 2017). The loss of PLA2G6 in *Drosophila* results in impaired retromer function, ceramide accumulation and lysosomal dysfunction, leading to age-dependent loss of neuronal activity (Lin et al. 2018). Dysfunction of mitochondria and increased lipid peroxidation have also been reported in *PLA2G6*-deficient *Drosophila*

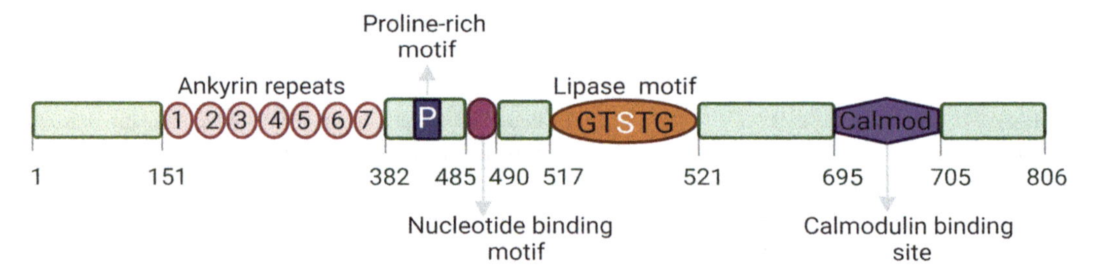

Fig. 1 Structure of PLA2G6 (iPLA2β) protein. The full-length protein is shown with seven ankyrin repeats (pink circles), a proline-rich motif (blue box), a glycine-rich nucleotide-binding motif (magenta), a lipase motif (orange with the active site highlighted), and a proposed C-terminal calcium-dependent calmodulin binding domain (purple). Numbers indicate amino acids

Table 2 List of PD-associated pathogenic mutations in *PLA2G6* (PARK14) gene loci

Mutation	Protein change
c.109 C > T	p.arg37-to-X (R37X)
c.216C > A	p.phe72-to-leu (F72L)
c.238 G > A	p.ala80-to-thr (A80T)
c.991G > T	p.asp331-to-tyr (D331Y)
c.1077 G > A	p.met358-llefsX6
c.1354C > T	p.gln452-to-X (Q452X)
c.1495G > A	p.ala499-to-thr (A499T)
c.1715C > T	p.thr572-to-ile (T572I)
c.1791delC	p.his597-fx69
c.1894C > T	p. arg632-to-trp (R632W)
c.1904G > A	p.arg635-to-gln (R635Q)
c.1976A > G	p.asn659-to-ser (N659S)
c.2215G > C	p.asp739-to-his (D739H)
c.2222G > A	p.argR741-to-gln (R741Q)
c.2239C > T	p.arg747-to-trp (R747W)

Mutations in the *PLA2G6* gene that are pathogenic (or likely pathogenic) and cause Parkinson's disease are documented

mimicking the human fibroblasts with R747W mutation (Kinghorn et al. 2015). Increased sensitivity to oxidative stress, progressive neurodegeneration and a severely reduced lifespan and impaired motor co-ordination is seen in *PLA2G6*-knockout flies (Iliadi et al. 2018). In yet another fly model, the loss of *PLA2G6* leads to shortening of phospholipid acyl chains, resulting in ER stress and impaired neuronal activity as well as formation of α-synuclein fibrils, demonstrating that phospholipid remodeling by *PLA2G6* is essential for DA neuron survival and function (Mori et al. 2019). Similarly, a rodent knockin model of D331Y *PLA2G6* mutation exhibited early degeneration of SNpc DA possibly via mitochondrial and ER stress, impaired autophagic mechanisms and gene expression changes (Chiu et al. 2019). Elevated levels of both α-synuclein and phosphorylated α-synuclein are seen in *PLA2G6* knockout mice models, facilitating the formation of LB and eventually death of affected DA neurons (Sumi-Akamaru et al. 2016). In an in vitro study examining the catalytic activity of PLA2G6 proteins, recombinant proteins containing the three mutations associated with dystonia-parkinsonism (R632W, R741Q and R747W) did not show any altered catalytic activity, whereas the mutations associated with INAD led to a significant loss of enzyme activity (Engel et al. 2010). In addition,

PLA2G6-PD mutants of SHSY5Y, a neuroblastoma cell line, failed to prevent rotenone-induced death of dopaminergic cells (Chiu et al. 2017). *PLA2G6*-PD does not present with a typical clinical scenario and continues to evolve with a wide phenotypic spectrum. It is safe to infer that the PD-relevant mutations do not significantly alter the catalytic activity of the enzyme but induce damage through parallel mechanisms like oxidative stress, mitochondrial dysfunction or even compromised lipid remodeling.

2.1.1 *PLA2G6* and Store-Operated Calcium Entry (SOCE)

Store-operated calcium entry (SOCE) is an arm of calcium signaling activated by depletion of ER stores that triggers influx of calcium across the plasma membrane (PM) brought about by the calcium sensor *STIM* and the PM pore channel *Orai* (Feske et al. 2005, 2006; Vig et al. 2006). Interestingly, in a genetic screening of *Drosophila*, not only *STIM1* and *Orai1*, but also a fly orthologue of *PLA2G6* encoded by the *CG6718* gene, were identified as SOCE activators (Vig et al. 2006). Many groups have from this time identified *PLA2G6* as an endogenous activator of SOCE (Smani et al. 2016; Schäfer et al. 2012; Singaravelu 2006; Bolotina and Orai 2008). The physiological relevance of neuronal SOCE is disputed (Lu and Fivaz 2016) and its role in DA neurons, particularly, is not known. In PD, abnormal calcium homeostasis triggers a cascade of downstream events that eventually leads to cell death (Michel et al. 2016). Interestingly, primary skin fibroblasts from idiopathic and PLA2G6-PD (R747W) patients revealed a significant deficit in endogenous SOCE and similarly, MEFs from the exon2-knockout mice exhibited deficient store-operated *PLA2G6*-dependent calcium signaling (Zhou et al. 2016a). This was also mirrored in the iPSC-derived DA neurons along with low ER calcium levels and deficient autophagic flux. The knockout mice also showed age-dependent loss of DA neurons. This study for the first time arrived at a causal relationship between *PLA2G6*-dependent SOCE, depleted stores, dysfunctional autophagy in DA neurons and a PD-like phenotype (Zhou et al. 2016a). Recently, in a patient-derived (D331Y) DA neuron model, imbalance of calcium homeostasis, markedly deficient SOCE, increase of UPR proteins, mitochondrial dysfunction, increase of ROS, and apoptosis was reported. Interestingly, the UPR modulator, azoramide rescued the phenotype of the mutant DA neurons, possibly via CREB signaling (Ke et al. 2020). These recent developments have opened new exciting areas to study the significant contributions of *PLA2G6* and SOCE in PD, and may involve new undiscovered molecules providing a yet unexplored arena for PD-drug discovery.

2.2 *Parkin*, *PINK1* and *DJ-1*

The E3 ubiquitin ligase parkin (PARK2) and the serine/threonine kinase PINK1 (phosphatase and tensin homolog-induced putative kinase1, PARK6), act together in a mitochondrial quality control pathway and promote the selective autophagy of depolarized mitochondria (mitophagy) (Narendra et al. 2012). PINK1 levels are low in healthy cells as it is continually cleaved inside the mitochondria in a sequential manner by proteases (Yamano and Youle 2013). The import of PINK1 into mitochondria is stopped when the organelle loses its inner membrane electrochemical gradient (depolarization), which leads to the stabilization of the protein on the mitochondrial outer membrane (Lin and Kang 2008). This accumulation of PINK1 kinase on the mitochondria triggers parkin recruitment and activation resulting in ubiquitination of various outer mitochondrial membrane proteins (Taanman and Protasoni 2017; Matsuda et al. 2010). The damaged mitochondria are eventually eliminated by autophagy. Pathogenic PD-associated mutations in either *Parkin* or *PINK1* causes accumulation of impaired mitochondria, increased ROS and neuronal cell death (Seirafi et al. 2015). More than 100 different *Parkin* mutations have been reported from PD patients, including deletions, insertions, multiplications, missense and truncating mutations, and over 40 point mutations and, rarely, large deletions, have been detected in *PINK1* (Lesage and Brice 2009). Clinically, both cause young-onset PD and show responsiveness to levodopa. The phenotype associated with the oncogene *DJ-1* mutations has been studied in a smaller number of patients but it is overall indistinguishable from that of the patients with *PINK1* or *Parkin* mutations (Bonifati et al. 2003). *DJ-1* is thought to be involved in the regulation of the integrity and calcium crosstalk between endoplasmic reticulum (ER) and mitochondria, and pathogenic mutations lead to impaired ER-mitochondria association in PD (Liu et al. 2019).

2.3 *ATP13A2, FBXO7, SPG11* and *POLG*

Mutations in *ATP13A2, FBXO7, spatacsin* and *POLG* cause juvenile-onset ARPD along with *PLA2G6* (Bonifati 2012). Mutations in *ATP13A2* or PARK9, were first identified in 2006 in a Chilean family and are associated with a juvenile-onset, levodopa-responsive type of parkinsonism called Kufor–Rakeb syndrome (KRS). KRS involves pyramidal degeneration, supranuclear palsy, and cognitive impairment (Ramirez et al. 2006). The *ATP13A2* gene encodes a large lysosomal protein, belonging to the P5-type ATPase family of transmembrane active transporters. Its substrate specificity remains unknown. It is suggested that *ATP13A2* recruits HDAC6 to lysosomes to promote autophagosome–lysosome fusion and maintain normal autophagic flux (Wang et al. 2019). This, in turn, is required for preventing α-synuclein aggregation in neurons. FBXO7, in turn is an adaptor protein in SCF^{FBXO7} ubiquitin E3 ligase complex that mediates degradative or non-degradative ubiquitination of substrates. *FBXO7* mutations aggravate protein aggregation in mitochondria and inhibit mitophagy (Zhou et al. 2018). *Parkin-* and *FBXO7*-linked PD have overlapping pathophysiologic mechanisms and clinical features. Wild-type FBXO7, but not PD-linked FBXO7 mutants, has been shown to rescue DA neuron degeneration in *Parkin* null *Drosophila* (Zhou et al. 2016b). Loss of activity of FBXO7 in seen in patients with PARK 15-PD and is therefore crucial for the maintenance of neurons (Zhao et al. 2011). *SPG11* or *spatacsin* mutations present with bilateral symmetric parkinsonism at an early age with rapid deterioration and development of spastic paraplegia and thinning of the corpus callosum on MRI, typical of spastic paraplegia 11 (Guidubaldi et al. 2011; Cao et al. 2013). An involvement of *POLG*, the mitochondrial DNA polymerase that is responsible for replication of the mitochondrial genome is considered in early-onset PD especially in the presence of additional symptoms, such as ophthalmoparesis, non-vascular white matter lesions and psychiatric comorbidity (Anett et al. 2020). A role of mitochondrial DNA defects in the pathogenesis of neurodegenerative parkinsonism with *POLG* mutations is speculated (Miguel et al. 2014).

3 Common Pathways in ARPD

There are many converging features seen at the molecular and clinical levels in ARPD that are discussed in the following section. Understanding these causal molecular mechanisms is crucial to identify common targets and devise therapeutic approaches. However, some fundamental underlying processes still remain unclear. The contributions of the intracellular organelle ER in ARPD pathology via the calcium signaling pathway, SOCE is poorly understood. The mitochondrion, which is the star player in PD pathogenesis, regulates SOCE activity possibly via sub-plasmalemmal calcium buffering, the generation of mediators, local ATP modulation and regulation of STIM1 (Malli and Graier 2017). In turn, SOCE-derived calcium significantly affects mitochondrial metabolism. Hence, the communication between SOCE and mitochondria is hypothesized to be interdependent and complex leading to fine-tuning of both SOCE and mitochondrial function (Spät and Szanda 2017). Though PD-relevant mitochondrial processes are studied extensively, SOCE and its role in PD are largely unexplored. The microbiota–gut–brain axis and its imbalance by alterations in the human microbiome also represent a risk factor for PD (Sampson et al. 2016). Such pathways that are not well-proven are omitted from this section, for ease of understanding.

3.1 Mitochondrial Pathways

The most compelling evidence for loss of mitochondrial fidelity comes from the genes *PINK1* and *Parkin*. As mentioned earlier, PINK1 accumulates on the outer membrane of damaged mitochondria and activates Parkin's E3 ubiquitin ligase activity. Parkin recruited to the damaged mitochondrion ubiquitinates the outer mitochondrial membrane proteins to trigger selective autophagy. In the late 1970s when accidental exposure to 1-methyl-4-phenyl-1,2,3,6-tetrahydropyridine (MPTP) was found to cause PD and neurodegeneration, the first causal mechanism speculated was mitochondrial dysfunction (William Langston et al. 1983). A specific defect of Complex I activity is also seen in the substantia nigra of patients with PD (Schapira et al. 1990). We now understand that the pathways included in mitochondrial quality control system are fission/fusion, mitochondrial transport, mitophagy and mitochondrial biogenesis (Scott et al. 2017). The precise mechanisms by which Parkin and PINK1 regulate fission and fusion is debated, but studies from *Drosophila* and mammalian culture systems, though contradictory, indicate unbalanced mitochondrial fission and fusion in PINK1 mutants (Scott et al. 2017; Chen and Chan 2009; Pryde et al. 2016; Scarffe et al. 2014; Yu et al. 2011). *DJ-1* (Irrcher et al. 2010) and *ATP13A2* (Park et al. 2014) mutants also show fragmented mitochondria. The combined effects of Parkin and PGC-1α in the maintenance of mitochondrial homeostasis in dopaminergic neurons is demonstrated (Zheng et al. 2017). PINK1 is also involved in mitochondrial motility along axons and dendrites of neurons. PINK1 interacts with Miro, a component of the motor/adaptor complex binding mitochondria to microtubules and allowing their movement to and from cellular processes (Brunelli et al. 2020). Miro is phosphorylated by PINK1 and ubiquitinated by parkin, leading to its degradation and halting mitochondrial transport promoting clearance of damaged mitochondria (Liu et al. 2012). Parkin/PINK1 is hence involved in mitochondrial trafficking (Scott et al. 2017; Weihofen et al. 2009).

The clearance of damaged mitochondria or mitophagy is a pathway common to mostly all ARPD-related genes. The role of Pink1-Parkin in mitophagy is well-established (Deas et al. 2011; Yamano and Youle 2020). Fbxo7 is also shown to induce mitophagy in response to mitochondrial depolarization in a common pathway with Parkin and PINK1, and PD-associated mutations interfere in this mechanism (Burchell et al. 2013). Parkin (Kuroda et al. 2006) and PINK1 (Pirooznia et al. 2020) is also linked to mitochondrial biogenesis, therefore probably being a part of mitochondrial transcription/replication.

3.2 Autophagy–Lysosomal Pathways

In addition to impaired mitophagy, protein degradation pathways, especially the autophagy–lysosomal pathway, are affected in PD. ATP13A2 is suggested as a regulator of the autophagy–lysosome pathway. ATP13A2 acts in concert with another PD-protein SYT11 and its loss of function results in dysfunctional autophagy–lysosomal pathway as seen in PD (Bento et al. 2016). α-synuclein-independent neurotoxicity due to endolysosomal dysfunction has also been demonstrated in ATP13A2 null mice (Kett et al. 2015). *Parkin* knockout neurons too show perturbed lysosomal morphology and mitochondrial stress (Okarmus et al. 2020). DJ-1 is associated to chaperone-mediated autophagy (CMA) and its deficiency aggravates α-synuclein aggregation by inhibiting CMA activation (Xu et al. 2017). Loss of DJ-1 could also lead to impaired autophagy and accumulation of dysfunctional mitochondria (Krebiehl et al. 2010). Autophagic defects are a mainstay in *PLA2G6*-PD. Genetic or molecular impairment of PLA2G6-dependent calcium signaling is a trigger for autophagic dysfunction, progressive loss of DA neurons and age-dependent L-DOPA-sensitive motor dysfunction in a mouse knockout model (Zhou et al. 2016a).

3.3 Cell Death and Oxidative Stress

Oxidative stress and apoptosis are frequently involved in ARPD pathogenesis. ROS accumulation plays a key role in the initiation and acceleration of cell death, compromising neuronal function and structural integrity (Schieber and Chandel 2014). The protein products of Parkin, PINK1 and DJ-1 are associated with disrupted oxidoreductive homeostasis in DA neurons. Impaired cell survival in part due to defective oxidative stress response is implicated in *PARK2* knockout neurons (Bogetofte et al. 2019). Further, transgenic overexpression of the parkin substrate, aminoacyl-tRNA synthetase complex interacting multifunctional protein-2 (AIMP2) leads to a selective, age-dependent progressive loss of dopaminergic neurons via activation of poly(ADP-ribose) polymerase-1 (PARP1) (Lee et al. 2013). Similarly, PINK1 is also shown to exert a neuroprotective effect by inhibiting ROS formation and maintaining normal mitochondrial membrane potential and morphology in cultured SN dopaminergic neurons (Wang et al. 2011). The profiles of oxidative damage in the whole brain and neurochemical metabolites in the striatum of *PINK1* knockout rats at different ages and genders were studied and oxidative damage revealed as a crucial factor for PD (Ren et al. 2019). Loss of PINK1 inhibits the mitochondrial Na(+)/Ca(2+) exchanger (NCLX), resulting in impaired mitochondrial calcium extrusion, which was, however, fully rescued by activation of the protein kinase A (PKA) pathway (Kostic et al. 2015). DJ-1 also has a role in cell death and combating oxidative stress. It suppresses PTEN activity, thereby promoting cell growth and promoting cellular defence against ROS through PI3K/Akt signaling (Chan and Chan 2015). Reduced anti-oxidative stress mechanisms have been reported in PD patients with mutant DJ-1 protein (Takahashi-Niki et al. 2004). It is also described that Daxx, the death-associated protein, translocated to the cytosol selectively in SNpc neurons due to MPTP-mediated down-regulation of DJ-1 after treatment with the neurotoxin in mouse models (Karunakaran et al. 2007). DJ-1 is also hypothesized to regulate the expression of UCP4 by oxidation and partially via NF-κB pathway in its protective response to oxidative stress (Xu et al. 2018). DJ-1, particularly in its oxidized form, is documented as a biomarker for many diseases including PD. DJ-1 may also work by increasing microRNA-221expression through the MAPK/ERK pathway, subsequently leading to repression of apoptotic molecules (Oh et al. 2018). Additionally, cell-permeable Tat-DJ-1 protein exerts neuroprotective effects in cell lines and mouse models of PD (Jeong et al. 2012). Data from the field indicate that DJ-1 may become activated in the presence of ROS or oxidative stress, but also as part of physiological receptor-mediated signal transduction and acts as a transcriptional regulator of antioxidative gene batteries (Kahle et al. 2009). ATP13A2, on the contrary, is thought to protect against hypoxia-induced oxidative stress (Xu et al. 2012). A recent study revealed a conserved neuroprotective mechanism that counters mitochondrial oxidative stress via ATP13A2-mediated lysosomal spermine export (Vrijsen et al. 2020). PLA2G6 protein is also indicated in oxidative stress-related pathways (Kinghorn et al. 2015; Ke et al. 2020).

4 Induced Pluripotent Stem Cells (iPSC) in Parkinson's Disease Research

Yamanaka's discovery in 2007 where key transcriptional factors (*Oct4*, *Sox2*, *Klf4* and *c-Myc*) were used to reprogramme adult cells to a de-differentiated, poised cell type called Induced pluripotent stem cells (iPSCs) revolutionized the field of human disease modeling (Takahashi et al. 2007). Reprogrammed iPSCs are similar to embryonic stem cells or ESCs, are pluripotent and can differentiate to multiple lineages. iPSCs when derived from a PD patient has the patient's complete genetic background and provides a valuable platform to study the impact of genetic mutations. Within a year of the discovery of iPSCs, PD-patient-derived iPSCs (Park et al. 2008) and DA neuron differentiation from iPSCs were reported (Soldner et al. 2009). iPSC models have successively been established from various sporadic and familial PD patients. To date, iPSCs are the most robust cellular system to understand PD and generate disease-relevant cell types for PD (Playne and Connor 2017).

iPSC studies typically involve few participants and random selection of cases and controls, which results in heterogeneous models in vitro. Consequently, a large sample size is required to increase statistical power and sample sizes of 10–30 individuals per iPSC study may be required to achieve a statistical power of 80% (Hoekstra et al. 2017; Tran et al. 2020). These are, in turn, labour-intensive and expensive; hence, it is unlikely that these requirements are met. A smaller number may be used if clinically and genotypically homogeneous subjects are used to reduce the variance in the cellular phenotypes. Hence, a preponderance of familial PD is seen in these studies. A recent report elegantly summarizes a meta-analysis of 385 iPSC-derived neuronal lines modeling mutations/deletions/triplications in *LRRK2, PRKN, PINK1, GBA* and *SNCA* (Tran et al. 2020). The authors discuss the importance of using the right controls in such studies. When healthy subjects are used as controls, differences in genetic background may give rise to variance in neuronal phenotypes studies that are not caused by disease mutations. Gene-editing techniques (TALEN, ZFN, CRISPR/Cas9) aid in the generation of isogenic lines that differ in only one single mutant gene, and this circumvents the issue of variance due to genetic background. CRISPR/Cas9 system, an RNA-based endonuclease, is the most common and effective tool used in the iPSC model for introducing the genetic changes seen in PD, including but not limited to knockout, knockin and gene correction (Arias-Fuenzalida et al. 2017; Qing et al. 2017; Soldner et al. 2016; Vermilyea et al. 2020). Disease-causing mutations, therefore, can be inserted in healthy ESC or iPSC lines or gene-correction of a single mutation can be performed in PD-lines to include comparative isogenic control lines (Tran et al. 2020).

Differentiation protocols for DA neurons mimic embryologic development in utero. Unlike cortical neurons, midbrain DA neurons are derived from the ventral floor plate of the neural tube (Ono et al. 2007). The molecular mechanisms that regulate the development of midbrain DA neurons in vivo, and how taking cues from this, one can generate in vitro human midbrain DA neurons from iPSCs was systematically reviewed previously (Arenas et al. 2015). Dual-SMAD inhibition along with modulation of sonic hedgehog (SHH) and WNT signaling by CHIR99021 (GSK3β inhibitor), and addition of FGF8 is routinely used to generate midbrain floor-plate precursors (Kirkeby et al. 2012; Kriks et al. 2011; Reinhardt et al. 2013). BDNF, GDNF, TGFβ3, dbcAMP, ascorbic acid, DAPT and ActivinA are used to enhance the purity and maturity of DA neurons which express the key marker TH (tyrosine hydroxylase) (Monzel et al. 2017a; Smits and Schwamborn 2020; Smits et al. 2019). A schematic and generalized diagram outlining the midbrain DA differentiation protocol from iPSCs is shown in Fig. 2 (the starting population can also be ESCs). It is important to note that, irrespective of the protocols used a heterogeneous cell population is attained, with neurons, glia and NSCs. To attain a high percentage of TH^+ DA neuron population, several strategies have been employed. Sorting of DA progenitors which are $CD184^{high}/CD44^-$(Suzuki et al. 2017) or $CORIN^+$(Kikuchi et al. 2017) is shown to increase the TH^+ DA neuron yield. CRISPR/ Cas9-based knockin of a fluorescent reporter to visually identify and purify TH^+ DA neurons has also been attempted (Calatayud et al. 2019). A monolayer-based neural differentiation protocol was described recently that reproducibly generates ~70–80% midbrain DA neurons (Stathakos et al. 2020; Stathakos et al. 2019). A higher concentration of 300 ng/ml SHH (100–200 ng/ml is used normally) in combination with a lower concentration of 0.6 μM CHIR99021 (0.8 μM–3 μM is used normally) and passaging and replating in the early differentiation and patterning stages maximized the yield of midbrain DA neurons as early as day 30. This monolayer platform is amenable to imaging and functional

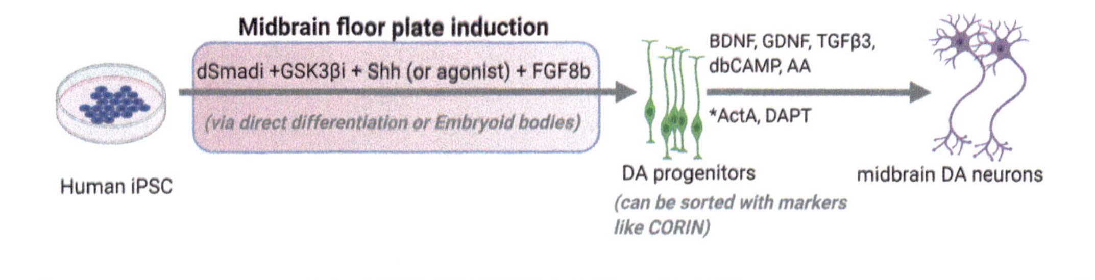

Fig. 2 A generalized schematic protocol for differentiation of human induced pluripotent stem cells (iPSCs) into midbrain dopaminergic (DA) neurons. Human iPSCs are treated with small molecules, such as GSK3β inhibitors (GSK3i) and SMAD inhibitors, along with Shh/FGF8b, to induce midbrain floor-plate formation and subsequent midbrain DA specification. This is done either by means of direct differentiation from iPSCs or through embryoid bodies (EBs). The DA progenitors can be sorted with a midbrain cell surface marker like CORIN to achieve higher purity of DA neurons via elimination of unwanted contaminant cells. Mature midbrain DA neurons are generated from these progenitors by addition of mentioned factors at the end of 40–70 DIV (days in vitro) in total. * indicates factors that may be used in the final differentiation step, but not compulsory. ActA, activin A; AA, ascorbic acid; BDNF, brain-derived neurotrophic factor; DAPT, γ-secretase inhibitor; dbcAMP, dibutyryl cyclic adenosine monophosphate; FGF8, fibroblast growth factor 8; GDNF, glial-cell-derived neurotrophic factor; Shh, sonic hedgehog; TGFβ3, transforming growth factor beta-3

assessments of autophagy/mitophagy (Stathakos et al. 2020). In an interesting study, autophagic dysfunction and premature aging was shown by PD-patient-derived NSCs (Zhu et al. 2019). One of these patients had early-onset PD with a novel mutation in *PLA2G6* gene. The authors hypothesize that developmental defects, and the subsequent depletion of NSC pool size could lead to lower DA neuron number and this impacts the onset and severity of the disease progression (Zhu et al. 2019). Hence, not only iPSC-derived DA neurons, but also the developmentally upstream NSCs could be a disease-relevant phenotype for prediction analyses and design of intervention therapies.

4.1 iPSC-Derived Two-Dimensional (2D) and Three-Dimensional (3D) Culture Models of ARPD

Mutations in the *PARK2* gene, encoding the protein parkin, have been identified as the most common cause of ARPD. Unsurprisingly, the limited in vitro iPSC-derived ARPD models primarily examine the cellular pathologies of this gene. Human iPSC-derived neurons with *PARK2* knockout is known to demonstrate severe mitochondrial dysfunction even in the absence of external stressors. *PARK2* patient iPSC-derived neurons showed increased oxidative stress, higher α-synuclein accumulation and enhanced activity of the nuclear factor erythroid 2-related factor 2 (Nrf2) pathway (Imaizumi et al. 2012). Interestingly, iPSC-derived neurons, but not fibroblasts or iPSCs, exhibited abnormal mitochondrial morphology and impaired mitochondrial homeostasis in their study. In a similar study, the loss of parkin significantly increased the spontaneous DA release independent of extracellular calcium and showed decreased dopamine uptake by reducing the total amount of correctly folded DAT along with DA-dependent oxidative stress. All these phenotypes could be rescued by overexpression of *parkin*, but not its PD-linked T240R mutant or GFP (Jiang et al. 2012). Mitochondrial dysfunction, elevated α-synuclein, synaptic dysfunction, DA accumulation, and increased oxidative stress and ROS have been reported in *PARK2*- and *PINK1*-patient-derived neurons in a floor-plate-based but not a neural-rosette-based directed differentiation strategy (Chung et al. 2016). Impairment of mitophagy via formation of S-nitrosylated PINK1 (SNO-PINK1) has also been shown in iPSC-derived parkin-mutant neurons (Oh et al. 2017). In a recent study, *PARK2* knockout neurons from isogenic lines exhibited lysosomal impairments and autophagic perturbations, suggesting an impairment of the autophagy–lysosomal pathway in parkin-deficient cells (Okarmus et al. 2020). The same group had earlier shown disturbances in oxidative stress defence, mitochondrial respiration and morphology, cell cycle control and cell viability of parkin-deficient neurons (Bogetofte et al. 2019).

Midbrain-specific 3D cultures are at present a powerful tool for modeling PD in vitro. The use of microwells by Tieng et al. was the very first attempt in this direction to generate embryoid bodies, which were then placed on an orbital shaker before being seeded and grown at air–liquid interface (Tieng et al. 2014). DA progenitor cells expressed FOXA2 and LMX1A as well as TH within a short span of 3 weeks. Subsequently, a number of reports have been published for midbrain organoids with neuromelanin expression seen in long-term cultures (Smits and Schwamborn 2020; Kim et al. 2019; Jo et al. 2016; Monzel et al. 2017b; Qian et al. 2016). However, PD modeling with midbrain organoids is largely focused on dominant mutations like *LRRK2* (Smits et al. 2019; Kim et al. 2019), *SNCA* (Jan et al. 2018) and also an only report on sporadic PD (Chlebanowska et al. 2020). A very recent study used CRISPR-Cas9 genome editing to develop isogenic loss-of-function 3D models of early-onset autosomal recessive PD (PARKIN$^{-/-}$, DJ-1$^{-/-}$, and ATP13A2$^{-/-}$) to identify common pathways (Ahfeldt et al. 2020). The DA neuronal population was markedly reduced in *PRKN*$^{-/-}$ organoids but no significant differences were observed in the other two cell lines. The death of newly differentiated TH$^+$ neurons and higher expression of VTA marker CALB1 in the *PRKN*$^{-/-}$ organoids were indicative of A9-like neurons being more severely affected than others. A dysregulation of the autophagy–lysosomal pathway and upregulated ROS in all cell lines and an upregulation of pathways associated with oxidative phosphorylation, mitochondrial dysfunction, and Sirtuin signaling, as well as a significant depletion of mitochondrial proteins were seen in the *PRKN−/−* DA neurons (Ahfeldt et al. 2020). Astrocytic pathologies in human *PRKN*-mutated iPSC-derived midbrain organoids were revealed for the first time, suggesting a non-autonomous cell death mechanism for dopaminergic neurons in brains of *PRKN*-mutated patients (Kano et al. 2020). Mutations in *PINK1* have also been reported to generate reduced TH$^+$ counts in midbrain organoids (Jarazo et al. 2019). Human midbrain organoid/spheroid cultures are a scalable and reproducible system to obtain DA neurons expressing markers of terminal differentiation along with neuromelanin production in a 3D environment that replicates the neuronal and glial cytoarchitecture of the human midbrain (Galet et al. 2020). They can hence provide a crucial platform to explore the molecular basis of ARPD, and also to delineate the associated cellular pathologies.

Cell Replacement Therapy with iPSC-Based DA Derivatives The various challenges pertaining to the safety and efficacy of stem-cell-based cell transplantations in PD have been elegantly reviewed and described (Fan et al. 2020). The right neural cell type for transplantation is of utmost importance. FGF8b inclusion in the differentiation protocols helps in acquisition of a caudal midbrain fate and promotes high dopaminergic graft volume, density and yield as evidenced by deep sequencing of more than 30 human ESC-derived midbrain tissues (Kirkeby et al. 2017). Dopaminergic precursors beyond the floor-plate progenitor stage but before formation of TH$^+$ dopaminergic neurons are found to be most efficient for graft survival, integration and function in animal models (Kikuchi et al. 2017; Nolbrant et al. 2017). Grafting of these precursors into the putamen area, where SNpc dopaminergic neurons innervate, is an approach most likely to succeed (Fan et al. 2020). The number of cells to be transplanted is still debated. Takahashi's group reported a minimum of 16,000 TH$^+$ cells in a primate model to see improvements in PD score and motor function (Kikuchi et al. 2017). The generation and implantation of iPSC-derived autologous dopaminergic progenitor cells in a patient with idiopathic PD is reported with clinical and imaging results suggesting possible benefit over a period of 24 months (Schweitzer et al. 2020). A global consortium, GForce-PD (http://www.gforce-pd.com), was set up in 2014, with major academic networks in Europe, the United States and Japan working on developing stem-cell-derived neural cell therapies for PD (Barker et al. 2015). The clinical trials using human ESCs are ongoing in Australia (NCT02452723) and China (NCT03119636), with their pre-clinical data published (Garitaonandia et al. 2016; Wang et al. 2018). A clinical trial (JMA-IIA00384, UMIN000033564) in Japan to treat PD patients by using iPSC-derived DA progenitors (DAPs) was started in 2018 by Takahashi and colleagues (Barker et al. 2017; Doi et al. 2020). The results of these trials are eagerly anticipated.

5 Limitations to iPSC-Based Disease Modeling of PD

Although iPSCs and their derivatives are currently in the forefront as PD models, there are many challenges which remain unaddressed. The most significant drawback of in vitro models is the absence of LB formation in PD iPSC-derived DA neurons. Increased levels of phosphorylated pS129 α-synuclein, however, have been observed (Lin et al. 2016). Additionally, the efficiency of generating DA neurons varies significantly between different methods and approximately 20–30% of the final cells are identified as DA neurons even with the most robust method such as the floor-plate induction protocol (Cui et al. 2016). Sorting of DA progenitors with markers such as CORIN seems to aid in a better yield of mature and functional DA neurons (Doi et al. 2020; Paik et al. 2018). Knocking in a reporter gene in the endogenous TH locus has been attempted to quantify the final yield of DA neurons (Cui et al. 2016) to understand the efficiency of different published protocols. However, no significant progress has been made to analyze if the DA neurons derived in vitro are similar to the SNpc neurons impacted in a PD patient. A TH^+ DA neuron is necessarily not a representation of the A9 or SNpc nuclei of the brain, though GIRK2/TH positivity and low Calbindin is considered as an A9 signature (Hartfield et al. 2014; Sánchez-Danés et al. 2012; Woodard et al. 2014). A reliable strategy would be multiplexing markers for reliable subtype identification (Kim et al. 2020). Another difficulty in modeling PD with iPSCs is the induction of 'aging' in a culture dish. Pharmacological inhibition of telomerase by the inhibitor BIBR1532 demonstrates moderate disease-relevant phenotypes in *PINK1* and *PARKIN* DA neurons (Vera et al. 2016). Progerin (the protein associated with premature aging) overexpression as a strategy to induce aging is also reported (Miller et al. 2013) but interpretation is complex as disease-relevant phenotypes and progerin-phenotypes are indistinguishable (Sison et al. 2018). Moreover, contrary to what is seen in PD pathology, an exogenous stressor is always necessary to observe disease phenotypes in an iPSC–DA system. In a PD-patient-derived iPSC model, DA neurons exhibit apoptosis only after exposure to stressors including hydrogen peroxidase, MG132 and 6-OHDA (Cooper et al. 2012). Lastly, 2D culture systems that are normally used to differentiate DA neurons do not mimic the complex in vivo environment. 3D organoids fill this gap by representing a more physiologically relevant model system. However, the tremendous progress seen in the field is largely limited to cortical or cerebral organoids. A few midbrain spheroid or organoid culture systems are nonetheless reported (Smits et al. 2019; Kim et al. 2019; Jo et al. 2016; Monzel et al. 2017b). Results from these studies indicate that 3D midbrain cultures are definitely an improvement over 2D cultures to model PD. A recent study describes the robust generation of midbrain organoids with homogeneous distribution of midbrain DA neurons along with other neuronal subtypes as well as functional glial cells, including astrocytes and oligodendrocytes (Kwak et al. 2020). Nevertheless, an overall low efficiency of generation and heterogeneity within the midbrain organoids along with its ethical considerations raises contentious questions towards a bench-to-bedside approach.

6 Conclusions

Human pluripotent stem cells, iPSCs in particular, are an invaluable tool to help us better understand PD pathology by generating functional DA neurons with A9-like identity and also reproducing the midbrain cell composition. Improvement in differentiation protocols and 3D culturing techniques combined with genome-editing technologies aids in better PD modeling studies. Additionally, these cultures exhibit key features of PD, such as α-syn accumulation, autophagic defects, oxidative stress and impairment of mitochondrial function. However, it may be advantageous to include other cell types like microglia in PD studies rather than focusing on midbrain-specific organoids to understand the disease pathology in a relevant way. The blood-brain barrier (BBB) and its dysfunction in PD should also be emphasized. Further, the future direction in investigating PD should make use of the organ-on-chip or organoids-on-chip model with a multi-organ configuration to study different cell types and involvement of various organs in PD progression and pathology. Lastly, ARPD genes other than *PRKN* and *PINK1*, though rare, may provide insights into the common molecular pathways of the monogenic disease forms and should be included in such detailed studies.

Acknowledgements This work was supported by the DBT/Wellcome Trust India Alliance Early Career Fellowship [IA/E/18/1/504319] awarded to the author. Figures were created with BioRender. Prof. Gaiti Hasan, National Centre for Biological Sciences (NCBS), TIFR provided critical inputs for the manuscript.

Conflict of Interests The author declares no competing interests.

Bibliography

Ahfeldt T et al (2020) Pathogenic pathways in early-onset autosomal recessive Parkinson's disease discovered using isogenic human dopaminergic neurons. Stem Cell Rep. https://doi.org/10.1016/j.stemcr.2019.12.005

Anett I et al (2020) Hereditary Parkinson's disease as a new clinical manifestation of the damaged POLG gene. Orv Hetil. https://doi.org/10.1556/650.2020.31724

Arenas E, Denham M, Villaescusa JC (2015) How to make a midbrain dopaminergic neuron. Dev. https://doi.org/10.1242/dev.097394

Arias-Fuenzalida J et al (2017) FACS-assisted CRISPR-Cas9 genome editing facilitates Parkinson's disease modeling. Stem Cell Reports 9:1423–1431

Badger JL, Cordero-Llana O, Hartfield EM, Wade-Martins R (2014) Parkinson's disease in a dish – using stem cells as a molecular tool. Neuropharmacology 76:88–96

Balsinde J, Balboa MA (2005) Cellular regulation and proposed biological functions of group VIA calcium-independent phospholipase A2 in activated cells. Cell Signal. https://doi.org/10.1016/j.cellsig.2005.03.002

Barker RA, Studer L, Cattaneo E, Takahashi J (2015) G-Force PD: a global initiative in coordinating stem cell-based dopamine treatments for Parkinson's disease. NPJ Park Dis. https://doi.org/10.1038/npjparkd.2015.17

Barker RA, Parmar M, Studer L, Takahashi J (2017) Human trials of stem cell-derived dopamine neurons for Parkinson's disease: dawn of a new era. Cell Stem Cell 21:569–573

Bento CF, Ashkenazi A, Jimenez-Sanchez M, Rubinsztein DC (2016) The Parkinson's disease-associated genes ATP13A2 and SYT11 regulate autophagy via a common pathway. Nat Commun. https://doi.org/10.1038/ncomms11803

Bogetofte H et al (2019) PARK2 mutation causes metabolic disturbances and impaired survival of human iPSC-derived neurons. Front Cell Neurosci. https://doi.org/10.3389/fncel.2019.00297

Bohlega SA et al (2016) Clinical heterogeneity of PLA2G6-related Parkinsonism: analysis of two Saudi families. BMC Res Notes. https://doi.org/10.1186/s13104-016-2102-7

Bolotina V, Orai M (2008) STIM1 and iPLA2 β: a view from a different perspective. J Physiol. https://doi.org/10.1113/jphysiol.2008.154997

Bonifati V (2012) Autosomal recessive parkinsonism. Park Relat Disord. https://doi.org/10.1016/s1353-8020(11)70004-9

Bonifati V (2014) Genetics of Parkinson's disease - state of the art, 2013. Park Relat Disord. https://doi.org/10.1016/S1353-8020(13)70009-9

Bonifati V et al (2003) Mutations in the DJ-1 gene associated with autosomal recessive early-onset parkinsonism. Science. https://doi.org/10.1126/science.1077209

Brunelli F, Valente EM, Arena G (2020) Mechanisms of neurodegeneration in Parkinson's disease: keep neurons in the PINK1. Mech Ageing Dev. https://doi.org/10.1016/j.mad.2020.111277

Burchell VS et al (2013) The Parkinson's disease-linked proteins Fbxo7 and Parkin interact to mediate mitophagy. Nat Neurosci. https://doi.org/10.1038/nn.3489

Calatayud C et al (2019) CRISPR/Cas9-mediated generation of a tyrosine hydroxylase reporter iPSC line for live imaging and isolation of dopaminergic neurons. Sci Rep. https://doi.org/10.1038/s41598-019-43080-2

Cao L et al (2013) Novel SPG11 mutations in Chinese families with hereditary spastic paraplegia with thin corpus callosum. Park Relat Disord. https://doi.org/10.1016/j.parkreldis.2012.10.007

Chan JYH, Chan SHH (2015) Activation of endogenous antioxidants as a common therapeutic strategy against cancer, neurodegeneration and cardiovascular diseases: a lesson learnt from DJ-1. Pharmacol Ther. https://doi.org/10.1016/j.pharmthera.2015.09.005

Chen H, Chan DC (2009) Mitochondrial dynamics-fusion, fission, movement, and mitophagy-in neurodegenerative diseases. Hum Mol Genet. https://doi.org/10.1093/hmg/ddp326

Chiu CC et al (2017) PARK14 PLA2G6 mutants are defective in preventing rotenoneinduced mitochondrial dysfunction, ROS generation and activation of mitochondrial apoptotic pathway. Oncotarget. https://doi.org/10.18632/oncotarget.20893

Chiu CC et al (2019) PARK14 (D331Y) PLA2G6 causes early-onset degeneration of substantia nigra dopaminergic neurons by inducing mitochondrial dysfunction, ER stress, mitophagy impairment and transcriptional dysregulation in a knockin mouse model. Mol Neurobiol. https://doi.org/10.1007/s12035-018-1118-5

Chlebanowska P, Tejchman A, Sułkowski M, Skrzypek K, Majka M (2020) Use of 3D organoids as a model to study idiopathic form of parkinson's disease. Int J Mol Sci. https://doi.org/10.3390/ijms21030694

Chu YT, Lin HY, Chen PL, Lin CH (2020) Genotype-phenotype correlations of adult-onset PLA2G6-associated neurodegeneration: case series and literature review. BMC Neurol. https://doi.org/10.1186/s12883-020-01684-6

Chung SY et al (2016) Parkin and PINK1 patient iPSC-derived midbrain dopamine neurons exhibit mitochondrial dysfunction and α-synuclein accumulation. Stem Cell Rep. https://doi.org/10.1016/j.stemcr.2016.08.012

Cooper O et al (2012) Pharmacological rescue of mitochondrial deficits in iPSC-derived neural cells from patients with familial Parkinson's disease. Sci Transl Med. https://doi.org/10.1126/scitranslmed.3003985

Cui J et al (2016) Quantification of dopaminergic neuron differentiation and neurotoxicity via a genetic reporter. Sci Rep. https://doi.org/10.1038/srep25181

de Lau L, Breteler MMMB, de Lau LML, Breteler MMMB (2006) Epidemiology of Parkinson's disease. Lancet Neurol 5:525–535

Deas E, Wood NW, Plun-Favreau H (2011) Mitophagy and Parkinson's disease: the PINK1-parkin link. Biochim Biophys Acta, Mol Cell Res. https://doi.org/10.1016/j.bbamcr.2010.08.007

Doi D et al (2020) Pre-clinical study of induced pluripotent stem cell-derived dopaminergic progenitor cells for Parkinson's disease. Nat Commun. https://doi.org/10.1038/s41467-020-17165-w

Engel LA, Jing Z, O'Brien DE, Sun M, Kotzbauer PT (2010) Catalytic function of PLA2G6 is impaired by mutations associated with infantile neuroaxonal dystrophy but not dystonia-parkinsonism. PLoS One 5

Fan Y, Winanto, Ng SY (2020) Replacing what's lost: A new era of stem cell therapy for Parkinson's disease. Transl Neurodegener. https://doi.org/10.1186/s40035-019-0180-x

Ferese R et al (2018) Heterozygous PLA2G6 mutation leads to iron accumulation within basal ganglia and Parkinson's disease. Front Neurol. https://doi.org/10.3389/fneur.2018.00536

Feske S, Prakriya M, Rao A, Lewis RS (2005) A severe defect in CRAC Ca2+ channel activation and altered K + channel gating in T cells from immunodeficient patients. J Exp Med 202:651–662

Feske S et al (2006) A mutation in Orai1 causes immune deficiency by abrogating CRAC channel function. Nature 441:179–185

Galet B, Cheval H, Ravassard P (2020) Patient-derived midbrain organoids to explore the molecular basis of Parkinson's disease. Front Neurol. https://doi.org/10.3389/fneur.2020.01005

Garitaonandia I et al (2016) Neural stem cell tumorigenicity and biodistribution assessment for phase I clinical trial in Parkinson's disease. Sci Rep. https://doi.org/10.1038/srep34478

Gregory A, Kurian MA, Maher ER, Hogarth P, Hayflick SJ (1993) PLA2G6-associated neurodegeneration. GeneReviews(®). NBK1675 [bookaccession]

Gui YX et al (2013) Four novel rare mutations of PLA2G6 in Chinese population with Parkinson's disease. Park Relat Disord. https://doi.org/10.1016/j.parkreldis.2012.07.016

Guidubaldi A et al (2011) Novel mutations in SPG11 cause hereditary spastic paraplegia associated with early-onset levodopa-responsive Parkinsonism. Mov Disord. https://doi.org/10.1002/mds.23552

Guo Y, Tang B, Guo J (2018) PLA2G6-associated neurodegeneration (PLAN): review of clinical phenotypes and genotypes. Front Neurol. https://doi.org/10.3389/fneur.2018.01100

Hartfield EM et al (2014) Physiological characterisation of human iPS-derived dopaminergic neurons. PLoS One 9

Hoekstra SD, Stringer S, Heine VM, Posthuma D (2017) Genetically-informed patient selection for iPSC studies of complex diseases may aid in reducing cellular heterogeneity. Front Cell Neurosci. https://doi.org/10.3389/fncel.2017.00164

Iliadi KG, Gluscencova OB, Iliadi N, Boulianne GL (2018) Mutations in the Drosophila homolog of human PLA2G6 give rise to age-dependent loss of psychomotor activity and neurodegeneration. Sci Rep. https://doi.org/10.1038/s41598-018-21343-8

Imaizumi Y et al (2012) Mitochondrial dysfunction associated with increased oxidative stress and α-synuclein accumulation in PARK2 iPSC-derived neurons and postmortem brain tissue. Mol Brain. https://doi.org/10.1186/1756-6606-5-35

Irrcher I et al (2010) Loss of the Parkinson's disease-linked gene DJ-1 perturbs mitochondrial dynamics. Hum Mol Genet. https://doi.org/10.1093/hmg/ddq288

Jan A et al (2018) Activity of translation regulator eukaryotic elongation factor-2 kinase is increased in Parkinson disease brain and its inhibition reduces alpha synuclein toxicity. Acta Neuropathol Commun. https://doi.org/10.1186/s40478-018-0554-9

Jankovic J (2008) Parkinson's disease: clinical features and diagnosis. J Neurol Neurosurg Psychiatry. https://doi.org/10.1136/jnnp.2007.131045

Jarazo J et al (2019) Parkinson's disease phenotypes in patient specific brain organoids are improved by HP-β-CD treatment. bioRxiv. https://doi.org/10.1101/813089

Jeong HJ et al (2012) Transduced tat-DJ-1 protein protects against oxidative stress-induced SH-SY5Y cell death and Parkinson disease in a mouse model. Mol Cells. https://doi.org/10.1007/s10059-012-2255-8

Jiang H et al (2012) Parkin controls dopamine utilization in human midbrain dopaminergic neurons derived from induced pluripotent stem cells. Nat Commun. https://doi.org/10.1038/ncomms1669

Jo J et al (2016) Midbrain-like organoids from human pluripotent stem cells contain functional dopaminergic and neuromelanin-producing neurons. Cell Stem Cell. https://doi.org/10.1016/j.stem.2016.07.005

Kahle PJ, Waak J, Gasser T (2009) DJ-1 and prevention of oxidative stress in Parkinson's disease and other age-related disorders. Free Radic Biol Med. https://doi.org/10.1016/j.freeradbiomed.2009.08.003

Kalinderi K, Bostantjopoulou S, Fidani L (2016) The genetic background of Parkinson's disease: current progress and future prospects. Acta Neurol Scand. https://doi.org/10.1111/ane.12563

Kano M et al (2020) Reduced astrocytic reactivity in human brains and midbrain organoids with PRKN mutations. NPJ Park Dis. https://doi.org/10.1038/s41531-020-00137-8

Kapoor S et al (2016) Genetic analysis of PLA2G6 in 22 Indian families with infantile neuroaxonal dystrophy, atypical late-onset neuroaxonal dystrophy and dystonia parkinsonism complex. PLoS One 11

Karimi-Moghadam A, Charsouei S, Bell B, Jabalameli MR (2018) Parkinson disease from Mendelian forms to genetic susceptibility: new molecular insights into the neurodegeneration process. Cell Mol Neurobiol. https://doi.org/10.1007/s10571-018-0587-4

Karunakaran S et al (2007) Activation of apoptosis signal regulating kinase 1 (ASK1) and translocation of death-associated protein, Daxx, in substantia nigra pars compacta in a mouse model of Parkinson's disease: protection by α-lipoic acid. FASEB J. https://doi.org/10.1096/fj.06-7580com

Ke M et al (2020) Azoramide protects iPSC-derived dopaminergic neurons with PLA2G6 D331Y mutation through restoring ER function and CREB signaling. Cell Death Dis. https://doi.org/10.1038/s41419-020-2312-8

Kett LR et al (2015) α-synuclein-independent histopathological and motor deficits in mice lacking the endolysosomal parkinsonism protein Atp13a2. J Neurosci. https://doi.org/10.1523/JNEUROSCI.0632-14.2015

Kikuchi T et al (2017) Human iPS cell-derived dopaminergic neurons function in a primate Parkinson's disease model. Nature. https://doi.org/10.1038/nature23664

Kim H et al (2019) Modeling G2019S-LRRK2 sporadic Parkinson's disease in 3D midbrain organoids. Stem Cell Rep. https://doi.org/10.1016/j.stemcr.2019.01.020

Kim TW, Koo SY, Studer L (2020) Pluripotent stem cell therapies for Parkinson disease: present challenges and future opportunities. Front Cell Dev Biol. https://doi.org/10.3389/fcell.2020.00729

Kinghorn KJ et al (2015) Loss of PLA2G6 leads to elevated mitochondrial lipid peroxidation and mitochondrial dysfunction. Brain. https://doi.org/10.1093/brain/awv132

Kirkeby A et al (2012) Generation of regionally specified neural progenitors and functional neurons from human embryonic stem cells under defined conditions. Cell Rep 1:703–714

Kirkeby A et al (2017) Predictive markers guide differentiation to improve graft outcome in clinical translation of hESC-based therapy for Parkinson's disease. Cell Stem Cell. https://doi.org/10.1016/j.stem.2016.09.004

Kostic M et al (2015) PKA phosphorylation of NCLX reverses mitochondrial calcium overload and depolarization, promoting survival of PINK1-deficient dopaminergic neurons. Cell Rep. https://doi.org/10.1016/j.celrep.2015.08.079

Krebiehl G et al (2010) Reduced basal autophagy and impaired mitochondrial dynamics due to loss of Parkinson's disease-associated protein DJ-1. PLoS One. https://doi.org/10.1371/journal.pone.0009367

Kriks S et al (2011) Dopamine neurons derived from human ES cells efficiently engraft in animal models of Parkinson's disease. Nature 480:547–551

Kuroda Y et al (2006) Parkin enhances mitochondrial biogenesis in proliferating cells. Hum Mol Genet. https://doi.org/10.1093/hmg/ddl006

Kwak TH et al (2020) Generation of homogeneous midbrain organoids with in vivo-like cellular composition facilitates neurotoxin-based Parkinson's disease modeling. Stem Cells. https://doi.org/10.1002/stem.3163
Lee Y et al (2013) Parthanatos mediates AIMP2-activated age-dependent dopaminergic neuronal loss. Nat Neurosci. https://doi.org/10.1038/nn.3500
Lesage S, Brice A (2009) Parkinson's disease: from monogenic forms to genetic susceptibility factors. Hum Mol Genet. https://doi.org/10.1093/hmg/ddp012
Lin W, Kang UJ (2008) Characterization of PINK1 processing, stability, and subcellular localization. J Neurochem. https://doi.org/10.1111/j.1471-4159.2008.05398.x
Lin L et al (2016) Molecular features underlying neurodegeneration identified through in vitro modeling of genetically diverse Parkinson's disease patients. Cell Rep. https://doi.org/10.1016/j.celrep.2016.05.022
Lin G et al (2018) Phospholipase PLA2G6, a parkinsonism-associated gene, affects Vps26 and Vps35, retromer function, and ceramide levels, similar to α-synuclein gain. Cell Metab. https://doi.org/10.1016/j.cmet.2018.05.019
Liu S et al (2012) Parkinson's disease-associated kinase PINK1 regulates miro protein level and axonal transport of mitochondria. PLoS Genet. https://doi.org/10.1371/journal.pgen.1002537
Liu Y et al (2019) DJ-1 regulates the integrity and function of ER-mitochondria association through interaction with IP3R3-Grp75-VDAC1. Proc Natl Acad Sci U S A. https://doi.org/10.1073/pnas.1906565116
Lu B, Fivaz M (2016) Neuronal SOCE: myth or reality? Trends Cell Biol. https://doi.org/10.1016/j.tcb.2016.09.008
Lu CS et al (2012) PLA2G6 mutations in PARK14-linked young-onset parkinsonism and sporadic Parkinson's disease. Am J Med Genet Part B Neuropsychiatr Genet. https://doi.org/10.1002/ajmg.b.32012
Lucking CB et al (2000) Association between early-onset Parkinson's disease and mutations in the parkin gene. French Parkinson's Disease Genetics Study Group. N Engl J Med
Malli R, Graier WF (2017) The role of mitochondria in the activation/maintenance of SOCE: the contribution of mitochondrial Ca2+ uptake, mitochondrial motility, and location to store-operated Ca2+ entry. In: Advances in experimental medicine and biology. https://doi.org/10.1007/978-3-319-57732-6_16
Marchetto MC, Brennand KJ, Boyer LF, Gage FH (2011) Induced pluripotent stem cells (iPSCs) and neurological disease modeling: Progress and promises. Hum Mol Genet. https://doi.org/10.1093/hmg/ddr336
Martínez-Morales PL, Liste I (2012) Stem cells as in vitro model of Parkinson's disease. Stem Cells Int. https://doi.org/10.1155/2012/980941
Matsuda N et al (2010) PINK1 stabilized by mitochondrial depolarization recruits Parkin to damaged mitochondria and activates latent Parkin for mitophagy. J Cell Biol. https://doi.org/10.1083/jcb.200910140
Michel PP, Hirsch EC, Hunot S (2016) Understanding dopaminergic cell death pathways in Parkinson disease. Neuron. https://doi.org/10.1016/j.neuron.2016.03.038
Miguel R et al (2014) POLG1-related levodopa-responsive parkinsonism. Clin Neurol Neurosurg. https://doi.org/10.1016/j.clineuro.2014.08.020
Miki Y et al (2017) PLA2G6 accumulates in Lewy bodies in PARK14 and idiopathic Parkinson's disease. Neurosci Lett 645:40–45
Miller JD et al (2013) Human iPSC-based modeling of late-onset disease via progerin-induced aging. Cell Stem Cell. https://doi.org/10.1016/j.stem.2013.11.006
Monzel AS et al (2017a) Derivation of human midbrain-specific organoids from neuroepithelial stem cells. Stem Cell Rep 8:1144–1154
Monzel AS et al (2017b) Derivation of human midbrain-specific organoids from neuroepithelial stem cells. Stem Cell Rep. https://doi.org/10.1016/j.stemcr.2017.03.010
Mori A et al (2019) Parkinson's disease-associated iPLA2-VIA/PLA2G6 regulates neuronal functions and α-synuclein stability through membrane remodeling. Proc Natl Acad Sci U S A. https://doi.org/10.1073/pnas.1902958116
Narendra D, Walker JE, Youle R (2012) Mitochondrial quality control mediated by PINK1 and Parkin: links to parkinsonism. Cold Spring Harb Perspect Biol. https://doi.org/10.1101/cshperspect.a011338
Nolbrant S, Heuer A, Parmar M, Kirkeby A (2017) Generation of high-purity human ventral midbrain dopaminergic progenitors for in vitro maturation and intracerebral transplantation. Nat Protoc. https://doi.org/10.1038/nprot.2017.078
O'Sullivan SS et al (2008) Nonmotor symptoms as presenting complaints in Parkinson's disease: a clinicopathological study. Mov Disord. https://doi.org/10.1002/mds.21813
Oh CK et al (2017) S-nitrosylation of PINK1 attenuates PINK1/Parkin-dependent mitophagy in hiPSC-based Parkinson's disease models. Cell Rep. https://doi.org/10.1016/j.celrep.2017.10.068
Oh SE et al (2018) The Parkinson's disease gene product DJ-1 modulates miR-221 to promote neuronal survival against oxidative stress. Redox Biol. https://doi.org/10.1016/j.redox.2018.07.021
Okarmus J et al (2020) Lysosomal perturbations in human dopaminergic neurons derived from induced pluripotent stem cells with PARK2 mutation. Sci Rep. https://doi.org/10.1038/s41598-020-67091-6
Ono Y et al (2007) Differences in neurogenic potential in floor plate cells along an anteroposterior location: midbrain dopaminergic neurons originate from mesencephalic floor plate cells. Development. https://doi.org/10.1242/dev.02879
Paik EJ, O'Neil AL, Ng SY, Sun C, Rubin LL (2018) Using intracellular markers to identify a novel set of

surface markers for live cell purification from a heterogeneous hIPSC culture. Sci Rep. https://doi.org/10.1038/s41598-018-19291-4

Paisan-Ruiz C et al (2009) Characterization of PLA2G6 as a locus for dystonia-parkinsonism. Ann Neurol 65:19–23

Paisán-Ruiz C et al (2010) Early-onset L-dopa-responsive Parkinsonism with pyramidal signs due to ATP13A2, PLA2G6, FBXO7 and Spatacsin mutations. Mov Disord. https://doi.org/10.1002/mds.23221

Paisán-Ruiz C et al (2012) Widespread Lewy body and tau accumulation in childhood and adult onset dystonia-parkinsonism cases with PLA2G6 mutations. Neurobiol Aging. https://doi.org/10.1016/j.neurobiolaging.2010.05.009

Park IH et al (2008) Disease-specific induced pluripotent stem cells. Cell. https://doi.org/10.1016/j.cell.2008.07.041

Park JS, Koentjoro B, Veivers D, Mackay-Sim A, Sue CM (2014) Parkinson's disease-associated human ATP13A2 (PARK9) deficiency causes zinc dyshomeostasis and mitochondrial dysfunction. Hum Mol Genet. https://doi.org/10.1093/hmg/ddt623

Pérez R, Melero R, Balboa MA, Balsinde J (2004) Role of group VIA calcium-independent phospholipase A2 in arachidonic acid release, phospholipid fatty acid incorporation, and apoptosis in U937 cells responding to hydrogen peroxide. J Biol Chem 279:40385–40391

Pirooznia SK et al (2020) PARIS induced defects in mitochondrial biogenesis drive dopamine neuron loss under conditions of parkin or PINK1 deficiency. Mol Neurodegener. https://doi.org/10.1186/s13024-020-00363-x

Playne R, Connor B (2017) Understanding Parkinson's disease through the use of cell reprogramming. Stem Cell Rev Rep. https://doi.org/10.1007/s12015-017-9717-5

Pryde KR, Smith HL, Chau KY, Schapira AHV (2016) PINK1 disables the anti-fission machinery to segregate damaged mitochondria for mitophagy. J Cell Biol. https://doi.org/10.1083/jcb.201509003

Qian X et al (2016) Brain-region-specific organoids using mini-bioreactors for modeling ZIKV exposure. Cell. https://doi.org/10.1016/j.cell.2016.04.032

Qing X et al (2017) CRISPR/Cas9 and piggyBac-mediated footprint-free LRRK2-G2019S knock-in reveals neuronal complexity phenotypes and α-synuclein modulation in dopaminergic neurons. Stem Cell Res. https://doi.org/10.1016/j.scr.2017.08.013

Ramanadham S et al (2015) Calcium-independent phospholipases A2 and their roles in biological processes and diseases. J Lipid Res. https://doi.org/10.1194/jlr.R058701

Ramirez A et al (2006) Hereditary parkinsonism with dementia is caused by mutations in ATP13A2, encoding a lysosomal type 5 P-type ATPase. Nat Genet. https://doi.org/10.1038/ng1884

Reinhardt P et al (2013) Derivation and expansion using only small molecules of human neural progenitors for neurodegenerative disease modeling. PLoS One 8

Ren X, Hinchie A, Swomley A, Powell DK, Butterfield DA (2019) Profiles of brain oxidative damage, ventricular alterations, and neurochemical metabolites in the striatum of PINK1 knockout rats as functions of age and gender: relevance to Parkinson disease. Free Radic Biol Med. https://doi.org/10.1016/j.freeradbiomed.2019.08.008

Sampson TR et al (2016) Gut microbiota regulate motor deficits and neuroinflammation in a model of Parkinson's disease. Cell. https://doi.org/10.1016/j.cell.2016.11.018

Sánchez-Danés A et al (2012) Efficient generation of A9 midbrain dopaminergic neurons by lentiviral delivery of LMX1A in human embryonic stem cells and induced pluripotent stem cells. Hum Gene Ther. https://doi.org/10.1089/hum.2011.054

Scarffe LA, Stevens DA, Dawson VL, Dawson TM (2014) Parkin and PINK1: much more than mitophagy. Trends Neurosci. https://doi.org/10.1016/j.tins.2014.03.004

Schäfer C, Rymarczyk G, Ding L, Kirber MT, Bolotina VM (2012) Role of molecular determinants of store-operated Ca(2+) entry (Orai1, phospholipase A2 group 6, and STIM1) in focal adhesion formation and cell migration. J Biol Chem 287:40745–40757

Schapira AHV et al (1990) Mitochondrial complex I deficiency in Parkinson's disease. J Neurochem. https://doi.org/10.1111/j.1471-4159.1990.tb02325.x

Schieber M, Chandel NS (2014) ROS function in redox signaling and oxidative stress. Curr Biol. https://doi.org/10.1016/j.cub.2014.03.034

Schweitzer JS et al (2020) Personalized iPSC-derived dopamine progenitor cells for Parkinson's disease. N Engl J Med. https://doi.org/10.1056/nejmoa1915872

Scott L, Dawson VL, Dawson TM (2017) Trumping neurodegeneration: targeting common pathways regulated by autosomal recessive Parkinson's disease genes. Exp Neurol. https://doi.org/10.1016/j.expneurol.2017.04.008

Seirafi M, Kozlov G, Gehring K (2015) Parkin structure and function. FEBS J. https://doi.org/10.1111/febs.13249

Shen T et al (2018) Genetic analysis of ATP13A2, PLA2G6 and FBXO7 in a cohort of Chinese patients with early-onset Parkinson's disease. Sci Rep. https://doi.org/10.1038/s41598-018-32217-4

Shen T et al (2019) Early-onset parkinson's disease caused by pla2g6 compound heterozygous mutation, a case report and literature review. Front Neurol. https://doi.org/10.3389/fneur.2019.00915

Sina F, Shojaee S, Elahi E, Paisán-Ruiz C (2009) R632W mutation in PLA2G6 segregates with dystonia-parkinsonism in a consanguineous Iranian family. Eur J Neurol. https://doi.org/10.1111/j.1468-1331.2008.02356.x

Singaravelu K (2006) Regulation of store-operated calcium entry by calcium-independent phospholipase A2 in rat cerebellar astrocytes. J Neurosci 26:9579–9592
Sison SL, Vermilyea SC, Emborg ME, Ebert AD (2018) Using patient-derived induced pluripotent stem cells to identify Parkinson's disease-relevant phenotypes. Curr Neurol Neurosci Rep. https://doi.org/10.1007/s11910-018-0893-8
Smani T, Dominguez-Rodriguez A, Callejo-Garcia P, Rosado JA, Avila-Medina J (2016) Phospholipase A2 as a molecular determinant of store-operated calcium entry. Adv Exp Med Biol 898:111–131
Smits LM, Schwamborn JC (2020) Midbrain organoids: a new tool to investigate Parkinson's disease. Front Cell Dev Biol. https://doi.org/10.3389/fcell.2020.00359
Smits LM et al (2019) Modeling Parkinson's disease in midbrain-like organoids. NPJ Park Dis. https://doi.org/10.1038/s41531-019-0078-4
Soldner F et al (2009) Parkinson's disease patient-derived induced pluripotent stem cells free of viral reprogramming factors. Cell. https://doi.org/10.1016/j.cell.2009.02.013
Soldner F et al (2016) Parkinson-associated risk variant in distal enhancer of α-synuclein modulates target gene expression. Nature. https://doi.org/10.1038/nature17939
Spät A, Szanda G (2017) The role of mitochondria in the activation/maintenance of SOCE: store-operated Ca2+ entry and mitochondria. In: Advances in experimental medicine and biology. https://doi.org/10.1007/978-3-319-57732-6_14
Stathakos P et al (2019) Imaging autophagy in hiPSC-derived midbrain dopaminergic neuronal cultures for parkinson's disease research. Methods Mol Biol. https://doi.org/10.1007/978-1-4939-8873-0_17
Stathakos P et al (2020) A monolayer hiPSC culture system for autophagy/mitophagy studies in human dopaminergic neurons. Autophagy. https://doi.org/10.1080/15548627.2020.1739441
Sumi-Akamaru H et al (2016) High expression of α-synuclein in damaged mitochondria with PLA2G6 dysfunction. Acta Neuropathol Commun. https://doi.org/10.1186/s40478-016-0298-3
Suzuki S et al (2017) Efficient induction of dopaminergic neuron differentiation from induced pluripotent stem cells reveals impaired mitophagy in PARK2 neurons. Biochem Biophys Res Commun. https://doi.org/10.1016/j.bbrc.2016.12.188
Taanman JW, Protasoni M (2017) Loss of PINK1 or parkin function results in a progressive loss of mitochondrial function. In: Autophagy: cancer, other pathologies, inflammation, immunity, infection, and aging, vol 12. Academic, London. https://doi.org/10.1016/B978-0-12-812146-7.00007-X
Takahashi K et al (2007) Induction of {pluripotent} {stem} {cells} from {adult} {human} {fibroblasts} by {defined} {factors}. Cell 131:861–872
Takahashi-Niki K, Niki T, Taira T, Iguchi-Ariga SMM, Ariga H (2004) Reduced anti-oxidative stress activities of DJ-1 mutants found in Parkinson's disease patients. Biochem Biophys Res Commun. https://doi.org/10.1016/j.bbrc.2004.05.187
Tieng V et al (2014) Engineering of midbrain organoids containing long-lived dopaminergic neurons. Stem Cells Dev. https://doi.org/10.1089/scd.2013.0442
Tolosa E, Wenning G, Poewe W (2006) The diagnosis of Parkinson's disease. Lancet Neurol. https://doi.org/10.1016/S1474-4422(05)70285-4
Tran J, Anastacio H, Bardy C (2020) Genetic predispositions of Parkinson's disease revealed in patient-derived brain cells. NPJ Park Dis. https://doi.org/10.1038/s41531-020-0110-8
Turk J, Ramanadham S (2004) The expression and function of a group VIA calcium-independent phospholipase A2 (iPLA2β) in β-cells. Can J Physiol Pharmacol. https://doi.org/10.1139/y04-064
van der Merwe C, Jalali Sefid Dashti Z, Christoffels A, Loos B, Bardien S (2015) Evidence for a common biological pathway linking three Parkinson's disease-causing genes: Parkin, PINK1 and DJ-1. Eur J Neurosci. https://doi.org/10.1111/ejn.12872
Vera E, Bosco N, Studer L (2016) Generating late-onset human iPSC-based disease models by inducing neuronal age-related phenotypes through telomerase manipulation. Cell Rep. https://doi.org/10.1016/j.celrep.2016.09.062
Vermilyea SC et al (2020) In vitro CRISPR/Cas9-directed gene editing to model LRRK2 G2019S Parkinson's disease in common marmosets. Sci Rep. https://doi.org/10.1038/s41598-020-60273-2
Vig M et al (2006) CRACM1 is a plasma membrane protein essential for store-operated Ca2+ entry. Science 312:1220–1223
Vrijsen S et al (2020) ATP13A2-mediated endo-lysosomal polyamine export counters mitochondrial oxidative stress. Proc Natl Acad Sci. https://doi.org/10.1073/pnas.1922342117
Wang HL et al (2011) PARK6 PINK1 mutants are defective in maintaining mitochondrial membrane potential and inhibiting ROS formation of substantia nigra dopaminergic neurons. Biochim Biophys Acta Mol Basis Dis. https://doi.org/10.1016/j.bbadis.2011.03.007
Wang YK et al (2018) Human clinical-grade parthenogenetic ESC-derived dopaminergic neurons recover locomotive defects of nonhuman primate models of Parkinson's disease. Stem Cell Rep. https://doi.org/10.1016/j.stemcr.2018.05.010
Wang R et al (2019) ATP13A2 facilitates HDAC6 recruitment to lysosome to promote autophagosome–lysosome fusion. J Cell Biol. https://doi.org/10.1083/jcb.201804165
Weihofen A, Thomas KJ, Ostaszewski BL, Cookson MR, Selkoe DJ (2009) Pink1 forms a multiprotein complex with miro and milton, linking Pink1 function to mitochondrial trafficking. Biochemistry. https://doi.org/10.1021/bi8019178
William Langston J, Ballard P, Tetrud JW, Irwin I (1983) Chronic parkinsonism in humans due to a product of

meperidine-analog synthesis. Science. https://doi.org/10.1126/science.6823561
Woodard CM et al (2014) IPSC-derived dopamine neurons reveal differences between monozygotic twins discordant for parkinson's disease. Cell Rep. https://doi.org/10.1016/j.celrep.2014.10.023
Xu Q et al (2012) Hypoxia regulation of ATP13A2 (PARK9) gene transcription. J Neurochem. https://doi.org/10.1111/j.1471-4159.2012.07676.x
Xu CY et al (2017) DJ-1 inhibits α-synuclein aggregation by regulating chaperone-mediated autophagy. Front Aging Neurosci. https://doi.org/10.3389/fnagi.2017.00308
Xu S, Yang X, Qian Y, Xiao Q (2018) Parkinson's disease-related DJ-1 modulates the expression of uncoupling protein 4 against oxidative stress. J Neurochem. https://doi.org/10.1111/jnc.14297
Yamano K, Youle RJ (2013) PINK1 is degraded through the N-end rule pathway. Autophagy. https://doi.org/10.4161/auto.24633
Yamano K, Youle RJ (2020) Two different axes CALCOCO2-RB1CC1 and OPTN-ATG9A initiate PRKN-mediated mitophagy. Autophagy. https://doi.org/10.1080/15548627.2020.1815457
Yu W, Sun Y, Guo S, Lu B (2011) The PINK1/Parkin pathway regulates mitochondrial dynamics and function in mammalian hippocampal and dopaminergic neurons. Hum Mol Genet. https://doi.org/10.1093/hmg/ddr235
Zhao T et al (2011) Loss of nuclear activity of the FBXO7 protein in patients with parkinsonian-pyramidal syndrome (PARK15). PLoS One. https://doi.org/10.1371/journal.pone.0016983
Zheng L et al (2017) Parkin functionally interacts with PGC-1α to preserve mitochondria and protect dopaminergic neurons. Hum Mol Genet. https://doi.org/10.1093/hmg/ddw418
Zhou Q et al (2016a) Impairment of PARK14-dependent Ca2+signalling is a novel determinant of Parkinson's disease. Nat Commun 7
Zhou ZD, Sathiyamoorthy S, Angeles DC, Tan EK (2016b) Linking F-box protein 7 and parkin to neuronal degeneration in Parkinson's disease (PD). Mol Brain. https://doi.org/10.1186/s13041-016-0218-2
Zhou ZD, Lee JCT, Tan EK (2018) Pathophysiological mechanisms linking F-box only protein 7 (FBXO7) and Parkinson's disease (PD). Mutat Res Rev Mutat Res. https://doi.org/10.1016/j.mrrev.2018.10.001
Zhu L et al (2019) Stress-induced precocious aging in PD-patient iPSC-derived NSCs may underlie the pathophysiology of Parkinson's disease. Cell Death Dis. https://doi.org/10.1038/s41419-019-1313-y

Adv Exp Med Biol - Cell Biology and Translational Medicine (2021) 14: 135–162
https://doi.org/10.1007/5584_2021_639

Published online: 13 May 2021

Stem Cell Applications in Lysosomal Storage Disorders: Progress and Ongoing Challenges

Sevil Köse, Fatima Aerts-Kaya, Duygu Uçkan Çetinkaya, and Petek Korkusuz

Abstract

Lysosomal storage disorders (LSDs) are rare inborn errors of metabolism caused by defects in lysosomal function. These diseases are characterized by accumulation of completely or partially degraded substrates in the lysosomes leading to cellular dysfunction of the affected cells. Currently, enzyme replacement therapies (ERTs), treatments directed at substrate reduction (SRT), and hematopoietic stem cell (HSC) transplantation are the only treatment options for LSDs, and the effects of these treatments depend strongly on the type of LSD and the time of initiation of treatment. However, some of the LSDs still lack a durable and curative treatment. Therefore, a variety of novel treatments for LSD patients has been developed in the past few years. However, despite significant progress, the efficacy of some of these treatments remains limited because these therapies are often initiated after irreversible organ damage has occurred.

Here, we provide an overview of the known effects of LSDs on stem cell function, as well as a synopsis of available stem cell-based cell and gene therapies that have been/are being developed for the treatment of LSDs. We discuss the advantages and disadvantages of use of hematopoietic stem cell (HSC), mesenchymal stem cell (MSC), and induced pluripotent stem cell (iPSC)-related (gene) therapies. An overview of current research data indicates that when stem cell and/or gene therapy applications are used in combination with existing therapies such as ERT, SRT, and chaperone therapies, promising results can be achieved, showing that these treatments may result in alleviation of existing symptoms

Authors Sevil Köse and Fatima Aerts-Kaya have equally contributed to this chapter.

S. Köse
Department of Medical Biology, Faculty of Medicine, Atilim University, Ankara, Turkey
e-mail: sevilarslan@hotmail.com

F. Aerts-Kaya
Department of Stem Cell Sciences, Hacettepe University Graduate School of Health Sciences, Ankara, Turkey

Hacettepe University Center for Stem Cell Research and Development (PEDI-STEM), Ankara, Turkey
e-mail: fatima.aerts@hacettepe.edu.tr; fatimaaerts@yahoo.com

D. Uçkan Çetinkaya
Hacettepe University Faculty of Medicine, Department of Pediatrics, Division of Hematology, Hacettepe University Center for Stem Cell Research and Development (PEDI-STEM), Ankara, Turkey

Department of Stem Cell Sciences, Hacettepe University Graduate School of Health Sciences, Ankara, Turkey
e-mail: duckan@hacettepe.edu.tr

P. Korkusuz (✉)
Department of Histology and Embryology, Hacettepe University Faculty of Medicine, Ankara, Turkey
e-mail: petek@hacettepe.edu.tr

and/or prevention of progression of the disease. All together, these studies offer some insight in LSD stem cell biology and provide a hopeful perspective for the use of stem cells. Further development and improvement of these stem cell (gene) combination therapies may greatly improve the current treatment options and outcomes of patients with a LSD.

Keywords

Lysosomal storage disorder · Lysosomal storage disease · Stem cell · Hematopoietic stem cell · Mesenchymal stem cell · Neural stem cell · Induced pluripotent stem cell · Gene therapy

Abbreviations

AAV	Adeno-associated vectors
ARSA	Arylsulfatase A
ASM	Acid sphingomyelinase
BBB	Brain-blood barrier
BM	Bone marrow
BMT	Bone marrow transplantation
CNS	Central nervous system
ER	Endoplasmic reticulum
ERT	Enzyme replacement therapy
GAG	Glycosaminoglycan
GD	Gaucher disease
GSD	Glycogen storage disease
GUSB	Beta-glucuronidase
GvHD	Graft-versus-host disease
HSCT	Hematopoietic stem cell transplantation
iPSCs	Induced pluripotent stem cells
LSD	Lysosomal storage disease/disorder
LV	Lentiviral vectors
MASCs	Multipotent adult stem cells
M-CSF	Macrophage colony-stimulating factor
MLD	Metachromatic leukodystrophy
MPS	Mucopolysaccharidosis
MPS-IH	Hurler's disease
MSC	Mesenchymal stem cell
NSC	Neural stem cells
PB	Peripheral blood
PCT	Pharmacological chaperone therapy
SIN	Self-inactivating
SRT	Substrate reduction therapy
UCB	Umbilical cord blood
UCBT	Umbilical cord blood transplantation

1 Introduction

Lysosomal storage diseases/disorders (LSDs) are a large group of hereditary diseases that lead to deficiency of specific soluble lysosomal enzymes responsible for breakdown of macromolecules in lysosomes (Leal et al. 2020). However, numerous defects in internal lysosomal membrane proteins can cause LSD. Although most LSDs are autosomal recessively inherited, some are X-linked. Besides the red blood cells, all cells in the body have lysosomes, and therefore these metabolic disorders can affect multiple organ systems simultaneously (Parenti et al. 2015). Stacking of completely or partially degraded substrates in the lysosomes causes numerous secondary alterations, such as endoplasmic reticulum (ER) and oxidative stress, disorders in autophagy, changes in calcium homeostasis, and a significant energy imbalance. Although individual LSDs are rare, when considered as a group, it is one of the most common hereditary diseases in childhood and affects about 1 in 5000 live births (Cox and Cachon-Gonzalez 2012). The most commonly observed LSDs include the combined groups of mucopolysaccharidoses, mucolipidoses, oligosaccharidoses, Pompe disease, Gaucher disease, Fabry disease, Niemann-Pick disease, and neuronal ceroid lipofuscinoses (Table 1, reviewed by (Sun 2018)). Generally, the clinical symptoms that characterize different diseases depend on the amount and type of the accumulated complex molecules, such as glycoproteins, glycosaminoglycans, sphingolipids, and glycogen (Ballabio and Gieselmann 2009). However, in the framework of this review, we will not discuss the clinical features of these syndromes in detail. Instead, we will focus on the effects of these diseases on stem cells and stem cell function and how (genetically

Table 1 Classification of lysosomal storage disorders and current treatment options

Disorder	Alternative name	Type	Affected gene	Affected protein	Involved organs/ tissues	Current treatment
Mucopolysaccharidosis (MPS) I	Hurler, Schreie syndrome	Mucopolysaccharidosis	*IDUA*	Alpha-L-iduronidase	Liver, spleen, heart, lungs, skeletal system, brain	ERT (laronidase), HSCT (<2 years of age)
MPS II	Hunter syndrome	Mucopolysaccharidosis	*IDS*	Iduronate-2-sulfatase protein	Liver, spleen, heart, lungs, skeletal system, brain	ERT (idursulfase)
MPS IIIA	Sanfilippo syndrome	Mucopolysaccharidosis	*SGSH*	Heparan N-sulfatase	Liver, spleen, brain	None
MPS IIIB	Sanfilippo syndrome	Mucopolysaccharidosis	*NAGLU*	N-Acetylglucosaminidase	Liver, spleen, brain	None
MPS IIIC	Sanfilippo syndrome	Mucopolysaccharidosis	*HGSNAT*	Acetyl-CoA glucosamine N-acetyltransferase	Liver, spleen, brain	None
MPS IIID	Sanfilippo syndrome	Mucopolysaccharidosis	*GNS*	N-Acetyl-glucosamine-6-sulfatase	Liver, spleen, brain	None
MPS IVA	Morquio syndrome A	Mucopolysaccharidosis	*GALNS*	N-Acetylgalactosamine-6-sulfate sulfatase	Skeletal system, lungs, heart	ERT (elosulfase alfa)
MPS IVB	Morquio syndrome B	Mucopolysaccharidosis	*GLB1*	β-Galactosidase	Skeletal system, lungs, heart	None
MPS VI	Maroteaux-Lamy syndrome	Mucopolysaccharidosis	*ARSB*	Arylsulfatase B	Skeletal system, liver, spleen, lungs	ERT (galsulfase)
MPS VII	Sly syndrome	Mucopolysaccharidosis	*GUSB*	β-Glucuronidase	Liver, spleen, skeletal system, lungs, heart	ERT (vestronidase alfa)
MPS IX		Mucopolysaccharidosis	*HYAL1*	Hyaluronidase	Skeletal system	None
Mucolipidosis I	Sialidosis	Mucolipidosis (glycoproteinosis)	*NEU1*	α-Neuraminidase	Skeletal system, brain	None
Mucolipidosis II/III		Mucolipidosis	*GNPTAB, GNPTG*	UDP-N-acetylglucosamine-1-phosphotransferase	Skeletal system, brain	None
Mucolipidosis IV		Mucolipidosis (gangliosidosis)	*MCOLN1*	Mucolipin-1	Skeletal system, brain	None

(continued)

Table 1 (continued)

Disorder	Alternative name	Type	Affected gene	Affected protein	Involved organs/ tissues	Current treatment
α-Mannosidosis		Oligosaccharidosis	*MAN2B1*	α-D-Mannosidase	Skeletal system, liver, spleen, neuromuscular system	HSCT before cognitive impairment, ERT tested in phase III
β-Mannosidosis		Oligosaccharidosis	*MANBA*	β-Mannosidase	Skeletal system, liver, spleen, neuromuscular system	None
Fucosidosis		Oligosaccharidosis	*FUCA1*	Fucosidase	Organomegaly	None
Galactosialidosis		Oligosaccharidosis	*CTSA*	Cathepsin A	Organomegaly	None
Salla disease		Oligosaccharidosis	*SLC17A5*	Sialin	Organomegaly	None
Infantile sialic acid storage disease		Oligosaccharidosis	*SLC17A5*	Sialin	Organomegaly	None
Aspartylglucosaminuria		Oligosaccharidosis	*AGA*	Aspartylglucosaminidase	Organomegaly	None
Schindler disease		Oligosaccharidosis	*NAGA*	α-N-Acetylgalactosaminidase	Organomegaly	None
GSD II	Pompe disease	Disorder of glycogenolysis	*GAA*	Acid α-glucosidase	Heart, muscles	ERT (alglucosidase alfa)
Gaucher disease type I, type II, type III		Disorder of sphingolipid metabolism	*GBA*	Glucocerebrosidase	Liver, spleen, and bone marrow, brain	ERT (imiglucerase, velaglucerase, taliglucerase) or SRT (miglustat, eliglustat)
Fabry disease		Disorder of sphingolipid metabolism	*GLA*	α-Galactosidase A	Gastrointestinal system, kidney, heart	ERT (fabrazyme, replagal) or SRT (lucerastat, venglustat, both phase III)
Acid sphingomyelinase deficiency	Niemann-Pick disease type A/B	Disorder of sphingolipid metabolism	*SMPD1*	Acid sphingomyelinase	Spleen, liver, lungs, BM, with (NP-A)/without (NP-B) brain involvement	None
Tay-Sachs disease	GM2 gangliosidosis	Disorder of sphingolipid metabolism	*HEXA*	β-Hexosaminidase A	Brain	None

Sandhoff disease	GM2 gangliosidosis	Disorder of sphingolipid metabolism	*HEXB*	β-Hexosaminidase B	Brain	None
GM1 gangliosidosis	GM1 gangliosidosis	Disorder of sphingolipid metabolism	*GLB1*	β-Galactosidase	Skeletal system, heart, brain	None
Globoid cell leukodystrophy	Krabbe disease	Disorder of sphingolipid metabolism	*GALC*	β-Galactosylceramidase (galactocerebrosidase, cerebroside β-galactosidase)	Central and peripheral nervous system	HSCT before onset of symptoms
Metachromatic leukodystrophy (MLD)		Disorder of sphingolipid metabolism	*ARSA*	Arylsulfatase A (ASA)	Central and peripheral nervous system	HSCT before onset of symptoms or in slowly progressing patients
Niemann-Pick disease type C (NP-C)		Cholesterolglyco-sphingolipidosis	*NPC1, NPC2*	NPC1 and NPC2 protein	Liver, spleen, lung, CNS	SRT (miglustat)566t7
Neuronal ceroid lipofuscinosis (CLN)	Batten disease	Neuronal ceroid lipofuscinoses	*CLN1/PPT1, CLN2/ TPP1, CLN3, CLN4/DNAJC5, CLN5, CLN6, CLN7/ MFSD8, CLN8, CLN10/ CTSD, CLN11/ GRN, CLN12/ATP13A2, CLN13/ CTSF, CLN14/ KCTD7*	Palmitoyl protein thioesterase 1, tripeptidyl peptidase 1, CLN3 transmembrane protein, cysteine string protein α, ceroid-lipofuscinosis neuronal protein 5, ceroid-lipofuscinosis neuronal protein 6, major facilitator superfamily domain-containing protein 8, CLN8, cathepsin D, progranulin, ATPase type 13A2, cathepsin F, potassium channel tetramerization domain-containing protein 7	Brain	Cerliponase alfa (Brineura) intracerebroventricular ERT for treatment of CLN2 only

modified) stem cell treatments/applications may be of benefit for the treatment of LSDs.

Novel stem cell therapies should not only be able to affect the disease globally, but should also be able to be directed toward adjusting levels of lysosomal storage, and on altering the chemical consequences of the destructive storage. Because different LSDs are caused by different molecular mechanisms, each disease requires specific attention and adaptation of treatments according to need (Ballabio and Gieselmann 2009). Significant progress in the management and treatment of LSDs has been obtained by increased knowledge of their molecular background and pathophysiology (Parkinson-Lawrence et al. 2010). For almost four decades, treatment of patients with LSDs was based largely on palliative or supportive medical treatments (Leal et al. 2020). The initial experiments leading to development of curative treatment strategies for correcting the main defect of these diseases came from in vitro studies demonstrating replacement of the enzyme defect in patient cells by exogenous supply of the missing enzyme through receptor-mediated endocytosis or by direct transfer from normal leukocytes in co-culture experiments (Olsen et al. 1981, 1982, 1983; Bou-Gharios et al. 1993).

These studies led to clinical studies involving intravenous injection of exogenously administered enzymes extracted from certain human tissues, infusion of plasma/plasma fractions and/or leukocytes, and implantation of skin fibroblasts or amniotic cells (Leal et al. 2020). Although theoretically these treatments could have shown potential, in practice these therapeutical interventions consistently led to poor clinical efficacy, and their use in patients was both impractical and time consuming. However, it has led to the development and use of advanced cellular therapies, such as HSCT for those LSDs where the missing enzyme is expressed in healthy donor hematopoietic cells and can be secreted in sufficient amounts to the tissues that are mostly affected by the deficiency (Zheng et al. 2003; Biffi et al. 2006, 2013; Biffi 2012; Langford-Smith et al. 2012; *Birth Defects Orig Artic Ser* 1986). These studies have also provided the basis for HSC gene therapy, where the lacking enzyme is overexpressed in the patient's own HSCs. In addition, several studies have now shown that where some therapeutic approaches provide no clinical benefit when used alone, when used in combination with other treatments, they may contribute to overall efficacy (Hawkins-Salsbury et al. 2011). In this chapter, we will focus on treatments that are currently approved for clinical use and on novel treatments that are being developed and show great promise for the treatment of LSDs.

2 Effects of LSDs on Stem Cell Function

Relatively few data are available on the direct effects of LSDs on stem cell function in general. However, in many LSD patients, in addition to specific organs, often skeletal development is affected. Although the severity of the skeletal involvement may vary between the different LSDs, it is highly likely that bone tissue and hematopoiesis can be affected by storage of deposited material in the bone marrow (BM) niche and BM stromal and hematopoietic stem cells in particular. Both multipotent mesenchymal stromal/stem cells (MSCs) and HSCs from Gaucher disease (GD) type 1 patients showed diminished activity of the lysosomal enzyme glucocerebrosidase (GBA). GD BM-MSCs revealed changes in cell size and cell cycle and decreased osteoblastic differentiation (Lecourt et al. 2013). In addition, these GD-MSCs were shown to secrete soluble factors that induced resorbing activities of osteoclasts and displayed a diminished capacity to support hematopoiesis in vitro. GD BM-MSCs also showed a significant rise in prostaglandin E2 (PGE2), cyclooxygenase-2 (COX-2), CCL2, and interleukin-8 (IL-8) secretion compared to normal controls (Campeau et al. 2009). These data indicated that the changed secretory profile, as well as the changes in differentiation and proliferation of GD-MSCs, may contribute to the observed skeletal problems and immune system failure in LSD patients. Similar to GD, also Hurler syndrome patients exhibit bone tissue abnormalities, and MSCs from these patients displayed an increase in their capacity to promote osteoclastogenesis compared to MSCs from healthy donors. The latter was correlated with an upregulation of the RANK/RANKL/OPG pathway in Hurler MSCs (Gatto et al. 2012). When the lysosomal enzymes GBA and alpha-galactosidase A (GLA), which cause Fabry disease, were silenced in BM-MSCs, these cells were shown to be prone to apoptosis and senescence due to impaired autophagy and DNA repair mechanisms (Squillaro et al. 2017). BM-MSCs derived from I-cell disease or mucolipidosis II patients showed a chondrogenic differentiation defect, as well as a preference for adipogenic differentiation, and may therefore contribute to the pathogenesis of I-cell disease through their effects on the BM niche (Kose et al. 2019). The glucose-6-phosphate transporter (G6PT) in healthy individuals is ubiquitously expressed in the ER membrane and facilitates transfer of cytoplasmic G6P into the lumen of the ER (Chou et al. 2010). Deficiency of the (*G6PT*) gene causes glycogen storage disease type Ib (GSD-Ib), which is characterized by disrupted glucose homeostasis and neutropenia. CRISPR/Cas9-mediated knockout of *G6PT* in human adipose tissue-derived MSCs resulted in enhanced proliferation and impaired adipogenic and osteogenic differentiation of these cells without apparent effects on the cells' morphology or immunophenotype (Sim et al. 2018). Furthermore, the absence of G6PT resulted in differences in metabolic activity and increased ER stress and mitochondrial oxidative stress in MSCs, similar to the situation as previously described in neutrophils (Kim et al. 2008). MLD is caused by mutations in the arylsulfatase A (ARSA) gene and causes accumulation of sulfatides, progressive demyelination, and neurological dysfunction. Patient-derived $ARSA^{-/-}$ BM-MSCs showed overall typical MSC morphology and a normal immunophenotype. The cells showed no specific deficit in adipogenic or osteogenic differentiation (Bohringer et al. 2017). In conclusion, although in most LSDs the effects of the disease on MSCs have not been studied, it appears that in most models, MSCs are affected in a manner similar to other cells. This may have several implications: firstly, MSCs have been shown to locally secrete several lysosomal enzymes (Jackson et al. 2015), and deficiency of these enzymes may further contribute to tissue pathology; and secondly, accumulation of proteins within the MSCs affects their regenerative and supportive function in the stem cell niches. However, the fact that healthy and genetically modified MSCs may offer the opportunity to serve as enzyme factories and cause cross-correction, as well as the notice that these cells have an intrinsic immunomodulatory and regenerative potential, underlines their potential for stem cell treatment of LSDs (discussed in detail below). Although in many LSDs the brain may be severely affected, the effect of storage of deposited material on neural stem cells (NSCs) has not been directly assessed. However, a study on the effect of aging on NSCs revealed that loss of lysosome function and the aggregate buildup in these cells during aging contributed to reduced NSC activity (Audesse and Webb 2018). Furthermore, stimulation of lysosomal activity in the aged NSCs increased the frequency of activated neural stem and progenitor cells in the NSC niche and restoration of healthy stem cell activity, indicating that involvement of LSDs in NSC activity and function is not only likely, but also that targeting affected NSCs may be helpful in decreasing LSD-related disease burden and symptoms (discussed below).

3 Hematopoietic Stem Cell Transplantation (HSCT) in LSDs

In addition to enzyme replacement therapies (ERTs) and treatments directed at substrate reduction (SRT), HSCT is the only intention curative treatment option for LSDs, such as sphingolipidosis and mucopolysaccharidosis (Hawkins-Salsbury et al. 2011; Poe et al. 2014). The purpose of HSCT is to provide a widespread, stable, and permanent endogenous supply of the missing enzyme through secretion of the missing enzyme by HSCs derived from a healthy donor or clinically unaffected carrier. The source of the HSCs may be

BM, mobilized peripheral blood (PB), or umbilical cord blood (UCB). After transplantation, the healthy donor HSCs rapidly migrate to the recipient's bone marrow where they repopulate the hematopoietic niche, replicate, and give rise to healthy, mature hematopoietic cells, which in turn migrate throughout the body and clear affected cells and tissues of deposited material (Tan et al. 2019; Prasad and Kurtzberg 2010a; Orchard et al. 2007; Boelens et al. 2010). This constitutes the first mechanism of treatment for LSDs by HSCT. Whereas ERT/SRT have been approved for the treatment of some of the LSDs (i.e., Gaucher disease, Fabry disease, lysosomal acid lipase deficiency, Hurler syndrome, Hunter syndrome, Maroteaux-Lamy syndrome, Morquio syndrome, and Pompe disease; see Table 1), successful treatment of LSDs with neurodegenerative aspects remains problematic because of the inability of the macromolecules to pass the brain-blood barrier (BBB) (Li 2018). Therefore, an important advantage of HSCT is the ability of some of the enzyme-producing donor-derived cells to cross the BBB and migrate to the brain (Tan et al. 2019; Neuwelt et al. 2008; Krivit et al. 1999). This action has the potential to improve neurocognitive function and standard of life, especially when applied early, before irreversible damage is caused by the disease. The basic mechanism of the therapeutic effect relies on the ability of circulating monocytes to egress from the vessels and migrate through tissues and organs as tissue macrophages. These macrophages locally secrete the missing enzyme that is then internalized by the affected cells. After uptake, the enzyme is transported to the lysosomes, where it aids in the clearance of stored, undigested material. For example, in the central nervous system (CNS), microglia cells have been shown to be partially replaced by donor cells and transport enzymes to affected neurons (Neuwelt et al. 2008; Krivit et al. 1999; Whitley et al. 1993; Peters and Steward 2003). A second, less important mechanism lies in the ability of the affected patient cells for uptake of functional lysosomal hydrolases released by donor cells to the bloodstream via an endocytosis-mediated mechanism. However, this approach is effective only in some LSDs, and its success in the treatment of advanced neurological disease is limited. The success of HSCT in LSDs is most evident in Hurler's disease (MPS-IH). HSCT significantly improves the clinical outcome by reducing visceromegaly, improving heart function and airway obstruction, improving neurocognitive outcome, and reducing early disease-related mortality and prolonging survival. However, the remaining disease manifestations particularly orthopedic problems require long-term follow-up and may require surgical intervention. HSCT has also been shown to ameliorate the clinical phenotype in other LSDs including patients with pre-symptomatic or late-onset Krabbe disease (globoid cell leukodystrophy), MPS VI, late-onset forms of MLD, and alpha mannosidosis (Neuwelt et al. 2008; Krivit et al. 1999; Whitley et al. 1993; Peters and Steward 2003). In 1991, the International Society for the Correction of Genetic Diseases by Transplantation (COGENT) established a guideline and proposed that only children under 2.5 years of age with a growth coefficient of 2.5 or smaller than 3 years should be offered an HSCT because of the expected benefits that rapidly decline after this age. In addition, the presence of a maximally HLA-compatible donor is recommended (Hobbs 1992). Although many years have passed since, this consensus is still generally accepted (de Ru et al. 2011). In the following years, HSCT has been shown to increase life expectancy and improve clinical symptoms when performed at an early age in children affected by mucopolysaccharidosis IH (or Hurler's disease) (Whitley et al. 1993; de Ru et al. 2011) and a wide range of other metabolic diseases with varying success (Boelens et al. 2010). However, in many centers, it remains common practice to use ERT until HSCT/during engraftment to assure the best health status of the patient pre-transplant and to improve the overall success rate of the procedure (Ghosh et al. 2016).

However, HSCT is often accompanied by significant morbidity and even mortality, as a result of rigid conditioning regimens and myeloablative protocols, insufficient engraftment or graft rejection, and the development of acute and chronic graft-versus-host disease (GvHD) (Talib and Shepard 2020; Aljurf et al. 2014; Prasad and Kurtzberg 2010b; Martin et al. 2013). Therefore, HSCT is usually restricted to LSDs that do not respond sufficiently to ERT/SRT (Krivit et al. 1999). In addition, the risk for the development of an immune response against the proteins delivered through ERT and SRT is considerable and causes a decrease in the effectiveness of these treatments. Since in contrast to the exogenous delivery of enzyme during ERT, in HSCT the enzyme is endogenously produced by the body's own cells, the risks of developing antibodies are much lower (Peters and Steward 2003; de Ru et al. 2011; Hoogerbrugge et al. 1995).

In the last two decades, UCB-derived HSCs have become the cell source of choice for the management of LSDs due to their rapid availability, low graft failure rates, improved engraftment potential, better tolerance of HLA mismatch, and reduced risk of GvHD (Jaing 2007; Mogul 2000; Aldenhoven and Kurtzberg 2015). Furthermore, recipients of UCB were shown to achieve full donor chimerism, normalization of enzyme levels, and superior long-term clinical prognosis compared to BM-HSC or PB-HSC transplantation (Aldenhoven and Kurtzberg 2015), resulting in prevention and/or delay of neurological damage, especially when administered to asymptomatic LSD patients (Prasad and Kurtzberg 2010b; Martin et al. 2013; Jaing 2007; Mogul 2000). Therefore, it appears that ERT/SRT and HSCT/UCBT can prolong and improve the quality of life for some LSDs patients, although none of these applications are fully curative and may have little impact on the developed skeletal malformations (Hoogerbrugge et al. 1995; Rappeport and Ginns 1984; Starer et al. 1987). Additional transplantation of allogeneic donor MSCs during/after HSCT may be considered to improve skeletal malformations in LSD patients at an early stage of disease development (Hawkins-Salsbury et al. 2011; Phinney and Isakova 2014). However, further assessment of the overall therapeutic effects of HSCT and UCBT is needed to be able to develop standardized criteria for patient selection, handling/management protocols, and posttreatment supportive care and follow-up (Talib and Shepard 2020; Aljurf et al. 2014).

4 Hematopoietic Stem Cell Gene Therapy for Treatment of LSDs

HSC transplantation has been used successfully for the treatment of some LSDs, such as Hurler, Krabbe, and metachromatic leukodystrophy in the early course of the disease (see Table 1), but has been found to be of little benefit for LSDs that severely affect the brain and result in loss of (previously acquired) neurocognitive skills. Furthermore, HSC transplantation is not useful for the treatment of LSDs if the enzyme is not expressed or secreted in sufficient quantities by HSCs to allow cross-correction of tissue damage. However, genetic engineering of autologous HSCs in order to constitutively overexpress the enzyme of interest has shown potential in several murine LSD models (Table 2), as well as in clinical trials.

Combining the use of autologous HSC together with gene therapy using integrating vectors avoids the risk of an immunological response, while at the same time, it allows cross-correction of affected cells due to overexpression and secretion of the enzyme throughout the body (Zheng et al. 2003; Biffi et al. 2006; Biffi 2012; Krall et al. 1994). Although initial studies used the gamma-retroviral MLV vectors to cure LSDs (Lutzko et al. 1999a, b), the insertional mutagenesis risks related to the use of these particular vectors have prompted researchers to pursue other means of gene transfer. The currently used third-generation self-inactivating (SIN) lentiviral vectors (LV) are highly effective in transferring the transgenes to the HSC genome and have been shown to alleviate LSD-related symptoms in many murine and larger animal models (Table 2). More recently, researchers have been exploring the possibilities of gene-editing methods, such as CRISPR/Cas9, rather than gene addition, to correct LSDs (Gomez-Ospina et al. 2019). Initial studies using RV vectors based on the MLV virus for the treatment of Niemann-Pick B disease showed the possibilities for efficient transfer of the missing enzyme throughout

Table 2 Effects of HSC-based gene therapy on animal models of lysosomal storage diseases

Cell type/dose	Vector	ROA	Disease	Animal model	Effects	References
1×10^6/g body weight murine Lin- HSCs	RV/MFG-hASM	i.v.	Niemann-Pick B disease (acid sphingomyelinase deficiency)	ASMKO mouse	High levels of ASM activity up to 10 months Extended life span up to 9 months Reduced sphingomyelin storage in the spleen, liver, and lung Despite increased Purkinje cells, ongoing brain damage	Erlich et al. (1999) and Miranda et al. (2000)
1×10^6 murine BM cells + 1×10^5 MSCs	RV/MFG-hASM	i.v. BM cells, intracerebral MSCs	Niemann-Pick B disease (acid sphingomyelinase deficiency)	ASMKO mouse	Normal levels of ASM in all tissues, including the brain Marked reduction in sphingomyelin Improved cerebellar function and increase in Purkinje cells Induction of anti-ASM antibodies	Jin and Schuchman (2003)
5×10^6–10^7 long-term murine marrow culture cells	MLV-LCαIDSN and MLV-M48αID	In utero	MPS-I (α-iduronidase deficiency)	MPS-I canine model	Despite engraftment, no amelioration of the disease Induction of anti-IDUA antibodies	Lutzko et al. (1999a, b)
1×10^6 murine Lin- HSCs	LV-IDUA	i.v.	MPS-I (α-iduronidase deficiency)	BL/6 idua$^{-/-}$ MPS-I mouse	Autologous, but not allogeneic, IDUA+ HSCs deliver functional enzyme to affected tissues T-cell depletion rescues engraftment of IDUA+ HSCs in animals with anti-IDUA CD8+ T-cells	Squeri et al. (2019) and Visigalli et al. (2010)
1×10^6 murine Lin- HSCs	LV-PGK-ARSA	i.v.	MLD	C57BL6/129 As2 (arsa)$^{-/-}$ MLD mouse	ARSA expression by all HSC progeny, CNS microglia, and PNS endoneurial macrophages Higher therapeutic impact than transplantation with WT HSCs	Biffi et al. (2006)
5×10^5 murine Lin- HSCs	LV-SF-GAA, LV-SF-GAAco	i.v.	Pompe disease (α-glucosidase deficiency)	FVB/N gaa$^{-/-}$ mouse model	High GAA activity in leukocytes (GAAco 20-fold > GAA) Reduced glycogen deposition in the heart, liver, lung, and muscle Ameliorated respiratory and normalized motor function	van Til et al. (2010) and Stok et al. (2020)

2 × 10^6 PB-CD34+ GUSB- HSPCs	SIN-LV-PGK-GUSB	i.v.	MPS-VII (B-glucuronidase (GUSB) deficiency)	NOD/SCID/ MPS-VII mice	Rapid engraftment and GUSB-expressing cells in the bone marrow, spleen, and liver Reduction in tissue pathology, reductions in GAGs and α-galactosidase activity	Hofling et al. (2004)
2 × 10^6 murine BM-MNCs	MLV-pUMFG-α-gal A	i.v.	Fabry disease (α-galactosidase A deficiency)	α-gal A-deficient mice	Engraftment of transduced cells Normalized α-galactosidase A activity in BM and PB-MNCs Correction of α-galactosidase A activity in all tissues, except the brain Lipid reduction in peripheral tissues	Takenaka et al. (1999, 2000)
1 × 10^6 murine Lin- BM cells	SIN-LV-IHK-coGALC	i.v.	Krabbe disease	C57BL/6 $GALC^{-/-}$ Twitcher mice	Efficient engraftment Increased life span of mice	Hu et al. (2016)

the body, but also pointed at the absence of significant effects on the central nervous system (Erlich et al. 1999; Miranda et al. 1998). In addition, previous ERT or high expression of previously lacking enzymes after gene transfer was shown to cause formation of antibodies against the transgene, causing suboptimal effects in mouse models of MPS-I (Lutzko et al. 1999a, b) and Niemann-Pick syndrome (Squeri et al. 2019). Whereas conditioning protocols (Hu et al. 2016), specific brain conditioning (Capotondo et al. 2012), or co-transplantation of genetically modified HSCs and MSCs was helpful in increasing brain engraftment of the transduced cells (Jin and Schuchman 2003), T-cell depletion greatly improved longevity of the treatments by preventing clearance of the enzyme by CD8+ T-cells (Squeri et al. 2019). Gene therapy using autologous HSCs further showed that overexpression of the lacking enzyme resulted in far greater efficacy of the treatment than transplantation of allogeneic WT-HSCs (Biffi et al. 2006; Visigalli et al. 2010). Further optimization of transgene constructs using SIN-LV vectors and codon optimization showed that increased transgene expression could efficiently clear deposited glycosaminoglycans (GAGs) throughout the body (liver, muscle, heart), but also from brain tissue in a mouse model of Pompe disease, indicating that a sufficiently high level of transgene expression may be helpful in preventing brain damage (van Til et al. 2010; Stok et al. 2020).

A clinical trial (NCT01560182) demonstrated that transplantation of autologous, lentivirally transduced HSCs from patients with MLD resulted in high-level stable bone marrow engraftment and above normal ARSA levels in hematopoietic cells and cerebrospinal fluid. Furthermore, disease-related symptoms were prevented in children treated before the symptomatic stage and halted in children in whom the disease had already progressed (Biffi et al. 2013; Sessa et al. 2016). Clinical studies for HSC gene therapy of Fabry disease, Pompe disease, and Krabbe disease are currently recruiting, but no clinical gene therapy data are available for the treatment of these diseases yet. Promising data on HSC gene therapy in clinical trials for the treatment of peroxisomal disorder X-linked adrenoleukodystrophy have been obtained and show that transplantation of autologous, genetically corrected *ABCD1* overexpressing CD34+ cells resulted in arrest of cerebral demyelination (Cartier et al. 2009, 2012). It is expected that similar results can be obtained with HSC gene therapy for LSDs.

Although clinical development of gene therapy for LSDs still has a long road to go, ex vivo HSC gene therapy as an alternative treatment option for treatment of LSDs has the advantage that in contrast to ERT a single treatment has the potential to be curative in essence (Yang et al. 2014). Although not discussed in this chapter, there is also a possibility to use in vivo gene therapy, where the viral vector is directly injected into the subject (Traas et al. 2007; Meyerrose et al. 2007). Depending on the target tissue or route of administration, adenoviral vectors (AdV), adeno-associated vectors (AAV), or LV vectors can be chosen. This approach has the advantage that within a tissue, different cell subsets can be reached, and that tissues, which are difficult to reach, such as the brain, can be treated through direct intracerebral injections.

Although HSC gene therapy has many advantages over allogeneic HSC transplantation, such as regulated gene expression by choosing, for example, tissue-specific promoters or co-expression of certain miRNAs and overexpression of the transgenes, resulting in better results due to cross-correction of affected cells, limitations and risks such as integrational mutagenesis, and the possibility for immunogenicity, should be carefully assessed.

5 Mesenchymal Stem Cell Applications in LSDs

Multipotent MSCs are non-hematopoietic, multipotent stem cells that have the capability to differentiate into mesodermal lineages such as adipocytes, chondrocytes, and osteocytes and some endodermal (hepatocytes) and ectodermal (neurons) lineages (Pittenger et al. 2019). Human

MSCs were firstly reported in BM, but have now been shown to reside in many other tissues, like umbilical cord (UC) Wharton's jelly, adipose tissue, endometrium, amniotic fluid, and dental tissues (Pittenger et al. 2019).

MSCs have been suggested to be suitable targets for the treatment of LSDs because of their capacity for multilineage differentiation and immunomodulation (Phinney and Isakova 2014). Since storage of deposited substances in the LSDs activates numerous inflammatory pathways, resulting in both local (e.g., neuroinflammation in the brain) and systemic inflammation (Rigante et al. 2017), MSCs may not only serve as source of the missing enzyme (Jackson et al. 2015; Muller et al. 2006) and support clearance of tissue damage but may also act through suppression of local inflammatory processes. For these reasons, MSCs might be a good alternative or addition to treatment with HSCs, particularly in LSDs with neurological involvement. Paracrine mechanisms, such as secretion of neurotrophins and other molecules that support neural cell recovery and neurite outgrowth during inflammation, may also contribute to the supportive effect of MSCs (Crigler et al. 2006; Joyce et al. 2010; Pisati et al. 2007; Qu et al. 2007). An overview of the use of MSCs for the treatment of mouse models with a neurodegenerative LSD has been provided in Table 3. Overall, the main source of MSCs used appears to be BM, although MSCs from umbilical cord (UC) and placental tissue also have been tested. The cells are generally used to treat the neurodegenerative effects of LSDs, such as different variants of the Niemann-Pick syndrome and are directly injected into different areas of the brain, including the cerebellum, ventricles, and cerebrum (hippocampus). Direct transplantation of culture-expanded BM-MSCs into the cerebellum of Niemann-Pick type C (NP-C) mice significantly alleviated the degree of microglial and astrocyte activation and reduced levels of the microglial activating cytokine macrophage colony-stimulating factor (M-CSF) (Bae et al. 2005a). Injection of UC-MSCs into the hippocampus of NP-C mice at an early stage of the disease was shown to improve motor skills due to increased proliferation and survival of neuronal cells (Seo et al. 2011). In addition, MSC transplantation in this model resulted in decreased cellular apoptosis and normalization of neurotransmitter homeostasis, further supporting recovery of the CNS. Intracerebral injection of BM-MSCs, co-cultured with embryonic NSCs from NP-C mice, resulted in enhanced NSC proliferation and neuronal differentiation in the subventricular zone of the brain of the NP-C mice (Lee et al. 2013). Similarly, co-culture of BM-MSCs with Schwann cells in the presence of psychosine, the toxic substrate deposited in Krabbe disease, induced proliferation and neurite outgrowth in the Twitcher mouse model. Combination treatment consisting of anti-BDNF antibodies together with MSCs was shown to further enhance the neuritogenic effect (Miranda et al. 2011). The effects of intrastriatal injection of BM-MSCs in Twitcher mice pups depended highly on age of the mouse and genotype. Although minor effects were observed on weight gain and severity of the twitching, as well as improved rotarod motor skills, overall the treatment was less effective than intraventricular injection. Injection of adipose-derived stromal cells in this mouse model failed to affect disease progression (Wicks et al. 2011).

Data from clinical trials using human MSCs are not available yet. According to the ClinicalTrials.gov website, only one clinical trial is currently assessing safety of MSC infusions (phase I) in LSD patients. However, in this study, the main aim is to use human placenta-derived stem cells for the treatment of immune-related issues after UCB transplantation for LSDs (ID: NCT01586455), rather than to use the cells as a source of enzyme.

MSC-derived extracellular vesicles (exosomes) have been widely studied, and, due to their content, they are candidates for use in the solution of many clinical problems. These exosomes are nano-sized extracellular vesicles that contain MSC-derived bioactive molecules (messenger RNA (mRNA), microRNAs (miRNAs)), chemokines, growth factors, cytokines, and enzymes) that modulate function and homing of immune cells and regulate survival

Table 3 Effects of unmodified MSCs in the treatment of murine neurodegenerative LSD models

Source	Dose/ROA	Disease	Mouse model	Effects	References
Bone marrow	1 × 10^6 MSCs into the cerebellum	Niemann-Pick disease type C1	BALB/c npc^{nih} (NP-C) mice	Upregulation of synaptic transmission-related genes following a BM-MSC/Purkinje cell fusion Improved morphological changes and cerebellar function	Bae et al. (2005a, b, 2007)
Bone marrow	1 × 10^5 MSCs into the cerebellum	Niemann-Pick disease type C1	BALB/c npc^{nih} (NP-C) mice	Correction of calcium homeostasis in Purkinje cells Increased sphingosine-1-phosphate levels Decreased sphingosine Inhibition of apoptosis	Lee et al. (2010)
Bone marrow	1 × 10^6 MSCs into the cerebrum	Niemann-Pick disease type C1	BALB/c npc^{nih} (NP-C) mice	Proliferation and neuronal differentiation of NSCs within the subventricular zone through release of CCL2 from BM-MSCs	Lee et al. (2013)
Umbilical cord	1 × 10^6 MSCs into the hippo-campus	Niemann-Pick disease type C1	BALB/c npc^{nih} (NP-C) mice	Recovery of motor function Normalized cholesterol homeostasis Enhanced neuronal cell survival and proliferation Reduced loss of Purkinje cells Decreased inflammation and apoptosis	Seo et al. (2011)
Bone marrow	0.5–1 × 10^5 MSCs, intraventricular injection	GM1 gangliosidosis	C57BL/6 β-galactosidase knockout (BKO) mice	Effective migration of cells through the whole brain Decreased GM1 ganglioside levels for up to 8 weeks	Sawada et al. (2009)
Bone marrow	20.000 MSCs intrastriatal injection	Krabbe disease	Twitcher mice (B6.CE-$Galc^{twi}$/J)	Age- and genotype-dependent weight gain, decrease in twitching severity, and improved motor skills	Wicks et al. (2011)

and proliferation of parenchymal cells (Toh et al. 2018). Because of the ability to carry enzymes and also their low immunogenic properties, these allogenic MSC-derived extracellular vesicles can be used for the treatment of LSDs (Joyce et al. 2010). Also extracellular vesicles derived from genetically modified MSC that secrete the deficient enzyme can be a good tool for the treatment of LSDs (Phinney and Isakova 2014). This practice can be considered as a permanent source of deficient enzyme due to engraftment of MSCs instead of ERT/SRT (Tancini et al. 2019).

Thus, the use of MSCs either by itself or as an adjuvant therapy in combination with HSCT appears to have many advantages, including rapid and easy isolation from several tissues, their capacity for multilineage differentiation, and a low risk for induction of immune reactions. In addition, their easy transduction offers the opportunity to serve as vectors for gene therapy as well (see below). However, the risks and/or possible side effects may include uncontrolled proliferation and differentiation and unregulated migration, which may be especially difficult to control after infusion into the brain. Nevertheless, current studies in animals have thus far not shown any indication of unwanted proliferation or differentiation and appear to be safe with little to no unwanted effects related to infusion.

6 MSC Gene Therapy for Treatment of LSDs

MSCs have a number of unique biological properties that make them well suitable to serve as cellular vectors. In addition, BM-MSCs have been shown to engraft and migrate throughout a large area of the brain when transplanted directly into the CNS of mice (Joyce et al. 2010).

Table 4 Effects of genetically modified MSCs on treatment of neurodegenerative mouse LSD models

Cell type/dose	Vector	ROA	Disease	Mouse model	Effects	References
4×10^6 BM-MSCs	LV-SB-IDUA	Intraperitoneal	Mucopolysaccharidosis type I (MPS-I)	C57bl/6 IDUA KO mouse	Migration of MSCs throughout the peritoneum and organs Temporary increase in IDUA levels Induction of anti-IDUA antibodies	Martin et al. (2014)
2×50.000 BM-MSCs	RV-MFG-hASM	Intracerebral (hippocampus) and cerebellar injection	Niemann-Pick disease A/B (acid sphingomyelinase deficiency)	ASMKO mice	Migration of MSCs throughout the brain High expression of ASM by MSCs Protective effect on Purkinje cells Extended the life span of treated mice Cross-correction of ASM levels in multiple neuronal cell types Immunological rejection of transduced MSCs causing only partial correction	Jin et al. (2002)
1×10^5–1×10^6 BM-MSCs (KUSA/A1)	RV-MND-HBG	Intraventricular injection	Sly syndrome (MPS-VII, β-glucuronidase deficiency)	MPS-VII mice	Increased GUSB activity throughout the brain Reduced levels of GAGs (near normal) 4 weeks after transplantation Markedly improved cognitive function	Sakurai et al. (2004)
1×10^6 hBM-MSCs	LV-MND-hGUSB	Intraperitoneal	Sly syndrome (MPS-VII, β-glucuronidase deficiency)	NOD-SCID MPS-VII mice	Increased GUSB serum levels up to 40% of normal levels Normalization of GAGs in several tissues Improved retinal function	Meyerrose et al. (2008)

Therefore, these cells have been proposed to serve as a potential target for gene therapy. Engineering MSCs to overexpress a variety of proteins has been shown to alleviate symptoms in mouse models of a wide range of neurological diseases, including Parkinson, Alzheimer, and Huntington, but also neurodegenerative diseases caused by LSDs (Phinney and Isakova 2014). Such studies have been undertaken in several different LSD mouse models and have been summarized in Table 4.

Intracranial transplantation of BM-MSCs, engineered to secrete acid sphingomyelinase (ASM) into ASM-deficient mice, revealed extensive engraftment throughout the CNS, delayed death of Purkinje cells, decreased levels of sphingomyelin, and extended animal survival (Jin et al. 2002). Combination therapy employing intracerebral and systemic application of ASM overexpressing BM-MSCs was shown to increase ASM activity in most peripheral tissues, as well as the brain (Jin and Schuchman 2003). Intracranial injection of murine BM-MSCs retrovirally modified to express human beta-glucuronidase (*HBG*) increased beta-glucuronidase (GUSB) enzyme levels in the striatum, cerebral cortex, and olfactory bulb, diminished GAG levels, and enhanced the cognitive function in the Sly syndrome mouse model (MPS VII) (Sakurai et al. 2004). Human BM-MSCs engineered to express GUSB using the MND/HBG lentiviral vector were injected intraperitoneally into the MPS-VII mice. The engineered cells migrated to diverse tissues within 4 months and secreted high levels of the deficient enzyme, decreasing storage of deposited substances in several tissues, including the retina (Meyerrose et al. 2008).

These data demonstrate that although MSCs can be suitable vectors to treat LSDs, their effects on neurologic defects may depend on the type of enzyme missing, disease progression, distribution of cells, as well as route of administration. In addition, data from the mouse studies indicate that the effects of genetically engineered MSCs may be temporary, and optimized methods to ensure prolongation of duration of the effects should be developed. Furthermore, studies exploring the mechanisms involved in MSC migration and engraftment as well as their regenerative potential in the treatment of LSD CNS involvement should be undertaken in order to be able to allow the development of clinically applicable treatment protocols.

7 Neural Stem Cell Applications in LSDs

The biggest challenge of HSCT in the treatment of LSDs is the limited number of cells that migrate, reach, and pass the BBB. During conventional HSCT, only a very small fraction of cells has been shown to differentiate into microglial cells or support CNS-resident cells (Kennedy and Abkowitz 1997). Neurocognitive symptoms, therefore, do not show any or very limited benefit from allogeneic BMT, HSCT, UCBT, or intravenously delivered MSC transplantations. These problems could be theoretically solved with intracerebral transplantation of MSCs (discussed above) or NSCs (Snyder et al. 2004).

NSCs are multipotent stem cells present in the specialized niches of the central nervous system that have self-renewal capacity and can produce new neurons as well as supporting cells, known as glial cells. Primary NSCs have been isolated from specific neurogenic niches in the brain (i.e., the subventricular zone and the subgranular zone), glial progenitors, and immortalized clonal cell lines. Alternatively, these cells can also be derived from embryonic stem cells (ESCs) or induced pluripotent stem cells (iPSCs) (Otsu et al. 2014). Activation of resident NSCs or their injection into defective areas of CNS can help the treatment of injury or age-related changes in animal models (Navarro Negredo et al. 2020). NSC-based applications could potentially be used as a treatment of LSDs defects through cell replacement, the release of trophic factors, and an anti-inflammatory effect. NSCs have been shown to express a wide range of lysosomal enzymes (Audesse and Webb 2018; Fukuhara et al. 2006), but can also be genetically engineered to overexpress the relevant enzyme that can then cross-correct the local defective niche (Navarro

Table 5 Results of studies using NSCs to treat mouse models of LSDs

Treatment	Construct	Disease	Mouse model	Effects	References
2–8 × 10^4 murine cerebellar C17.2 NSCs into mouse ventricles	None	Sly syndrome (MPS-VII, β-glucuronidase deficiency)	MPS VII mice	Donor cell migration to the parenchyma Diffuse engraftment without conditioning GUSB expression by NSCs Lysosomal storage decreased or was absent in neurons and glial cells	Snyder et al. (1995)
5 × 10^5 murine cerebellar C17.2 NSCs or human HFT13 NSCs into lateral ventricles	None	Niemann-Pick syndrome type A (NP-A)	ASMKO mice	Engraftment without conditioning Decrease in neuronal and glial vacuolation and cholesterol accumulation in the neocortex, hippocampus, striatum, and cerebellum Cross-correction of brain tissue	Sidman et al. (2007)
4 × 10^4 murine cerebellar C17.2 NSCs into mouse cerebellum	None	Niemann-Pick syndrome type C (NP-C)	BALB/c *Npc1*$^{-/-}$ mice	Widespread migration through the cerebellum Increased life span No effect on numbers of Purkinje cells No clear effect on ataxia	Ahmad et al. (2007)
5 × 10^4 C17.2 mNSCs, primary Rosa E10.5 mNSCs, primary human fetal NSCs, hESC-derived NSCs into cerebral ventricles	None	Sandhoff disease (β-hexosaminidase deficiency)	*Hexb*$^{-/-}$ mice	Prolonged life and improved motor skills NSCs engraft widely and generate neurons Increased Hex activity and diminished gangliotriaosylceramide (GA2) and monosialoganglioside (GM2) Reduced microglial activation SRT synergizes with NSC transplantation	Lee et al. (2007)
2.5–5 × 10^4 murine fetal neurospheres into mouse lateral ventricles	None	Sly syndrome (MPS-VII, β-glucuronidase deficiency)	B6.C-H-2^{bml}/*ByBir-gus*^{mps}/*gus*^{mps} mice	Increased GUSB activity in the brain Cell migration throughout the brain Decrease in lysosomal storage in the hippocampus, cortex, and ependyma No indication for any tumor formation	Fukuhara et al. (2006)
3 × 10^5 murine neurosphere cells in lateral ventricles	LV-GFP	Metachromatic leukodystrophy, ARSA deficiency	C56BL/6 MLD mice	Successful engraftment Increased ARSA activity and improved sulfatide metabolism Cross-correction of cells throughout the brain Improved locomotor performance	Givogri et al. (2008)

(continued)

Table 5 (continued)

Treatment	Construct	Disease	Mouse model	Effects	References
10^5 murine neuronal progenitor cells intracerebral at multiple sites	MLV-hASM	Niemann-Pick syndrome type A (NP-A)	ASMKO mice	Transplanted NPCs survived, migrated, and showed region-specific differentiation In vivo low ASM activity, but sufficient for cross-correction of host cells Reversal of lysosomal pathology and clearance of sphingomyelin and cholesterol storage No correction of pathology in mice transplanted with WT-NPCs	Shihabuddin et al. (2004)
1 × 10^5 human embryonal NSCs into cerebral lateral ventricle	RV-LHC-HBG	Sly syndrome (MPS-VII, **β-glucuronidase deficiency)**	B6.C-H-2^{bml}/ *ByBir-gus*^{mps}/ *gus*^{mps} **mice**	Human NSC engraftment High-level β-glucuronidase production **Reduction in substrates of β-glucuronidase** **Widespread clearance of lysosomal storage** **Short-term effect due to apoptosis of NSCs**	Meng et al. (2003)

Negredo et al. 2020). In the past, fetal neural tissues have been used successfully for the treatment of Parkinson-related symptoms (Lindvall et al. 1990; Freed et al. 1992). However, expansion of NSCs in vitro has proven difficult, and from an ethical point of view, the use of human NSCs derived from multiple human fetuses in order to obtain sufficient amounts of cells may be not suitable. Nevertheless, the use of autologous and/or culture-expanded NSCs may prove feasible to ameliorate symptoms of LSDs related to advanced brain storage. Preclinical animal studies performed in in particular the Sly disease (MPS-VII) and Niemann-Pick mouse models have shown promising features (see Table 5), although several problems, such as expansion of NSCs, longevity of the effects, and viability of NSCs upon transplantation, cell migration, etc., still need to be addressed. Another promising feature of NSC treatment is that thus far no tumor formation related to intracranial injection of NSCs has been observed.

In an early report using the Sly syndrome MPS-VII model mouse, NSCs were engineered to overexpress GUSB and transplanted into the lateral ventricles of the mice. Donor-derived cells were found throughout the brain, with correction of β-glucuronidase levels, resulting in overall correction of lysosomal storage in neurons and glia of the MPS-VII mice (Snyder et al. 1995). Similarly, intracranial administration of fetal murine neurospheres to MPS-VII mice raised brain GUSB activity, resulting in diminished lysosomal storage (Fukuhara et al. 2006). In the ASMKO mouse model of Niemann-Pick disease type A, injection of either mouse or human NSCs into neonatal mice lateral ventricles resulted in a decrease in neuronal and glial storage and cholesterol accumulation throughout the brain (Sidman et al. 2007). Injection of NSCs in the cerebellum of neonatal Niemann-Pick type C $npc1^{-/-}$ mice resulted in widespread migration through the cerebellum and prolonged life span. However, no effect was seen on number of Purkinje cells, or ataxia (Ahmad et al. 2007), indicating that route of administration (cerebellar vs cerebrum) and cells' dose may be critical factors. Comparison of immortalized (C17.2) murine NSCs (mNSCs), primary Rosa E10.5 embryonic mNSCs, fetal human NSCs, and hESC-derived NSCs for the treatment of a mouse model of Sandhoff disease showed that all sources of NSCs improved CNS function, increased β-hexosaminidase (Hex) levels, and resulted in a decrease of glycosphingolipids. In addition, SRT synergized with transplantation of NSCs (Lee et al. 2007). Whereas in the absence of functional enzyme, the impact of SRT for Sandhoff disease is only limited, in combination with a constant (low-level expression) of β-hexosaminidase provided by the transplanted NSCs, SRT use may in fact be very effective, and in this mouse model, combination therapy resulted in a doubled life span (Lee et al. 2007). In addition, at least part of the effects of NSC transplantation has been contributed to the anti-inflammatory effects of NSCs. Recent studies have been focused on improving the effects of NSC transplantation by genetically modifying the cells to overexpress the missing enzyme. Using this approach, a better clearance of ASM in the brains of Niemann-Pick type A ASMKO mice could be achieved than with transplantation of WT neuronal progenitor cells (NPCs) (Shihabuddin et al. 2004). Similarly, Meng et al. showed the feasibility of treating Sly syndrome MPS-VII mice with NSCs retrovirally transduced to overexpress the human β-glucuronidase gene.

Thus, neuronal progenitors and stem cells provide a unique and powerful tool for the treatment of the neurodegenerative component of LSDs. However, (ethical) issues concerning procurement, optimized production and expansion, as well as transplantation-related risks need to be further explored before clinical application is possible.

8 Induced Pluripotent Stem Cell Applications in LSDs

Induced pluripotent stem cells can be generated by reprogramming adult somatic cells (Takahashi et al. 2007; Takahashi and Yamanaka 2006). Use of human iPSCs has been explored for the development of patient-specific cell therapies and research models for inherited or acquired diseases. Disease-specific iPSCs can be generated from subjects with a genetic disease (Rowe and Daley 2019), and iPSC clones can now be generated from patients with currently untreatable diseases from whom it is otherwise difficult to obtain stem cells or stem cells in sufficient amounts and can be used to study the pathophysiology of the disease in vitro, to test novel diagnostic procedures, and to enable drug development and are therefore a very new and promising tool for modeling of human diseases and development of new treatment strategies.

Using iPSC modeling, researchers are able to generate a functional model of a disease directly from patient-specific cells. These disease-specific iPSCs are pluripotent, can be used without ethical restraints, and are suitable for use in combination with gene therapy and can be expanded or differentiated without exhaustion. In addition, in case these cells would be used in the future, the risk of immunorejection of these cells would be expected to be negligible since autologous patient-derived cells can be used (Yousefi et al. 2020), and the only foreseen immune response could be a response against the transgenic protein that the corrected patient-derived iPSCs may express. Since the first publication describing reprogramming of somatic cells, many iPSC models have been generated from a wide range of different LSDs, including Fabry (Kawagoe et al. 2013); Gaucher (Tiscornia et al. 2013); Pompe (Higuchi et al. 2014; Raval et al. 2015; Sato et al. 2015); Niemann-Pick type A (Long et al. 2016) and type C (Maetzel et al. 2014; Trilck et al. 2017); GM1 gangliosidosis (Son et al. 2015) and GM2 gangliosidosis (Trilck et al. 2017); MPS-I (Tolar et al. 2011), MPS-II (Kobolak et al. 2019), MPS-III (Lemonnier et al. 2011; Canals et al. 2015a), and MPS-VII (Griffin et al. 2015); MLD (Doerr et al. 2015); and even neuronal ceroid lipofuscinoses (Lojewski et al. 2014; Chandrachud et al. 2015), and have been shown to largely recapitulate the disease phenotype and can be differentiated into hematopoietic (stem) cells, neuronal (stem/progenitor) cells, endothelial cells, cardiomyocytes, hepatocytes, and macrophages (Kido et al. 2020; Xu et al. 2016). Using Gaucher disease patient-derived skin fibroblasts, iPSCs were generated and differentiated into macrophages and dopaminergic neurons. These cells were then used to assess the ability of the newly developed non-inhibitory enzyme glucocerebrosidase chaperone, to correct the cellular phenotype (Tiscornia et al. 2013; Mazzulli et al. 2011; Aflaki et al. 2016). iPSCs obtained from skin fibroblasts of Fabry patients (Itier et al. 2014), and both infantile-type and late-onset-type Pompe patients could be successfully differentiated into cardiomyocytes. Using lentiviral transduction to overexpress α-glucosidase in the Pompe iPSCs, researchers were further able to decrease glycogen storage in cells and increase enzyme activity (Sato et al. 2015, 2016). Furthermore, they used the Pompe iPSC model to assess whether oxidative stress or a damaged anti-oxidative stress response mechanism might contribute to the molecular pathology of late-onset Pompe disease by using metabolomics (Sato et al. 2016). Considering the devastating effect of many LSDs on neurocognitive functions, many groups have tried to differentiate LSD-derived iPSCs into self-renewing neuronal progenitor and/or stem cells (Son et al. 2015; Lemonnier et al. 2011; Lojewski et al. 2014; Meneghini et al. 2017), astroglial progenitors, and neuroepithelial stem cells (Doerr et al. 2015). Furthermore, when ARSA overexpressing NPCs/NSCs derived from MLD-iPSCs were transplanted into ARSA-deficient mice, mouse brains displayed decreased sulfatide levels (Doerr et al. 2015). Hepatocyte-like cells, neural progenitors (Soga et al. 2015), and neurons (Lee et al. 2014) derived from the iPSC lines generated from Niemann-Pick type C (NPC) patient-derived fibroblasts or multipotent adult stem cells (MASCs) (Bergamin et al. 2013) recapitulated the disease and displayed damaged intracellular transport of glycolipids and cholesterol. This model was then used to test the efficacy of 2-hydroxypropyl-γ-cyclodextrin (Soga et al. 2015) and vascular endothelial growth factor (VEGF) (Lee et al. 2014) and showed diminished lipid accumulation in the NPC patient-derived iPSCs. iPSCs derived from Gaucher patients and MPS-I (Hurler) patients were both successfully differentiated into hematopoietic

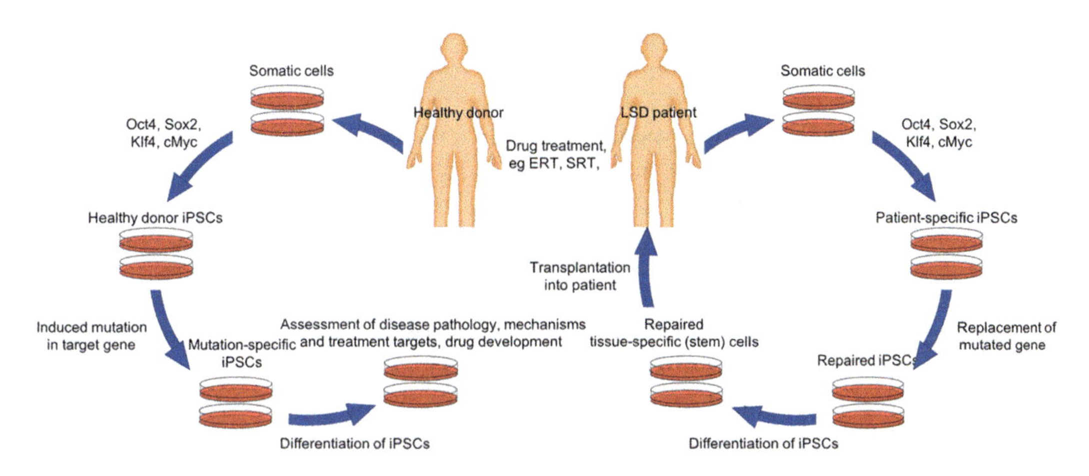

Fig. 1 Use of iPSC modeling of LSDs to develop new treatment options. Whereas patient-specific iPSCs can be used to model the disease and/or test new treatment options. Cells may be used in the future after repair, differentiation/expansion, and quality control for the treatment of the patient him/herself. In contrast, healthy donor iPSCs can be used to specifically knockout genes to assess mutation-specific effects and used to assess disease pathophysiology in vitro

progenitor and stem cells. In addition, gene-corrected MPS-I iPSCs using viral transfer of α-L-iduronidase were successfully differentiated into both non-hematopoietic and hematopoietic cells, showing the potential of patient-specific iPSCs to obtain genetically corrected autologous HSCs (Tolar et al. 2011).

Nevertheless, notwithstanding the many advantages of iPSC modeling, prolonged culture expansion of iPSCs may occasionally cause (epi)-genetic changes, which may lead to DNA mutations and/or modified gene expression, and it is possible that these changes may affect the in vitro disease phenotype and cell function of the disease-derived iPSCs (Lund et al. 2012). Even more, the development of iPSC-based LSD models is relatively costly and labor-intensive, and the efficacy of development may be directly related to the metabolic defects that cause the disease (Borger et al. 2017). Thus far, iPSC modeling of LSDs has been largely used to assess disease pathophysiology (neuropathology in specific) (Kobolak et al. 2019) and testing/development of drugs. CRISPR/Cas9 technology is increasingly being used to develop LSD models from healthy donor iPSCs by inducing mutations into genes that are thought to be related to the development of LSDs (Maguire et al. 2019). In addition, iPSCs have been made from LSD mouse models, as well as from human patients. Although iPSC technology has certainly advanced understanding of LSDs, its potential for the development of new therapeutic treatment options for LSD patients remains to be yet fully explored. Nevertheless, using patient- or disease-specific iPSCs that can be modified, differentiated and expanded, in combination with healthy donor-derived iPSCs, that can be used to specifically knockout genes, (thought to be) related to the development of LSDs, will tremendously accelerate the development of new treatment options and allow a better understanding of disease pathophysiology (Fig. 1).

9 Challenges and Final Remarks: Two May Be Better than One

In this chapter, some of the newest developments in the field of stem cell treatments for LSDs have been discussed in detail. However, many more types of treatment that have the potential to treat or cure LSDs have not been discussed here. Among these are the direct in vivo gene therapeutics (e.g., using adenoviral or adeno-associated viral constructs) and, for example, personalized chemical or PCT, which may be an option for some patients that carry chaperone-sensitive or chaperone-specific mutations (Haneef and Doss 2016). Also of interest is the development of a completely drug-free approach, where researchers attempt to selectively suppress genes responsible for the production of the depositing substrates, using RNA interference (RNAi), also known as genetic SRT (gSRT) (Diaz-Font et al. 2006; Dziedzic et al. 2010; Canals et al. 2015b). However, for now, it appears that instead of focusing on a single treatment, using patient-specific combination treatments that may consist of a HSCT (either allogeneic or after genetic modification of autologous cells, if available) with or without supportive treatment (e.g., anti-inflammatory agents or chaperones that may decrease secondary symptoms related to increased oxidative or endoplasmic reticulum stress) and the help of ERT/SRT or ERT with SRT to reduce disease load and in case of neurocognitive impairment additional treatment with (genetically modified) MSCs (Fig. 2) may have several advantages. Using ERT before and during HSCT/HSC-GT allows the patient to maintain a better physical and/or cognitive condition with fewer lysosomal deposits, which may in turn allow better engraftment and accelerated recovery. Interestingly, some therapeutic approaches have been shown to provide virtually no clinical benefit when used alone, but may still contribute to an increased overall efficacy when used in combination with other treatments. For example, in the mouse model of Sandhoff disease, the effects of treatment with SRT were limited, whereas a combination of SRT together with infusion of NSCs or BMT was highly effective (Lee et al. 2007; Jeyakumar et al. 2001). Similarly, suppression of inflammation in the same mouse model using NSAIDs in combination with SRT improved survival of the mice (Jeyakumar et al. 2004). Another example of the advantage of combination treatments was shown using the globoid cell leukodystrophy Twitcher mouse model, where a combination of BMT and SRT resulted in a synergistic effect and a doubled life span of these mice (Biswas and LeVine 2002).

It appears that combination of treatments (ERT with or without SRT, HSCT, MSCs as supportive therapy, NSCs, direct gene transfer, cross-correction gene therapy strategies, etc.) may be required to alleviate the main symptoms of (especially neurodegenerative) LSDs in the long term. Meanwhile, other supportive treatments may

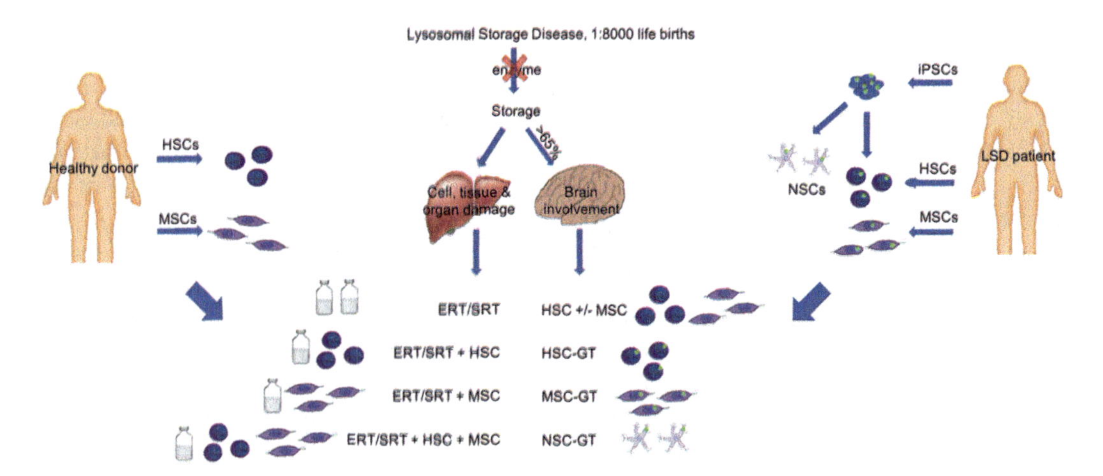

Fig. 2 Future treatment strategies and combination therapies for LSDs. Depending on the type of LSD, ERT/SRT may be used in combination with donor/third-party HSC or MSCs in order to treat the patient. Especially in patients with neurodegenerative LSDs, strategies involving infusion of autologous, gene-corrected (iPSC-derived) HSCs and/or local therapy with gene-corrected autologous or healthy donor MSCs and/or NSCs may offer symptom relief or even cure

address other complexing factors of the disease, thus ensuring better quality of life and prolonged survival. Further development and improvement of stem cell therapies and gene therapy vectors continues to offer different treatment strategies, and the field is constantly evolving. Eventually, results from ongoing and pending clinical trials involving stem cell and gene therapy approaches will greatly improve our understanding of treatments and improve treatment outcomes in patients with LSD.

Acknowledgments This study was supported by a grant from the Turkish Ministry of Development, PediSTEM nr. 2006-K120640, and the Scientific and Technological Research Council of Turkey (TUBİTAK), project nr. 219S675.

Conflict of Interest The authors declare that they have no conflict of interest.

Ethical Approval The authors declare that this article does not contain any studies with human participants or animals.

References

Aflaki E, Borger DK, Moaven N, Stubblefield BK, Rogers SA, Patnaik S et al (2016) A new glucocerebrosidase chaperone reduces alpha-synuclein and glycolipid levels in iPSC-derived dopaminergic neurons from patients with gaucher disease and parkinsonism. J Neurosci 36(28):7441–7452

Ahmad I, Hunter RE, Flax JD, Snyder EY, Erickson RP (2007) Neural stem cell implantation extends life in Niemann-Pick C1 mice. J Appl Genet 48(3):269–272

Aldenhoven M, Kurtzberg J (2015) Cord blood is the optimal graft source for the treatment of pediatric patients with lysosomal storage diseases: clinical outcomes and future directions. Cytotherapy 17 (6):765–774

Aljurf M, Rizzo JD, Mohty M, Hussain F, Madrigal A, Pasquini MC et al (2014) Challenges and opportunities for HSCT outcome registries: perspective from international HSCT registries experts. Bone Marrow Transplant 49(8):1016–1021

Audesse AJ, Webb AE (2018) Enhancing Lysosomal Activation Restores Neural Stem Cell Function During Aging. J Exp Neurosci 12:1179069518795874

Bae JS, Furuya S, Ahn SJ, Yi SJ, Hirabayashi Y, Jin HK (2005a) Neuroglial activation in Niemann-Pick Type C mice is suppressed by intracerebral transplantation of bone marrow-derived mesenchymal stem cells. Neurosci Lett 381(3):234–236

Bae JS, Furuya S, Shinoda Y, Endo S, Schuchman EH, Hirabayashi Y et al (2005b) Neurodegeneration augments the ability of bone marrow-derived mesenchymal stem cells to fuse with Purkinje neurons in Niemann-Pick type C mice. Hum Gene Ther 16 (8):1006–1011

Bae JS, Han HS, Youn DH, Carter JE, Modo M, Schuchman EH et al (2007) Bone marrow-derived mesenchymal stem cells promote neuronal networks with functional synaptic transmission after transplantation into mice with neurodegeneration. Stem Cells 25 (5):1307–1316

Ballabio A, Gieselmann V (2009) Lysosomal disorders: from storage to cellular damage. Biochim Biophys Acta 1793(4):684–696

Bergamin N, Dardis A, Beltrami A, Cesselli D, Rigo S, Zampieri S et al (2013) A human neuronal model of Niemann Pick C disease developed from stem cells isolated from patient's skin. Orphanet J Rare Dis 8:34

Biffi A (2012) Genetically-modified hematopoietic stem cells and their progeny for widespread and efficient protein delivery to diseased sites: the case of lysosomal storage disorders. Curr Gene Ther 12(5):381–388

Biffi A, Capotondo A, Fasano S, del Carro U, Marchesini S, Azuma H et al (2006) Gene therapy of metachromatic leukodystrophy reverses neurological damage and deficits in mice. J Clin Invest 116 (11):3070–3082

Biffi A, Montini E, Lorioli L, Cesani M, Fumagalli F, Plati T et al (2013) Lentiviral hematopoietic stem cell gene therapy benefits metachromatic leukodystrophy. Science 341(6148):1233158

Biswas S, LeVine SM (2002) Substrate-reduction therapy enhances the benefits of bone marrow transplantation in young mice with globoid cell leukodystrophy. Pediatr Res 51(1):40–47

Boelens JJ, Prasad VK, Tolar J, Wynn RF, Peters C (2010) Current international perspectives on hematopoietic stem cell transplantation for inherited metabolic disorders. Pediatr Clin North Am 57(1):123–145

Bohringer J, Santer R, Schumacher N, Gieseke F, Cornils K, Pechan M et al (2017) Enzymatic characterization of novel arylsulfatase A variants using human arylsulfatase A-deficient immortalized mesenchymal stromal cells. Hum Mutat 38(11):1511–1520

Borger DK, McMahon B, Roshan Lal T, Serra-Vinardell J, Aflaki E, Sidransky E (2017) Induced pluripotent stem cell models of lysosomal storage disorders. Dis Model Mech 10(6):691–704

Bou-Gharios G, Abraham D, Olsen I (1993) Lysosomal storage diseases: mechanisms of enzyme replacement therapy. Histochem J 25(9):593–605

Campeau PM, Rafei M, Boivin MN, Sun Y, Grabowski GA, Galipeau J (2009) Characterization of Gaucher disease bone marrow mesenchymal stromal cells reveals an altered inflammatory secretome. Blood 114 (15):3181–3190

Canals I, Soriano J, Orlandi JG, Torrent R, Richaud-Patin-Y, Jimenez-Delgado S et al (2015a) Activity and high-order effective connectivity alterations in sanfilippo C patient-specific neuronal networks. Stem Cell Rep 5 (4):546–557

Canals I, Beneto N, Cozar M, Vilageliu L, Grinberg D (2015b) EXTL2 and EXTL3 inhibition with siRNAs as a promising substrate reduction therapy for Sanfilippo C syndrome. Sci Rep 5:13654

Capotondo A, Milazzo R, Politi LS, Quattrini A, Palini A, Plati T et al (2012) Brain conditioning is instrumental for successful microglia reconstitution following hematopoietic stem cell transplantation. Proc Natl Acad Sci U S A 109(37):15018–15023

Cartier N, Hacein-Bey-Abina S, Bartholomae CC, Veres G, Schmidt M, Kutschera I et al (2009) Hematopoietic stem cell gene therapy with a lentiviral vector in X-linked adrenoleukodystrophy. Science 326 (5954):818–823

Cartier N, Hacein-Bey-Abina S, Bartholomae CC, Bougneres P, Schmidt M, Kalle CV et al (2012) Lentiviral hematopoietic cell gene therapy for X-linked adrenoleukodystrophy. Methods Enzymol 507:187–198

Chandrachud U, Walker MW, Simas AM, Heetveld S, Petcherski A, Klein M et al (2015) Unbiased cell-based screening in a neuronal cell model of batten disease highlights an interaction between Ca2+ homeostasis, autophagy, and CLN3 protein function. J Biol Chem 290(23):14361–14380

Chou JY, Jun HS, Mansfield BC (2010) Neutropenia in type Ib glycogen storage disease. Curr Opin Hematol 17(1):36–42

Cox TM, Cachon-Gonzalez MB (2012) The cellular pathology of lysosomal diseases. J Pathol 226 (2):241–254

Crigler L, Robey RC, Asawachaicharn A, Gaupp D, Phinney DG (2006) Human mesenchymal stem cell subpopulations express a variety of neuro-regulatory molecules and promote neuronal cell survival and neuritogenesis. Exp Neurol 198(1):54–64

de Ru MH, Boelens JJ, Das AM, Jones SA, van der Lee JH, Mahlaoui N et al (2011) Enzyme replacement therapy and/or hematopoietic stem cell transplantation at diagnosis in patients with mucopolysaccharidosis type I: results of a European consensus procedure. Orphanet J Rare Dis 6:55

Diaz-Font A, Chabas A, Grinberg D, Vilageliu L (2006) RNAi-mediated inhibition of the glucosylceramide synthase (GCS) gene: a preliminary study towards a therapeutic strategy for Gaucher disease and other glycosphingolipid storage diseases. Blood Cells Mol Dis 37(3):197–203

Doerr J, Bockenhoff A, Ewald B, Ladewig J, Eckhardt M, Gieselmann V et al (2015) Arylsulfatase A overexpressing human iPSC-derived neural cells reduce cns sulfatide storage in a mouse model of metachromatic leukodystrophy. Mol Ther 23(9):1519–1531

Dziedzic D, Wegrzyn G, Jakobkiewicz-Banecka J (2010) Impairment of glycosaminoglycan synthesis in mucopolysaccharidosis type IIIA cells by using siRNA: a potential therapeutic approach for Sanfilippo disease. Eur J Hum Genet 18(2):200–205

Erlich S, Miranda SR, Visser JW, Dagan A, Gatt S, Schuchman EH (1999) Fluorescence-based selection of gene-corrected hematopoietic stem and progenitor cells from acid sphingomyelinase-deficient mice: implications for Niemann-Pick disease gene therapy and the development of improved stem cell gene transfer procedures. Blood 93(1):80–86

Freed CR, Breeze RE, Rosenberg NL, Schneck SA, Kriek E, Qi JX et al (1992) Survival of implanted fetal dopamine cells and neurologic improvement 12 to 46 months after transplantation for Parkinson's disease. N Engl J Med 327(22):1549–1555

Fukuhara Y, Li XK, Kitazawa Y, Inagaki M, Matsuoka K, Kosuga M et al (2006) Histopathological and behavioral improvement of murine mucopolysaccharidosis type VII by intracerebral transplantation of neural stem cells. Mol Ther 13(3):548–555

Gatto F, Redaelli D, Salvade A, Marzorati S, Sacchetti B, Ferina C et al (2012) Hurler disease bone marrow stromal cells exhibit altered ability to support osteoclast formation. Stem Cells Dev 21(9):1466–1477

Ghosh A, Miller W, Orchard PJ, Jones SA, Mercer J, Church HJ et al (2016) Enzyme replacement therapy prior to haematopoietic stem cell transplantation in Mucopolysaccharidosis Type I: 10 year combined experience of 2 centres. Mol Genet Metab 117 (3):373–377

Givogri MI, Bottai D, Zhu HL, Fasano S, Lamorte G, Brambilla R et al (2008) Multipotential neural precursors transplanted into the metachromatic leukodystrophy brain fail to generate oligodendrocytes but contribute to limit brain dysfunction. Dev Neurosci 30 (5):340–357

Gomez-Ospina N, Scharenberg SG, Mostrel N, Bak RO, Mantri S, Quadros RM et al (2019) Human genome-edited hematopoietic stem cells phenotypically correct Mucopolysaccharidosis type I. Nat Commun 10 (1):4045

Griffin TA, Anderson HC, Wolfe JH (2015) Ex vivo gene therapy using patient iPSC-derived NSCs reverses pathology in the brain of a homologous mouse model. Stem Cell Rep 4(5):835–846

Haneef SA, Doss CG (2016) Personalized pharmacoperones for lysosomal storage disorder: approach for next-generation treatment. Adv Protein Chem Struct Biol 102:225–265

Hawkins-Salsbury JA, Reddy AS, Sands MS (2011) Combination therapies for lysosomal storage disease: is the whole greater than the sum of its parts? Hum Mol Genet 20(R1):R54–R60

Higuchi T, Kawagoe S, Otsu M, Shimada Y, Kobayashi H, Hirayama R et al (2014) The generation of induced pluripotent stem cells (iPSCs) from patients with infantile and late-onset types of Pompe disease and the effects of treatment with acid-alpha-glucosidase in Pompe's iPSCs. Mol Genet Metab 112(1):44–48

Hobbs JR (1992) Bone marrow transplants in genetic diseases. Eur J Pediatr 151(Suppl 1):S44–S49

Hofling AA, Devine S, Vogler C, Sands MS (2004) Human CD34+ hematopoietic progenitor cell-directed lentiviral-mediated gene therapy in a xenotransplantation model of lysosomal storage disease. Mol Ther 9 (6):856–865

Hoogerbrugge PM, Brouwer OF, Bordigoni P, Ringden O, Kapaun P, Ortega JJ et al (1995) Allogeneic bone marrow transplantation for lysosomal storage diseases. The European Group for Bone Marrow Transplantation. Lancet 345(8962):1398–1402

Hu P, Li Y, Nikolaishvili-Feinberg N, Scesa G, Bi Y, Pan D et al (2016) Hematopoietic Stem cell transplantation and lentiviral vector-based gene therapy for Krabbe's disease: Present convictions and future prospects. J Neurosci Res 94(11):1152–1168

Itier JM, Ret G, Viale S, Sweet L, Bangari D, Caron A et al (2014) Effective clearance of GL-3 in a human iPSC-derived cardiomyocyte model of Fabry disease. J Inherit Metab Dis 37(6):1013–1022

Jackson M, Derrick Roberts A, Martin E, Rout-Pitt N, Gronthos S, Byers S (2015) Mucopolysaccharidosis enzyme production by bone marrow and dental pulp derived human mesenchymal stem cells. Mol Genet Metab 114(4):584–593

Jaing TH (2007) Umbilical cord blood transplantation: application in pediatric patients. Acta Paediatr Taiwan 48(3):107–111

Jeyakumar M, Norflus F, Tifft CJ, Cortina-Borja M, Butters TD, Proia RL et al (2001) Enhanced survival in Sandhoff disease mice receiving a combination of substrate deprivation therapy and bone marrow transplantation. Blood 97(1):327–329

Jeyakumar M, Smith DA, Williams IM, Borja MC, Neville DC, Butters TD et al (2004) NSAIDs increase survival in the Sandhoff disease mouse: synergy with N-butyldeoxynojirimycin. Ann Neurol 56(5):642–649

Jin HK, Schuchman EH (2003) Ex vivo gene therapy using bone marrow-derived cells: combined effects of intracerebral and intravenous transplantation in a mouse model of Niemann-Pick disease. Mol Ther 8 (6):876–885

Jin HK, Carter JE, Huntley GW, Schuchman EH (2002) Intracerebral transplantation of mesenchymal stem cells into acid sphingomyelinase-deficient mice delays the onset of neurological abnormalities and extends their life span. J Clin Invest 109(9):1183–1191

Joyce N, Annett G, Wirthlin L, Olson S, Bauer G, Nolta JA (2010) Mesenchymal stem cells for the treatment of neurodegenerative disease. Regen Med 5(6):933–946

Kawagoe S, Higuchi T, Otaka M, Shimada Y, Kobayashi H, Ida H et al (2013) Morphological features of iPS cells generated from Fabry disease skin fibroblasts using Sendai virus vector (SeVdp). Mol Genet Metab 109(4):386–389

Kennedy DW, Abkowitz JL (1997) Kinetics of central nervous system microglial and macrophage engraftment: analysis using a transgenic bone marrow transplantation model. Blood 90(3):986–993

Kido J, Nakamura K, Era T (2020) Role of induced pluripotent stem cells in lysosomal storage diseases. Mol Cell Neurosci 108:103540

Kim SY, Jun HS, Mead PA, Mansfield BC, Chou JY (2008) Neutrophil stress and apoptosis underlie myeloid dysfunction in glycogen storage disease type Ib. Blood 111(12):5704–5711

Kobolak J, Molnar K, Varga E, Bock I, Jezso B, Teglasi A et al (2019) Modelling the neuropathology of lysosomal storage disorders through disease-specific human induced pluripotent stem cells. Exp Cell Res 380(2):216–233

Kose S, Aerts Kaya F, Kuskonmaz B, Uckan CD (2019) Characterization of mesenchymal stem cells in mucolipidosis type II (I-cell disease). Turk J Biol 43 (3):171–178

Krall WJ, Challita PM, Perlmutter LS, Skelton DC, Kohn DB (1994) Cells expressing human glucocerebrosidase from a retroviral vector repopulate macrophages and central nervous system microglia after murine bone marrow transplantation. Blood 83(9):2737–2748

Krivit W, Peters C, Shapiro EG (1999) Bone marrow transplantation as effective treatment of central nervous system disease in globoid cell leukodystrophy, metachromatic leukodystrophy, adrenoleukodystrophy, mannosidosis, fucosidosis, aspartylglucosaminuria, Hurler, Maroteaux-Lamy, and Sly syndromes, and Gaucher disease type III. Curr Opin Neurol 12(2):167–176

Langford-Smith A, Wilkinson FL, Langford-Smith KJ, Holley RJ, Sergijenko A, Howe SJ et al (2012) Hematopoietic stem cell and gene therapy corrects primary neuropathology and behavior in mucopolysaccharidosis IIIA mice. Mol Ther 20(8):1610–1621

Leal AF, Espejo-Mojica AJ, Sanchez OF, Ramirez CM, Reyes LH, Cruz JC et al (2020) Lysosomal storage diseases: current therapies and future alternatives. J Mol Med (Berl) 98(7):931–946

Lecourt S, Mouly E, Freida D, Cras A, Ceccaldi R, Heraoui D et al (2013) A prospective study of bone marrow hematopoietic and mesenchymal stem cells in type 1 Gaucher disease patients. PLoS One 8(7):e69293

Lee JP, Jeyakumar M, Gonzalez R, Takahashi H, Lee PJ, Baek RC et al (2007) Stem cells act through multiple mechanisms to benefit mice with neurodegenerative metabolic disease. Nat Med 13(4):439–447

Lee H, Lee JK, Min WK, Bae JH, He X, Schuchman EH et al (2010) Bone marrow-derived mesenchymal stem cells prevent the loss of Niemann-Pick type C mouse Purkinje neurons by correcting sphingolipid metabolism and increasing sphingosine-1-phosphate. Stem Cells 28(4):821–831

Lee H, Kang JE, Lee JK, Bae JS, Jin HK (2013) Bone-marrow-derived mesenchymal stem cells promote proliferation and neuronal differentiation of Niemann-Pick type C mouse neural stem cells by upregulation and secretion of CCL2. Hum Gene Ther 24 (7):655–669

Lee H, Lee JK, Park MH, Hong YR, Marti HH, Kim H et al (2014) Pathological roles of the VEGF/SphK pathway in Niemann-Pick type C neurons. Nat Commun 5:5514
Lemonnier T, Blanchard S, Toli D, Roy E, Bigou S, Froissart R et al (2011) Modeling neuronal defects associated with a lysosomal disorder using patient-derived induced pluripotent stem cells. Hum Mol Genet 20(18):3653–3666
Li M (2018) Enzyme replacement therapy: a review and its role in treating lysosomal storage diseases. Pediatr Ann 47(5):e191–e1e7
Lindvall O, Brundin P, Widner H, Rehncrona S, Gustavii B, Frackowiak R et al (1990) Grafts of fetal dopamine neurons survive and improve motor function in Parkinson's disease. Science 247(4942):574–577
Lojewski X, Staropoli JF, Biswas-Legrand S, Simas AM, Haliw L, Selig MK et al (2014) Human iPSC models of neuronal ceroid lipofuscinosis capture distinct effects of TPP1 and CLN3 mutations on the endocytic pathway. Hum Mol Genet 23(8):2005–2022
Long Y, Xu M, Li R, Dai S, Beers J, Chen G et al (2016) Induced pluripotent stem cells for disease modeling and evaluation of therapeutics for Niemann-Pick disease type A. Stem Cells Transl Med 5(12):1644–1655
Lund RJ, Narva E, Lahesmaa R (2012) Genetic and epigenetic stability of human pluripotent stem cells. Nat Rev Genet 13(10):732–744
Lutzko C, Omori F, Abrams-Ogg AC, Shull R, Li L, Lau K et al (1999a) Gene therapy for canine alpha-L-iduronidase deficiency: in utero adoptive transfer of genetically corrected hematopoietic progenitors results in engraftment but not amelioration of disease. Hum Gene Ther 10(9):1521–1532
Lutzko C, Kruth S, Abrams-Ogg AC, Lau K, Li L, Clark BR et al (1999b) Genetically corrected autologous stem cells engraft, but host immune responses limit their utility in canine alpha-L-iduronidase deficiency. Blood 93(6):1895–1905
Maetzel D, Sarkar S, Wang H, Abi-Mosleh L, Xu P, Cheng AW et al (2014) Genetic and chemical correction of cholesterol accumulation and impaired autophagy in hepatic and neural cells derived from Niemann-Pick Type C patient-specific iPS cells. Stem Cell Rep 2(6):866–880
Maguire JA, Cardenas-Diaz FL, Gadue P, French DL (2019) Highly efficient CRISPR-Cas9-mediated genome editing in human pluripotent stem cells. Curr Protoc Stem Cell Biol 48(1):e64
Martin HR, Poe MD, Provenzale JM, Kurtzberg J, Mendizabal A, Escolar ML (2013) Neurodevelopmental outcomes of umbilical cord blood transplantation in metachromatic leukodystrophy. Biol Blood Marrow Transplant 19(4):616–624
Martin PK, Stilhano RS, Samoto VY, Takiya CM, Peres GB, da Silva Michelacci YM et al (2014) Mesenchymal stem cells do not prevent antibody responses against human alpha-L-iduronidase when used to treat mucopolysaccharidosis type I. PLoS One 9(3): e92420
Mazzulli JR, Xu YH, Sun Y, Knight AL, McLean PJ, Caldwell GA et al (2011) Gaucher disease glucocerebrosidase and alpha-synuclein form a bidirectional pathogenic loop in synucleinopathies. Cell 146 (1):37–52
Meneghini V, Frati G, Sala D, De Cicco S, Luciani M, Cavazzin C et al (2017) Generation of human induced pluripotent stem cell-derived bona fide neural stem cells for ex vivo gene therapy of metachromatic leukodystrophy. Stem Cells Transl Med 6(2):352–368
Meng XL, Shen JS, Ohashi T, Maeda H, Kim SU, Eto Y (2003) Brain transplantation of genetically engineered human neural stem cells globally corrects brain lesions in the mucopolysaccharidosis type VII mouse. J Neurosci Res 74(2):266–277
Meyerrose TE, De Ugarte DA, Hofling AA, Herrbrich PE, Cordonnier TD, Shultz LD et al (2007) In vivo distribution of human adipose-derived mesenchymal stem cells in novel xenotransplantation models. Stem Cells 25(1):220–227
Meyerrose TE, Roberts M, Ohlemiller KK, Vogler CA, Wirthlin L, Nolta JA et al (2008) Lentiviral-transduced human mesenchymal stem cells persistently express therapeutic levels of enzyme in a xenotransplantation model of human disease. Stem Cells 26(7):1713–1722
Miranda SR, Erlich S, Friedrich VL Jr, Haskins ME, Gatt S, Schuchman EH (1998) Biochemical, pathological, and clinical response to transplantation of normal bone marrow cells into acid sphingomyelinase-deficient mice. Transplantation 65(7):884–892
Miranda SR, Erlich S, Friedrich VL Jr, Gatt S, Schuchman EH (2000) Hematopoietic stem cell gene therapy leads to marked visceral organ improvements and a delayed onset of neurological abnormalities in the acid sphingomyelinase deficient mouse model of Niemann-Pick disease. Gene Ther 7(20):1768–1776
Miranda CO, Teixeira CA, Liz MA, Sousa VF, Franquinho F, Forte G et al (2011) Systemic delivery of bone marrow-derived mesenchymal stromal cells diminishes neuropathology in a mouse model of Krabbe's disease. Stem Cells 29(11):1738–1751
Mogul MJ (2000) Unrelated cord blood transplantation vs matched unrelated donor bone marrow transplantation: the risks and benefits of each choice. Bone Marrow Transplant 25(Suppl 2):S58–S60
Muller I, Kustermann-Kuhn B, Holzwarth C, Isensee G, Vaegler M, Harzer K et al (2006) In vitro analysis of multipotent mesenchymal stromal cells as potential cellular therapeutics in neurometabolic diseases in pediatric patients. Exp Hematol 34(10):1413–1419
Navarro Negredo P, Yeo RW, Brunet A (2020) Aging and rejuvenation of neural stem cells and their niches. Cell Stem Cell 27(2):202–223
Neuwelt E, Abbott NJ, Abrey L, Banks WA, Blakley B, Davis T et al (2008) Strategies to advance translational research into brain barriers. Lancet Neurol 7(1):84–96

Olsen I, Dean MF, Harris G, Muir H (1981) Direct transfer of a lysosomal enzyme from lymphoid cells to deficient fibroblasts. Nature 291(5812):244–247
Olsen I, Dean MF, Muir H, Harris G (1982) Acquisition of beta-glucuronidase activity by deficient fibroblasts during direct contact with lymphoid cells. J Cell Sci 55:211–231
Olsen I, Muir H, Smith R, Fensom A, Watt DJ (1983) Direct enzyme transfer from lymphocytes is specific. Nature 306(5938):75–77
Orchard PJ, Blazar BR, Wagner J, Charnas L, Krivit W, Tolar J (2007) Hematopoietic cell therapy for metabolic disease. J Pediatr 151(4):340–346
Otsu M, Nakayama T, Inoue N (2014) Pluripotent stem cell-derived neural stem cells: From basic research to applications. World J Stem Cells 6(5):651–657
Parenti G, Andria G, Ballabio A (2015) Lysosomal storage diseases: from pathophysiology to therapy. Annu Rev Med 66:471–486
Parkinson-Lawrence EJ, Shandala T, Prodoehl M, Plew R, Borlace GN, Brooks DA (2010) Lysosomal storage disease: revealing lysosomal function and physiology. Physiology (Bethesda) 25(2):102–115
Peters C, Steward CG (2003) National Marrow Donor P, International Bone Marrow Transplant R, Working Party on Inborn Errors EBMTG. Hematopoietic cell transplantation for inherited metabolic diseases: an overview of outcomes and practice guidelines. Bone Marrow Transplant 31(4):229–239
Phinney DG, Isakova IA (2014) Mesenchymal stem cells as cellular vectors for pediatric neurological disorders. Brain Res 1573:92–107
Pisati F, Bossolasco P, Meregalli M, Cova L, Belicchi M, Gavina M et al (2007) Induction of neurotrophin expression via human adult mesenchymal stem cells: implication for cell therapy in neurodegenerative diseases. Cell Transplant 16(1):41–55
Pittenger MF, Discher DE, Peault BM, Phinney DG, Hare JM, Caplan AI (2019) Mesenchymal stem cell perspective: cell biology to clinical progress. NPJ Regen Med 4:22
Poe MD, Chagnon SL, Escolar ML (2014) Early treatment is associated with improved cognition in Hurler syndrome. Ann Neurol 76(5):747–753
Prasad VK, Kurtzberg J (2010a) Transplant outcomes in mucopolysaccharidoses. Semin Hematol 47(1):59–69
Prasad VK, Kurtzberg J (2010b) Cord blood and bone marrow transplantation in inherited metabolic diseases: scientific basis, current status and future directions. Br J Haematol 148(3):356–372
Qu R, Li Y, Gao Q, Shen L, Zhang J, Liu Z et al (2007) Neurotrophic and growth factor gene expression profiling of mouse bone marrow stromal cells induced by ischemic brain extracts. Neuropathology 27(4):355–363
Rappeport JM, Ginns EI (1984) Bone-marrow transplantation in severe Gaucher's disease. N Engl J Med 311 (2):84–88
Raval KK, Tao R, White BE, De Lange WJ, Koonce CH, Yu J et al (2015) Pompe disease results in a Golgi-based glycosylation deficit in human induced pluripotent stem cell-derived cardiomyocytes. J Biol Chem 290(5):3121–3136
Rigante D, Cipolla C, Basile U, Gulli F, Savastano MC (2017) Overview of immune abnormalities in lysosomal storage disorders. Immunol Lett 188:79–85
Rowe RG, Daley GQ (2019) Induced pluripotent stem cells in disease modelling and drug discovery. Nat Rev Genet 20(7):377–388
Sakurai K, Iizuka S, Shen JS, Meng XL, Mori T, Umezawa A et al (2004) Brain transplantation of genetically modified bone marrow stromal cells corrects CNS pathology and cognitive function in MPS VII mice. Gene Ther 11(19):1475–1481
Sato Y, Kobayashi H, Higuchi T, Shimada Y, Era T, Kimura S et al (2015) Disease modeling and lentiviral gene transfer in patient-specific induced pluripotent stem cells from late-onset Pompe disease patient. Mol Ther Methods Clin Dev 2:15023
Sato Y, Kobayashi H, Higuchi T, Shimada Y, Ida H, Ohashi T (2016) TFEB overexpression promotes glycogen clearance of Pompe disease iPSC-derived skeletal muscle. Mol Ther Methods Clin Dev 3:16054
Sawada T, Tanaka A, Higaki K, Takamura A, Nanba E, Seto T et al (2009) Intracerebral cell transplantation therapy for murine GM1 gangliosidosis. Brain Dev 31 (10):717–724
Seo Y, Yang SR, Jee MK, Joo EK, Roh KH, Seo MS et al (2011) Human umbilical cord blood-derived mesenchymal stem cells protect against neuronal cell death and ameliorate motor deficits in Niemann Pick type C1 mice. Cell Transplant 20(7):1033–1047
Sessa M, Lorioli L, Fumagalli F, Acquati S, Redaelli D, Baldoli C et al (2016) Lentiviral haemopoietic stem-cell gene therapy in early-onset metachromatic leukodystrophy: an ad-hoc analysis of a non-randomised, open-label, phase 1/2 trial. Lancet 388(10043):476–487
Shihabuddin LS, Numan S, Huff MR, Dodge JC, Clarke J, Macauley SL et al (2004) Intracerebral transplantation of adult mouse neural progenitor cells into the Niemann-Pick-A mouse leads to a marked decrease in lysosomal storage pathology. J Neurosci 24 (47):10642–10651
Sidman RL, Li J, Stewart GR, Clarke J, Yang W, Snyder EY et al (2007) Injection of mouse and human neural stem cells into neonatal Niemann-Pick A model mice. Brain Res 1140:195–204
Sim SW, Park TS, Kim SJ, Park BC, Weinstein DA, Lee YM et al (2018) Aberrant proliferation and differentiation of glycogen storage disease type Ib mesenchymal stem cells. FEBS Lett 592(2):162–171
Snyder EY, Taylor RM, Wolfe JH (1995) Neural progenitor cell engraftment corrects lysosomal storage throughout the MPS VII mouse brain. Nature 374 (6520):367–370
Snyder EY, Daley GQ, Goodell M (2004) Taking stock and planning for the next decade: realistic prospects for stem cell therapies for the nervous system. J Neurosci Res 76(2):157–168
Soga M, Ishitsuka Y, Hamasaki M, Yoneda K, Furuya H, Matsuo M et al (2015) HPGCD outperforms HPBCD

as a potential treatment for Niemann-Pick disease type C during disease modeling with iPS cells. Stem Cells 33(4):1075–1088

Son MY, Kwak JE, Seol B, Lee DY, Jeon H, Cho YS (2015) A novel human model of the neurodegenerative disease GM1 gangliosidosis using induced pluripotent stem cells demonstrates inflammasome activation. J Pathol 237(1):98–110

Squeri G, Passerini L, Ferro F, Laudisa C, Tomasoni D, Deodato F et al (2019) Targeting a Pre-existing Anti-transgene T Cell Response for Effective Gene Therapy of MPS-I in the Mouse Model of the Disease. Mol Ther 27(7):1215–1227

Squillaro T, Antonucci I, Alessio N, Esposito A, Cipollaro M, Melone MAB et al (2017) Impact of lysosomal storage disorders on biology of mesenchymal stem cells: Evidences from in vitro silencing of glucocerebrosidase (GBA) and alpha-galactosidase A (GLA) enzymes. J Cell Physiol 232(12):3454–3467

Starer F, Sargent JD, Hobbs JR (1987) Regression of the radiological changes of Gaucher's disease following bone marrow transplantation. Br J Radiol 60 (720):1189–1195

Stok M, de Boer H, Huston MW, Jacobs EH, Roovers O, Visser TP et al (2020) Lentiviral hematopoietic stem cell gene therapy corrects murine Pompe disease. Mol Ther Methods Clin Dev 17:1014–1025

Sun A (2018) Lysosomal storage disease overview. Ann Transl Med 6(24):476

Takahashi K, Yamanaka S (2006) Induction of pluripotent stem cells from mouse embryonic and adult fibroblast cultures by defined factors. Cell 126(4):663–676

Takahashi K, Tanabe K, Ohnuki M, Narita M, Ichisaka T, Tomoda K et al (2007) Induction of pluripotent stem cells from adult human fibroblasts by defined factors. Cell 131(5):861–872

Takenaka T, Qin G, Brady RO, Medin JA (1999) Circulating alpha-galactosidase A derived from transduced bone marrow cells: relevance for corrective gene transfer for Fabry disease. Hum Gene Ther 10 (12):1931–1939

Takenaka T, Murray GJ, Qin G, Quirk JM, Ohshima T, Qasba P et al (2000) Long-term enzyme correction and lipid reduction in multiple organs of primary and secondary transplanted Fabry mice receiving transduced bone marrow cells. Proc Natl Acad Sci U S A 97 (13):7515–7520

Talib S, Shepard KA (2020) Unleashing the cure: Overcoming persistent obstacles in the translation and expanded use of hematopoietic stem cell-based therapies. Stem Cells Transl Med 9(4):420–426

Tan EY, Boelens JJ, Jones SA, Wynn RF (2019) Hematopoietic Stem Cell Transplantation in Inborn Errors of Metabolism. Front Pediatr 7:433

Tancini B, Buratta S, Sagini K, Costanzi E, Delo F, Urbanelli L et al (2019) Insight into the role of extracellular vesicles in lysosomal storage disorders. Genes (Basel) 10(7)

Tiscornia G, Vivas EL, Matalonga L, Berniakovich I, Barragan Monasterio M, Eguizabal C et al (2013) Neuronopathic Gaucher's disease: induced pluripotent stem cells for disease modelling and testing chaperone activity of small compounds. Hum Mol Genet 22 (4):633–645

Toh WS, Lai RC, Zhang B, Lim SK (2018) MSC exosome works through a protein-based mechanism of action. Biochem Soc Trans 46(4):843–853

Tolar J, Park IH, Xia L, Lees CJ, Peacock B, Webber B et al (2011) Hematopoietic differentiation of induced pluripotent stem cells from patients with mucopolysaccharidosis type I (Hurler syndrome). Blood 117 (3):839–847

Traas AM, Wang P, Ma X, Tittiger M, Schaller L, O'Donnell P et al (2007) Correction of clinical manifestations of canine mucopolysaccharidosis I with neonatal retroviral vector gene therapy. Mol Ther 15 (8):1423–1431

Trilck M, Peter F, Zheng C, Frank M, Dobrenis K, Mascher H et al (2017) Diversity of glycosphingolipid GM2 and cholesterol accumulation in NPC1 patient-specific iPSC-derived neurons. Brain Res 1657:52–61

van Til NP, Stok M, Aerts Kaya FS, de Waard MC, Farahbakhshian E, Visser TP et al (2010) Lentiviral gene therapy of murine hematopoietic stem cells ameliorates the Pompe disease phenotype. Blood 115 (26):5329–5337

Visigalli I, Delai S, Politi LS, Di Domenico C, Cerri F, Mrak E et al (2010) Gene therapy augments the efficacy of hematopoietic cell transplantation and fully corrects mucopolysaccharidosis type I phenotype in the mouse model. Blood 116(24):5130–5139

Whitley CB, Belani KG, Chang PN, Summers CG, Blazar BR, Tsai MY et al (1993) Long-term outcome of Hurler syndrome following bone marrow transplantation. Am J Med Genet 46(2):209–218

Wicks SE, Londot H, Zhang B, Dowden J, Klopf-Eiermann J, Fisher-Perkins JM et al (2011) Effect of intrastriatal mesenchymal stromal cell injection on progression of a murine model of Krabbe disease. Behav Brain Res 225(2):415–425

Xu M, Motabar O, Ferrer M, Marugan JJ, Zheng W, Ottinger EA (2016) Disease models for the development of therapies for lysosomal storage diseases. Ann N Y Acad Sci 1371(1):15–29

Yang B, Li S, Wang H, Guo Y, Gessler DJ, Cao C et al (2014) Global CNS transduction of adult mice by intravenously delivered rAAVrh.8 and rAAVrh.10 and nonhuman primates by rAAVrh.10. Mol Ther 22 (7):1299–1309

Yousefi N, Abdollahii S, Kouhbanani MAJ, Hassanzadeh A (2020) Induced pluripotent stem cells (iPSCs) as game-changing tools in the treatment of neurodegenerative disease: Mirage or reality? J Cell Physiol

Zheng Y, Rozengurt N, Ryazantsev S, Kohn DB, Satake N, Neufeld EF (2003) Treatment of the mouse model of mucopolysaccharidosis I with retrovirally transduced bone marrow. Mol Genet Metab 79 (4):233–244

Adv Exp Med Biol - Cell Biology and Translational Medicine (2021) 14: 163–181
https://doi.org/10.1007/5584_2021_648

Published online: 22 July 2021

Mechanisms of Drug Resistance and Use of Nanoparticle Delivery to Overcome Resistance in Breast Cancers

Huseyin Beyaz, Hasan Uludag, Doga Kavaz, and Nahit Rizaner

Abstract

Breast cancer is the leading cancer type diagnosed among women in the world. Unfortunately, drug resistance to current breast cancer chemotherapeutics remains the main challenge for a higher survival rate. The recent progress in the nanoparticle platforms and distinct features of nanoparticles that enhance the efficacy of therapeutic agents, such as improved delivery efficacy, increased intracellular cytotoxicity, and reduced side effects, hold great promise to overcome the observed drug resistance. Currently, multifaceted investigations are probing the resistance mechanisms associated with clinical drugs, and identifying new breast cancer-associated molecular targets that may lead to improved therapeutic approaches with the nanoparticle platforms. Nanoparticle platforms including siRNA, antibody-specific targeting and the role of nanoparticles in cellular processes and their effect on breast cancer were discussed in this article.

Keywords

Breast cancer · Drug resistance · Therapeutic targets/delivery (new discoveries)

Abbreviations

ABC	ATP Binding Cassette
AIs	Aromatase Inhibitors
ALDH	Aldehyde Dehydrogenase
AuNPs	Gold Nanoparticles
CA	Carbonate Apatite
CKAP4	Cytoskeleton-Associated Protein 4
CSC	Cancer Stem Cell
DKK1	Dickkopf-1
DOX	Doxorubicin
EGFR	Epidermal Growth Factor Receptor
EMT	Epithelial-Mesenchymal Transition
EPR	Enhanced Permeability and Retention Effect

H. Beyaz (✉)
Bioengineering Department, Faculty of Engineering, Cyprus International University, Nicosia, Turkey
e-mail: hbeyaz@ciu.edu.tr

H. Uludag
Department of Chemical and Materials Engineering, Faculty of Engineering, University of Alberta, Edmonton, AB, Canada

Faculty of Pharmacy and Pharmaceutical Sciences, University of Alberta, Edmonton, AB, Canada

Department of Biomedical Engineering, Faculty of Medicine and Dentistry, University of Alberta, Edmonton, AB, Canada

D. Kavaz and N. Rizaner
Bioengineering Department, Faculty of Engineering, Cyprus International University, Nicosia, Turkey

Biotechnology Research Center, Cyprus International University, Nicosia, Turkey

ER	Estrogen Receptor
ERα and Erβ	Estrogen Receptors, Alpha and Beta
FZD	Frizzled Receptor
GLI	Glioma-Associated Oncogene
HER2	Human Endothelial Growth Factor Receptor 2
Hh	Hedgehog
HSA	Human Serum Albumin
ICG	Indocyanine Green
IGF1R	IGF1 Receptor
IGF1R	Insulin-Like Growth Factor 1 Receptor
LRP5/6	Low-Density Lipoproteins 5/6
MAPK	Mitogen-Activated Protein Kinase
MDR1	Multidrug Resistance 1
MPA	Medroxyprogesterone Acetate
NIR	Near-Infrared
NQC	Quinacrine Nanoparticles
OGEO	*Ocimum gratissimum* Plant Essential Oils
PDA	Polydopamine
PEG	Polyethylene Glycol
P-gp	P-Glycoprotein
PHB-CMCh	Poly(3-hydroxybutyrate)-carboxymethyl Chitosan
PI3K	Phosphoinositide 3-Kinase
PR	Progesterone
RNAi	RNA Interference
SEM	Selective Estrogen Modulators
SERDs	Selective Estrogen Receptor Downregulators
siRNA	Short Interfering RNA
SPRMs	Selective PR Modulators
SUR	Surfactin
T-DM1	Ado-trastuzumab Emtansine
TNBC	Triple-Negative Breast Cancer
TPA	Telapristone Acetate
WHO	World Health Organization
ZnO	Zinc Oxide

1 Introduction

Globally, cancer is one of the leading causes of morbidity and mortality. It is considered as one of the most complex diseases due to the involvement of multitude of genetic disorders and cellular abnormalities (Das et al. 2009; Mohanty et al. 2011). According to the statistics from the World Health Organization (WHO), ~13.1 million people are estimated to die from cancer by 2030 (Boyle and Levin 2008). During the previous decades, efforts have been made to improve early detection, prevention and treatment efficacy of cancer (Zhang et al. 2017; Hassanpour and Dehghani 2017). However, the complexity of signalling mechanisms associated with transformed cells, tumour heterogeneity and plasticity as well as metastasis remained as the major challenges to overcome (Saraswathy and Gong 2014; Endoh and Ohtsuki 2009).

Surgery is the foremost treatment method which is followed by other methods such as radiotherapy, hormonal therapy, chemotherapy or induction therapy. In the case of nodal dissection for locoregional and sentinel lymph nodes, surgery is the first line of treatment strategy. In order to improve the efficacy of surgery, single or combination treatment with chemotherapy or endocrine therapy can be utilized (Matter et al. 2000). Surgery increases the overall survival and decreases the breast cancer mortality rate by reducing complications and metastasis due to resection of primary tumour and/or tumour-spread tissues such as lung, ovaries, and liver (Matter et al. 2000; Matsunaga et al. 2016). Radiation therapy is typically applied after surgery and mastectomy. However, 7–12.6% of patients develop tumour relapse and resistance against to radiation therapy in 5 years. Therefore, together with radiation therapy, endocrine therapy is used to improve the treatment efficacy (Feys et al. 2015; Yu et al. 2015; Murphy et al. 2015). Endocrine therapy is considered as an effective and systematic therapy in estrogen receptor (ER)-positive tumours at early- and late-stage breast cancer (Bartsch et al. 2012). It can also be used to reduce toxicity related with other treatments. To reduce toxicity, endocrine therapy can be applied pre-operatively (neoadjuvant) or post-operatively (adjuvant) or during metastatic breast cancer stage (i.e. palliative treatment; Shioi et al. 2014, Palmieri et al. 2014).

Chemotherapy is one of the main approaches used to combat against breast cancer, especially the metastasized disease. Chemotherapy induces tumour cell death and decrease tumour size (Abdullah and Chow 2013). However, 90% of drug failures is associated with chemoresistance in metastatic cancers; thereby, chemotherapy is unable to completely eliminate cancer cells on its own (Longley and Johnston 2005). Chemoresistance has been attributed as the main reason of cancer relapse and metastasis development. Therefore, it is important to understand the molecular mechanism(s) leading to chemoresistance, otherwise named drug resistance, and identify alternative therapeutic targets for breast cancer therapy (Brasseur et al. 2017; Lu and Shervington 2008). Chemoresistance occurs when the transformed cells become insensitive to the applied chemical drugs (Nikolaou et al. 2018). Depending on the time of resistance development, it is categorized as pre-existing (intrinsic) or drug-induced (acquired) resistance (Lippert et al. 2008). Intrinsic resistance arises from the inability of tumour response to initial therapy, whereas acquired resistance arises from development of resistance over the therapy due to the exposure of transformed cells to the chemotherapeutic agent (Giaccone and Pinedo 1996; Goldie 1983).

The accumulation of information regarding drug resistance and recent developments in medicinal technologies have directed cancer therapy towards *targeted drug delivery* (Huang et al. 2016). New discoveries in the nanotechnology and targeted delivery system fields had already provided extensive knowledge for improved therapeutic efficacy for breast cancer based on these emerging technologies (Rosenblum et al. 2018). Nanotechnology has become a powerful tool in medical sciences, and nanoparticles have been promising in detection, prevention and treatment of a multitude of diseases due to their unique characteristics including small size and special coating capability which enables targeted delivery of anti-cancer drugs to the desired target. Nanoparticle formulations are extensively explored in both early-stage and metastatic breast cancers in targeting tumour cells together with chemotherapeutic agents to enhance treatment efficacy and reduce the side effects of anti-cancer drugs (Hussain et al. 2018). In fact, several FDA-approved nanoparticle-based drugs such as Doxil, DaunoXome, Marqibo and Abraxane are already present in the market and are available for patient care (Dawidczyk et al. 2014). This article will review (a) the molecular background of breast cancer and (b) association of drug resistance with breast cancer and (c) discuss the progress of the new discoveries in targeted drug delivery system with a perspective on nanoparticle usage for targeted drug delivery.

2 Breast Cancer and Drug Resistance

Breast cancer is the most frequently diagnosed cancer type among women in the world (Tang et al. 2016). Nearly 30% of patients who are diagnosed with early stage of the disease progress to metastatic breast cancer (O'Shaughnessy 2005). Additionally, 30–40% of patients who have been treated previously display recurrent local and/or metastatic breast cancer (Table 1) (Garcia-Saenz et al. 2015; Yousefi et al. 2018; Sharma et al. 2017).

2.1 Breast Cancer-Associated Receptors

Breast cancer is traditionally classified into four categories, luminal A, luminal B, human endothelial growth factor receptor 2 (HER2) and triple-negative breast cancer (TNBC), depending on the expression of three receptors, estrogen (ER), progesterone (PR) and HER2 (Perou et al. 2000; Sotiriou et al. 2003; Samadi et al. 2018). Luminal A and B subtypes are characterized by high expression of ER and PR but absence of HER2 receptors (Garcia-Saenz et al. 2015; Yousefi et al. 2018). HER2 protein is highly expressed in nearly 20% of breast cancer patients (Elster et al. 2015). TNBC is characterized by the absence of ER, PR and HER2 (Ismail-Khan and Bui 2010) and accounts for 10–20% of breast cancer patients (Neophytou et al. 2018). Different therapies have been developed to target each breast cancer subtype such as the specific endocrine therapy, chemotherapy and biologics therapy (McArthur et al. 2011; Liedtke and Kolberg 2016) and immunotherapy together with chemotherapy (Bernard-Marty et al. 2004). The decision regarding the treatment strategy is determined by menopausal status, tumour status, presence of metastasis, co-morbidities and duration of treatment failure (Bernard-Marty et al. 2004).

2.1.1 Estrogen Receptor

Estrogen receptors, alpha and beta (ERα and ERβ), play an important role in breast cancer development and progression (Normanno et al. 2005; Leclercq et al. 2006). Almost 70% of breast cancer patients are ER positive (Lumachi et al. 2013). Therefore, ERs are considered as an important therapeutic target in breast cancer. ERα mediates angiogenesis via transcriptional regulation of genes, proliferation and metastasis (Normanno et al. 2005, Leclercq et al. 2006). Endocrine therapy is extensively used to treat ER-dependent breast cancer patients. Several therapeutic agents, such as tamoxifen and fulvestrant, are currently in the market whose aim is to suppress ER signalling and enhance the overall survival rate (Lumachi et al. 2013). Unfortunately, systematic administration of these agents results in the development of endocrine therapy resistance (Lumachi et al. 2015). Tamoxifen is one of the most widely used drug in ER(+) breast cancer cells but Erα(+) breast cancer cells are unresponsive to tamoxifen. The development of resistance to tamoxifen leads to re-development of tumour (Huang et al. 2015). Therefore, there is an immediate need for new targeting strategies to overcome the ER-associated limitations of the therapy.

2.1.2 Progesterone Receptor

Progesterone receptor (PR) is mediated by PR-A and PR-B receptor isoforms, which are located in multiple organs in the body (Lange and Yee 2008). PR isoforms are expressed in response to ERα-dependent transcriptional events and also independently (Hewitt and Korach 2000). So far, ER has been implicated as the main steroid hormone inducing breast cancer (Goepfert et al. 2000). The complex PR function had been studied by Carnevale et al. (2007); the involvement of mitogen-activated protein kinase (MAPK) and phosphoinositide 3-kinase (PI3K) signalling mechanisms was documented based on blocking the PR function, which acts as an activator of signalling events to prevent growth as well as metastasis in PR-positive breast cancer (Carnevale et al. 2007).

2.1.3 Human Epithelial Receptor 2

Almost 20% of breast cancer patients overexpresses HER2 protein (Elster et al. 2015), a member of the HER family protein, including epidermal growth factor receptor (EGFR) EGFR/HER1, c-erb/HER2, HER3 and HER4. HER2 functions as a co-receptor for other family members. In case of HER2 over-expression or amplification, tumour growth is enhanced and invasion and survival of transformed cells are elevated via activation of MAPK and PI3K signalling mechanisms (Spector and Blackwell 2009; Graus-Porta et al. 1997; Moasser 2007). These signalling mechanisms are involved in resistance development to trastuzumab, an anti-HER2-specific antibody (Wang and Xu 2019). Trastuzumab resistance mechanism involves decreased HER2-antibody binding capability due to increased signalling via RTK (HER2 family receptor tyrosine kinases), increased PI3K/Akt activity and insulin-like growth factor 1 receptor (IGF1R) signalling (Elster et al. 2015).

2.1.4 Triple-Negative Breast Cancer

TNBC accounts for 10–20% of breast cancer cases (Neophytou et al. 2018). TNBC is the most aggressive form of breast cancer (O'Toole et al. 2013) and can metastasize typically to lungs, liver and bones (Weigelt et al. 2005). TNBC is characterized by the lack expression for ER, PR and HER2 receptors thereof, so alternative signalling mechanisms gained an importance for the identification of new therapeutic target (O'Toole et al. 2013). Due to the lack of expression of ER, PR and HER2 receptors, it is not possible to treat TNBC with endocrine therapy or HER2 targeting strategies (McGee 2010). Unfortunately, there is no approved targeted therapy for TNBC yet

(Corkery et al. 2009). At the present time, the optimal TNBC treatment strategy is based on broad spectrum chemotherapy due to the lack of alternative strategies for this aggressive subtype of breast cancer (Crown et al. 2012). Studies have documented the advantages of chemotherapy usage in the neoadjuvant, adjuvant and metastatic TNBC conditions. Depending on tumour profile, different chemotherapeutic agents can be used to treat TNBC such as platinum compounds (cisplatin, carboplatin), taxanes (docetaxel), capecitabine (fluorouracil) and an anthracycline (doxorubicin) (Isakoff 2010). Furthermore, TNBC-specific EGFR over-expression enhances resistance to conventional chemotherapeutic therapies (McGee 2010), whose silencing has the potential to improve TNBC treatment efficacy (Al-Mahmood et al. 2018).

3 Therapeutic Targets and Drug Resistance

Receptors and signalling mechanisms have been important molecular targets, and detailed studies are launched to develop and/or improve available therapeutic agents (Ariazi et al. 2006). A major drug resistance mechanism in breast cancer is mediated by ATP binding cassette (ABC) transporters (Kuo 2007). ABC transporter superfamily has 47 genes (Allikmets et al. 1996) including ABCB1, also named as P-glycoprotein (P-gp) or multidrug resistance 1 (MDR1) gene (Chen et al. 1986), and ABCB5 (Frank et al. 2003). ABC transporters are capable of pumping intracellular chemotherapeutic drugs out of the cells. More than 80% of drug efflux is facilitated by these proteins, which are readily over-expressed in breast cancer cells. Hence, targeting drug efflux-associated transporters had become a promising therapeutic approach to enhance intracellular drug accumulation in breast cancer (Kuo 2007). ABCB1 and ABCB5 enhance chemotherapeutic drug efflux from tumour, thereby increasing drug resistance (Luo et al. 2012). Liao et al. (2019) documented the importance of ABCB1 as a therapeutic target to overcome chemoresistance in breast cancer. Knock-down of ABCB1 expression resulted in enhanced intracellular chemotherapeutic drug presence in cancer cells that lead to an improved treatment efficacy for breast cancer (Liao et al. 2019). Yao et al. (2017) had targeted ABCB5 protein, which is over-expressed in breast cancer including metastatic tissues that promote metastasis and increase epithelial-mesenchymal transition (EMT). They showed a prevention of metastatic breast cancer-related activities by inhibition of ABCB5 expression (Yao et al. 2017). Tumour suppressor protein p53 downregulates ABCB1 expression by binding to its promoter region (Chin et al. 1992). A mutant p53 enhanced the ABCB1 expression that led to drug resistance in colon cancer (Thottassery et al. 1997). The p73 gene belonging to the p53 family encodes N-terminally truncated isoforms, collectively named as ΔNp73 (Di et al. 2013). The ΔNp73 knock-out resulted in increased drug sensitivity in embryonic fibroblasts (Wilhelm et al. 2010) and ΔNp73 over-expression was associated with reduced drug response in breast cancer (Di et al. 2013). Sakil et al. (2017) showed ΔNp73 to be associated with increased expression of ABC transporter in breast cancer samples, while ΔNp73 knocked-down resulted in a decrease of ABCB1 and ABCB5 expression. In addition, reduced expression of ΔNp73 led to better intracellular retention of DOX and reduced cell proliferation in MDA-MB-231 and MCF7 cells. The results indicate the regulatory role of ΔNp73 expression in drug resistance via ABCB1 and ABCB5 proteins. The expression of p73ΔEx2/3, which is an isoform of ΔNp73, was associated with ABCB5 expression in metastasis in melanoma patients and melanoma-derived cell line SK-MEL-28 (Sakil et al. 2017). However, the role of p73ΔEx2/3 on ΔNp73 expression and resistance against the chemotherapeutics has not been addressed yet.

3.1 Breast Cancer-Associated Receptors and Inhibitors

Since ER, PR and HER2 receptors are closely involved in breast cancer development and progression (Perou et al. 2000; Sotiriou et al. 2003; Samadi et al. 2018; Longley and Johnston 2005), many studies are investigating new inhibitors for ER, PR and HER2 receptors. Endocrine therapies, a treatment option for ER+ breast cancer patients, are classified into three groups which are aromatase inhibitors (AIs), selective estrogen receptor downregulators (SERDs) and selective estrogen modulators (SEM). Letrozole, anastrozole and exemestane are used to reduce the estrogen production via inhibition of aromatase enzyme in ER+ breast cancer patients (Hertz et al. 2017; Ali et al. 2016; Sharma et al. 2018). Aromatase enzyme catalyzes the final step in estrogen biosynthesis, and AIs are effectively being used in adjuvant and first-line metastatic treatment (Chumsri et al. 2011). The SERD fulvestrant abolishes the estradiol binding to ER, which impairs receptor dimerization and results in the blockage of nuclear localization of the receptor (Osborne et al. 2004). The SEM tamoxifen specifically binds to ER and displaces estrogen that leads to the inhibition of estrogen function in breast cancer cells. These conformational alterations leading to blockage of ER and co-activator protein interaction inhibit the activation of genes associated with cell proliferation. Tamoxifen had decreased mortality rate by 30% in breast cancer patients, but the presence of acquired resistance to tamoxifen is a major limitation to overcome (Ali et al. 2016).

Selective PR modulators (SPRMs) are synthetic steroids specific for PR (Chabbert-Buffet et al. 2018). Mifepristone, ulipristal, proellex, onapristone, asoprisnil and lonaprisan are in clinical development as SPRMs (Wagenfeld et al. 2016). Lee et al. (2016) had shown that telapristone acetate (TPA) diminished tumour incidence, burden, cell proliferation and angiogenesis in mouse mammary glands (Lee et al. 2016). Clinical trials had revealed that combination hormone replacement therapy (ER and progestins) had an enhanced risk for breast cancer recurrence compared to ER or placebo-receiving patients. Medroxyprogesterone acetate (MPA), a component of synthetic progestins, was shown to enhance breast cancer tumour growth and metastasis. Furthermore, MPA increased CD44 expression and aldehyde dehydrogenase (ALDH) activity, which are cancer stem cell (CSC) markers. Liang et al. (2017) had evaluated the effect of RO48-8071, a cholesterol synthesis inhibitor, in hormone-dependent human breast cancer cell lines, namely, T47-D and BT-474. RO48-8071 decreased the PR-dependent MPA-medicated CD44 expression and MPA-induced CSC expansion by downregulating the expression of PR protein in T47-D and BT-474 cell lines (Liang et al. 2017).

Trastuzumab, pertuzumab, lapatinib and ado-trastuzumab emtansine (T-DM1) are currently being used as anti-HER2 agents (Wang and Xu 2019). However, with the development of recurrence and acquired resistance, studies have focused on the investigation of resistance mechanisms and identification of new inhibitors for HER2 (Schroeder et al. 2014). Among the anti-HER2 inhibitors, abemaciclib (LY2835219) is a promising CDK4/6 inhibitor due to its high selectivity for CDK4/6 complexes. The studies had indicated that abemaciclib inhibited phosphorylation of the retinoblastoma tumour suppressor via G1 cell cycle arrest in tumour cells. Abemaciclib also induced a dose-dependent anti-tumour activity, leading to ~70% reduction in tumour volume in vivo (Gelbert et al. 2014). Furthermore, abemaciclib enhanced anti-tumour immunity via promoting tumour antigen presentation and suppression of regulatory T cell activity (Goel et al. 2017).

3.2 Signalling Mechanisms and Drug Resistance

The main goals of breast cancer therapy are to inhibit proliferation, eradicate metastasis and induce apoptosis in transformed cells. Molecular approaches such as the physiologic alterations in signalling mechanisms are important for the modulation of apoptosis (Oltersdorf et al. 2005). Alterations in signalling mechanisms can cause aberrant regulation of apoptosis and are particularly associated with the development of drug resistance. PI3K/AKT pathway, for example, is activated by ER and HER family receptors.

Activation of PI3K/AKT pathway with increasing phosphatidylinositol-3,4,5-triphosphate leads to the activation of signalling mechanism involving AKT/PKB pathway. Activation of AKT prevented apoptotic alterations by activating NF-kappa B. NF-kB activation results in phosphorylation of Bad which is a pro-survival gene (Datta et al. 1997; Shimamura et al. 2003). The relation between Bad protein and AKT signalling was mediated via Bcl-2 protein. The anti-apoptotic protein Bag-1 is over-expressed on breast cancer cell lines and is involved in the formation of B-Raf, C-Raf and AKT complexes in breast cancer cells, which resulted in phosphorylation of pro-apoptotic Bad for apoptosis inhibition (Kizilboga et al. 2019).

The identification and emergence CSCs have been an important therapeutic consideration for breast cancer therapy (Yang et al. 2020). CSCs possess self-renewal capacity and play a major role in differentiation, metastasis, cancer recurrence, resistance to drugs and radiation (Yang et al. 2020). Therefore, strategies leading to clearance of CSCs is an important therapeutic approach for metastatic breast cancer (Zuo et al. 2016). Main breast CSC associated signalling pathways are Wnt, Notch and Hedgehog signalling pathway (Yang et al. 2020).

3.2.1 Hedgehog Signalling Pathway

The Hedgehog (Hh) signalling mechanism plays a significant role in embryogenic development (Skoda et al. 2018). The main factor behind basal cell carcinoma is dysregulation of Hh mechanism so that the components of this mechanism had been important therapeutic targets. Hh signalling activates glioma-associated oncogene (GLI) transcription factors. GLI1 over-expression is associated with poor prognosis in breast cancer and also GLI1 activation leading to ER exposure enhances breast CSC proliferation and EMT (Bhateja et al. 2019). Estrogen is related to the enhancement of CSCs and EMT via GLU1 in ERα(+) breast cancer cells. GLI1 depletion increased tamoxifen cytotoxicity and reduced ERα protein expression levels due to diminished ERα signalling activity in tamoxifen-resistant and sensitive cells (Diao et al. 2016). Targeting GLI transcription factor family may provide valuable information about the modulation of Hh mechanism and its effect on breast cancer therapy.

3.2.2 Notch Signalling Pathway

Notch signalling mechanism involves Notch receptors and DSL ligands (Ranganathan et al. 2011). Notch receptor interaction with DSL ligands initiates Notch signalling mechanism that undergoes cleavage processes (Cohen et al. 2010; Leong et al. 2007). Notch signalling interferes in the regulation of cellular processes (Miele et al. 2006). However, dysregulation of the Notch signalling mechanism is involved in endocrine resistance in ERα(+) breast cancer (Bai et al. 2020). Notch1 receptor silencing reversed the EMT and inhibited the growth of xenografts and metastatic invasion in MCF-7 and MBA-MD-231 breast cancer models (Shao et al. 2015). Similar results were documented by a recent study of Xiao et al. (2019); Notch1 receptor silencing inhibited Akt pathway, decreased the EMT and enhanced sensitivity to cisplatin and DOX in MDA-MB-231DDPR cells (Xiao et al. 2019).

3.2.3 Wnt Signalling Pathway

Aberrant activation of Wnt signalling was related to breast cancer progression (Yin et al. 2018; Pohl et al. 2017). Wnt ligands bind to frizzled receptor (FZD) and co-receptors low-density lipoproteins 5/6 (LRP5/6), which initiates β-catenin-dependent (canonical) and β-catenin-independent (non-canonical) signalling pathways (Yin et al. 2018). Dickkopf-1 (DKK1) protein is important for the regulation of Wnt signalling (Yin et al. 2018). The increased expression of DKK1 and cytoplasm/nuclear-β-catenin was observed in breast cancer patients and those without lymph node metastasis (Sun et al. 2019). Further, Sada et al. (2019) addressed the association between DKK1 and AKT signalling; DKK1 induces depalmitoylation of cytoskeleton related protein 4 (CKAP4) and LRP5/6 receptors which are needed to activate PI3K-AKT pathway. LRP5/6 knock-down reduced DKK1-mediated AKT activities and tumour proliferation via CKAP4 (Sada et al. 2019). The role of DKK1 on T-DM1 resistance was reported by Li et al. (2018). The

DKK1 and canonical Wnt signalling-dependent MMP7 activation resulted in the induction of T-DM1 resistance and poor gastric adenocarcinoma prognosis, whereas the reverse effects on DKK1 or canonical Wnt signalling led to a reduction in T-DM1 resistance via reduced expression of MMP7 in gastric adenocarcinoma (Li et al. 2018).

3.3 Drug Resistance in TNBC

Neither hormonal therapy nor anti-HER2 therapies are effective in TNBC treatment due to the lack of the relevant targets. Currently, broad spectrum chemotherapy is the only treatment option for TNBC (Crown et al. 2012). However, chemoresistance limits the success of the agents against TNBC since majority of the agents fail due to the development of either intrinsic or acquired resistance to chemotherapeutic agents (Longley and Johnston 2005). The underlying chemoresistance mechanisms in TNBC can be listed as expression of ABC transports, mutations DNA repair enzymes and activation of chemoresistance-associated signalling mechanisms (Abdullah and Chow 2013). In addition, Wnt/β-catenin, Notch and Hedgehog signalling mechanisms are involved in progression and exacerbation of TNBC (O'Toole et al. 2013).

4 Targeted Drug Delivery

Targeted drug delivery is an approach to administer single or multiple therapeutic compounds to achieve beneficial effects against a particular target (i.e. cells bearing the target), while displaying little effects on other (non-relevant) tissues or cells. To reach this aim, several drug delivery systems have been developed and analysed (Tiwari et al. 2012). Among these systems, nanoparticles represent a great success in the application of targeted drug delivery due to their unique features (Singh and Vyas 1996).

4.1 Targeted Drug Delivery with Nanoparticles

Nanoparticles are nano-scaled particles ranging in size <100 nm in at least one dimension. The increased interests in nanoparticles have been greatly facilitated with the prospect of developing nanomedicines (Hussain et al. 2018; Tang et al. 2017) and their facilitated delivery in the tumour vicinity (Ould-Ouali et al. 2005; Kipp 2004), increased specificity to tumour tissue and bioavailability of chemotherapeutics at the site of the disease (Fonseca et al. 2002; Koziara et al. 2006). The final efficacy of nanomedicines is based on their distinct features such as the optimal size, reduced drug toxicity at other sites and controlled drug release at the site of tumours. Targeted delivery of anti-cancer drugs into tumour cells can overcome drug resistance (Koziara et al. 2006; Koziara et al. 2004) if the drugs are packaged in nanoparticles. The development of drug resistance avoids the induction of cytotoxicity of chemotherapeutics in cancer cells (Cho et al. 2008). Huang et al. (2018) had formulated surfactin (SUR)-based nanoparticles loaded with doxorubicin (DOX) that induced cytotoxicity to DOX-resistant MCF-7/ADR. This study had shown enhanced cellular uptake, tumour accumulation and decreased cellular efflux of DOX in vitro. Tumour inhibition was shown without changing the body weight of the preclinical model along with reduced cardiotoxicity, which is one of the main limitations of DOX. These results show the biosafety and targeted delivery capability of nanoparticles to tumours (Huang et al. 2018). A study on HER2 targeting by Mondal et al. (2019) explored nanoparticle formulations with an incorporated anti-HER2 antibody and PLGA nanoparticles loaded with DOX (8.5% w/w). Using human breast cancer cell lines SKRB-3, MCF-7 and MDA-MB-231, an increased accumulation of designed nanoparticle formulations was shown in tumour cells, which led to a reduction in tumour volume and decreased cardiotoxicity (Mondal et al. 2019). Another study targeted over-expressed HER2 receptors (Li et al. 2017) in breast CSC using polymer-lipid hybrid nanoparticles loaded with an antibody against HER2 receptor and salinomycin (Li et al. 2017), an anti-breast CSC drug (Gupta et al. 2009; Magnifico et al. 2009; Naujokat and Steinhart 2012). The result of the study showed an increased cytotoxic effect, reduction in tumour growth and decreased breast tumour sphere formation rate (Li et al. 2017).

4.2 Properties of Nanoparticles

Nanocarriers can be used for diagnostic and therapeutic purposes. Polymers such as polyethylene glycol (PEG) are coated onto nanoparticles so that they cannot be detected by the immune system and undergo prolonged circulation time in the body. With respect to functionalization, ligands are used to target over-expressed receptors in tumour cells. Internalization of nanoparticles by endocytotic pathways may overcome the major challenge of drug resistance. Nanoparticles can be designed to be sensitive to intracellular stimuli and enable controlled intracellular drug release (Ferrari 2005; Duncan 2006). The importance of nanoparticle structure in breast cancer therapy was shown by Tiash and Chowdhury (2019). Anti-cancer drug DOX and siRNA were loaded into pH-sensitive carbonate apatite (CA) nanoparticles to target ABC transporter genes. Low polydispersity nanoparticles at ~200 nm were taken up more efficiently and decreased tumour size at low dose of DOX in 4 T1 breast cancer cells (Tiash and Chowdhury 2019). The importance of pH was shown by Onyebuchi and Kavaz (2019), who used *Ocimum gratissimum* plant essential oils (OGEO) encapsulated by chitosan and N,N,N-trimethyl chitosan nanoparticles. The results indicated higher drug release at pH 3 compared to 7.4 in vitro with MDA-MB-231 breast cancer cell line (Onyebuchi and Kavaz 2019). Furthermore, nanoparticle size and structure were important for cellular uptake efficacy. Akbal et al. (2017) had analysed human serum albumin (HSA) and poly(3-hydroxybutyrate)-carboxymethyl chitosan (PHB-CMCh) nanocarriers. HSA nanoparticles had smaller size and higher drug entrapment efficacy compared to PHB-CMCh nanoparticles. Furthermore, HSA nanoparticles represented cytotoxic and high cellular uptake efficacy in MCF-7 cells (Akbal et al. 2017). These outcomes remain to be translated to other cell types.

4.3 Enhanced Permeability and Retention Effect

Enhanced permeability and retention effect (EPR) is a unique characteristic of solid tumours associated with anatomical and pathophysiological differences from normal tissue (Fang et al. 2011). Blood vessels of solid tumours have defective 'leaky' structures. Cancer cells over-produce angiogenic and vascular permeability factors, thereby stimulating vascular permeability to supply nutrients and oxygen to tumour tissue for rapid growth. Due to this unique pathophysiology, this leaky vasculature enables transportation and accumulation of macromolecules into the tumour tissue (Greish 2007; Matsumura and Maeda 1986). In fact, this unique phenomenon of nanoparticles provides opportunities to target solid tumours through EPR effect (Iyer et al. 2006). The underlying principle of EPR effect is targeted delivery and accumulation capability of nanoparticles at the tumours by either active targeting or passive targeting (Park et al. 2009). However, tumour size-dependent heterogeneity of tumour tissue and the variety of macromolecular accumulation within the tumour tissue are the main challenges to overcome (Maki et al. 1985; Nagamitsu et al. 2009).

Active targeting strategy, whereby anti-cancer drugs are transported specifically to target cells (Duncan 2006; Ferrari 2005), relies on the interaction between ligands on nanoparticle surface and target cells. The biological ligands are incorporated into nanoparticles to bind to specific receptors on the target cell. This resulted in an enhanced uptake of nanoparticle formulations into the cell (Fig. 1) (Byrne et al. 2008; Muhamad et al. 2018). Passive targeting strategy, whereby nanoparticles accumulate in tumour cells by the EPR effect (Huang et al. 2016), relies on the

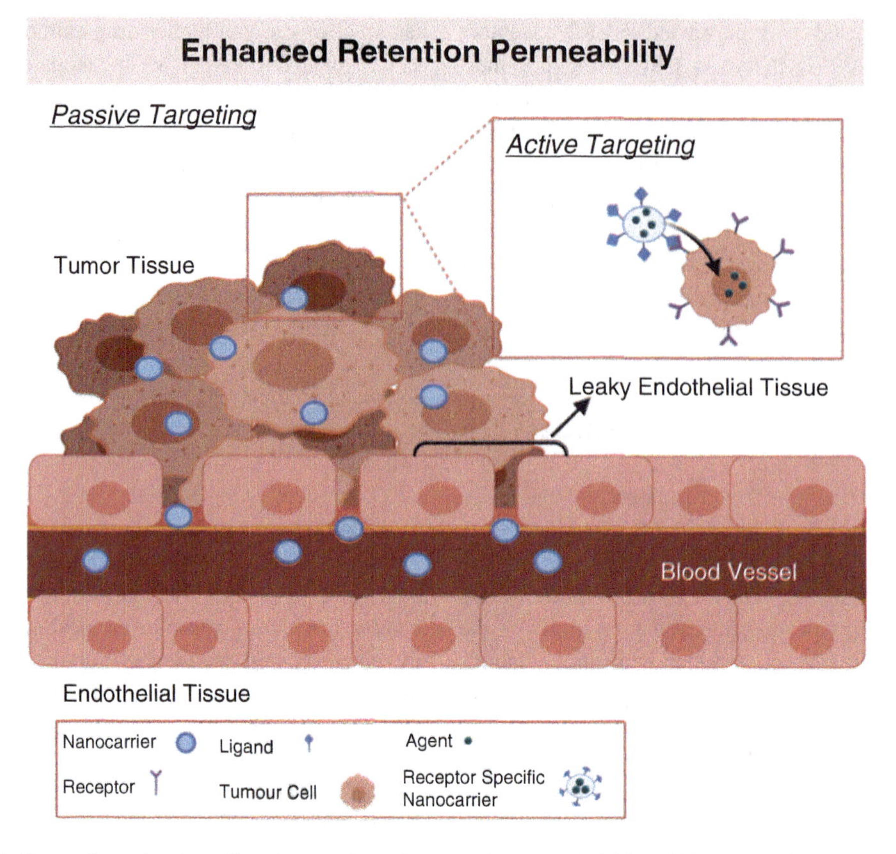

Fig. 1 **Active and passive targeting strategy by using nanoparticles to target tumours via enhanced retention permeability (EPR) effect**. Nanoparticles are used as nanocarriers which include formulations to inhibit tumour activities. The leaky structure of blood vessels enables nanocarriers to accumulate into the tumour and exert anti-tumour activities. This process is called as passive targeting. Conversely, active targeting relies on ligand-receptor interaction. Receptor-specific ligands are loaded to nanocarriers and these nanocarriers specifically bind to the receptor and abolish tumour activities

unique physiological structure of solid tumours such as the leaky vasculature and defective lymphatic drainage (Fig. 1) (Rosenblum and Peer 2014; Matsumura and Maeda 1986). Doxil and Abraxane are some of passively targeted nanodrugs which are currently used in a clinical setting (Shi et al. 2017).

4.4 Targeted Therapy Strategies for Breast Cancer-Associated Receptors

Targeted delivery strategies have gained importance (Peer et al. 2007) and nanoparticle formulations have extensively been used to this end for breast cancers. Verderio et al. (2014) reported inhibition of cellular growth in ER-dependent MCF-7 breast cancer cells by using ASC-J9-loaded PLGA nanoparticles. The release of ASC-J9 in loaded PLGA nanoparticles was associated with time- and dose-dependent drug activity. This nanoparticle blocked G2/M cell cycle in the cytosol of MCF-7 cells (Verderio et al. 2014). Kubota et al. (2018) used anti-HER2 antibody trastuzumab-coated gold nanoparticles (AuNPs) and showed an increased cytotoxicity in trastuzumab-resistant gastric cancer cells. Similar effects have also been observed in in vivo studies (Kubota et al. 2018).

HER2-based targeting was also employed by Gu et al. (2018) that created HER2-specific antibody-bearing nanoparticles with siRNA against HER2(+) breast cancer cells displaying lapatinib acquired resistance. The cells were treated for 7 months to investigate the long-term effect of formulation. Even after the removal of formulation, long-term treated cells grow slower than cells which did not receive any treatment. Furthermore, treated cells did not undergo EMT or had tumour-initiating cell enrichment. HER2 ablation with this formulation inhibited HER2 signalling reactivation that is present in lapatinib-resistant cells. The formulation had successfully prevented trastuzumab and lapatinib resistance development in HER2 BT474 breast cancer cell line (Gu et al. 2018).

Another study by Zhang et al. (2020) developed nontoxic transformable peptide nanoparticles to target HER2 receptor cells in vivo. The design used micelles in aqueous solution that transformed into nanofibers to disrupt HER2 dimerization and downstream signalling events and results in apoptosis of cancer cells. Disrupted HER2 dimerization blocked cell proliferation and cell survival signalling at the tumour tissue in mouse xenografts (Zhang et al. 2020).

EGFR is another important target for breast cancers due to its over-expression in ~15% of metastatic breast cancers (Gallardo et al. 2012; DiGiovanna et al. 2005). Covarrubias et al. (2019) targeted EGFR by designing dual-ligand nanoparticles with DOX to target EFGR and $\alpha_v \beta_3$ integrin in vitro and in vivo. This treatment prolonged survival time under metastatic conditions and delayed outgrowth of metastasis. Additionally, DOX-loaded nanoparticles did not lead to a reduction in weight loss, minimizing the well-known side effects of DOX (Covarrubias et al. 2019). Since co-expression of EGFR and HER2 is associated with poor prognosis and reduced survival (Gallardo et al. 2012, DiGiovanna et al. 2005), Houdaihed et al. (2020) had targeted both HER2 and EGFR receptors using polymeric nanoparticles that encapsulated paclitaxel and everolimus drugs. Paclitaxel has been extensively used in breast cancer treatment, but like other therapeutics, its usage is limited with chemoresistance (Ajabnoor et al. 2012; Murray et al. 2012) and unacceptable side effects (Gelderblom et al. 2001). The results indicated stronger induction of cytotoxicity and enhanced cellular uptake in EGFR-/HER2-targeting nanoparticles, as compared to single and non-targeted nanoparticle formulations in breast cancer cells (Houdaihed et al. 2020).

4.5 Nanoparticles and siRNA

RNA interference (RNAi) technology is a promising gene regulation technology to treat pathologic diseases such as cardiovascular diseases and cancer by silencing the genes of interest. Despite the success of preclinical RNAi studies, clinical RNAi application is not yet possible due to the limitations of delivery system. Short interfering RNA (siRNA) enables scientists to regulate expression levels of target genes. The siRNA has gained importance as a therapeutic agent (Dana et al. 2017). A study from Jafari et al. (2019) has proved the importance and therapeutic efficacy of using nanoparticles for drug targeting (Jafari et al. 2019). IGF1 receptor (IGF1R) is required for growth and survival in normal cells, whereas IGF1R signalling plays a major role in tumour growth, development and metastasis. Additionally, IGF1R signalling cross talks with other signalling mechanisms (Baserga et al. 2003), leading to the activation of several transcription factors, thereby upregulating a multitude of genes responsible from invasion to angiogenesis (Banerjee and Resat 2016). Simultaneous silencing of IGF1R using siRNA and targeted delivery of docetaxel-loaded chitosan nanoparticles with an anti-mucin1 aptamer resulted in increased cellular uptake of nanoparticles and a decrease in cell viability and reduction in gene expression, which led to tumour progression and metastasis (Jafari et al. 2019).

Nanoparticles can overcome inherent limitations and enable targeted delivery of naked siRNA to the desired site of action (Chen et al. 2018). TNBC, the most aggressive subtypes of breast cancer, undergoes metastasis with the activation of EMT. Since TGF-β upregulates β_3-integrin, which is required for EMT and metastasis, Parvani et al. (2015) targeted β_3-integrin using ECO/siRNA nanoparticles. Blocking of β_3 integrin resulted in the attenuation of TGFβ-mediated EMT and metastasis in vitro and a reduction of primary tumour burden and metastasis in vivo (Parvani et al. 2015).

4.6 Nanoparticles and Signalling Mechanisms

A nanoparticle formulation had been used to target Hh signalling components. GLI1 had been targeted with GANT61, a Hh inhibitor (Koike et al. 2017), and curcumin-loaded polymeric nanoparticles in the MCF-7 breast cancer cell line. This nanoparticle formulation successfully induced cytotoxic effects at a mid-minimal dosage via autophagy and apoptosis, and a reduction in self-renewal capability of CSC (Borah et al. 2020). Another study targeted Hh signalling component GLI1 by Nayak et al. (2016); quinacrine nanoparticles (NQC) reduced GLI1 mRNA expression dose-dependently in cervical CSC. NQC reduced tumour size and induced apoptosis through GLI1 inhibition in Hh-GLI cascade (Nayak et al. 2016). However, the effect of NQC on GLI1 has not yet been tested in breast cancer cell lines.

TNBC cells typically over-expresses Notch 1 receptor and Bcl-2 anti-apoptotic protein, which can be inhibited by ABT-737 to initiate apoptotic signalling. Valcourt et al. (2020) formulated PLGA nanoparticles bearing Notch-1 specific antibodies. This formulation enhanced TNBC-specific binding and Notch signalling inhibition through the blockage of the Notch 1 receptor. Notch inhibition via ABT-737 further induced apoptosis to TNBC cells. In vivo studies showed accumulation of nanoparticles in TNBC xenograft, which resulted in a reduction of tumour burden and extended animal survival (Valcourt et al. 2020). Mamaeva et al. (2011) had also prepared nanoparticle formulations to overcome drug resistance. The formulation promoted Notch signalling in breast cancer by enhancing CSC self-renewal capability. Mesoporous silica nanoparticles with glucose (since CSCs display glycolytic activity) were loaded with Notch signalling inceptor, γ-secretase inhibitor. This formulation is accumulated both in transformed and CSCs in vitro and in vivo, which also reduced the CSC population in vivo (Mamaeva et al. 2011).

Metallic nanoparticles have gained an increased interest due to nontoxic and stable chemical structures (Albrecht et al. 2006). Both positively and negatively charged metallic gold nanoparticles (AuNPs) induced cytotoxicity via oxidative stress and resulted in alterations in Wnt signalling pathway in TNBC cells. Surapaneni et al. (2018) extensively studied the AuNP mechanism of action on TNBC cells. The results indicated the positively and negatively charged AuNPs to exert their cytotoxic effects via different mechanisms. Negatively charged AuNPs slowed cell death process, whereas positively charged AuNPs resulted in abrupt destruction of MDA-MB-231 cell line by enhanced phosphorylation of histone H3 ser 10 (Surapaneni et al. 2018). Thymidylate synthetase over-expression was associated with anti-cancer drug 5-FU resistance (Zhang et al. 2008). Both positively and negatively charged AuNP treatments reduced the expression of thymidylate synthetase which made MDA-MB-231 cells more sensitive to 5-FU (Surapaneni et al. 2018). This study showed the importance of signalling mechanism in the induction of cytotoxicity and sensitivity difference between differentially charged AuNPs. Cytotoxic effects of another metallic nanoparticle, zinc oxide (ZnO), were studied by Umar et al. (2018) on MDA-MB-231 and MCF-7 cells. Their results indicated highest cytotoxic effect to be observed at concentration that corresponded to the highest zinc ion concentration. Therefore, the increased presence of zinc ion was suggested as the reason of antioxidant activity of ZnO nanoparticles (Umar et al. 2018).

5 Conclusion and Perspectives

Gene therapy with nanoparticles is expected to provide enhanced treatment efficacy. Nanoparticle platform allows a sophisticated delivery of a multitude of therapeutic agents and modalities. As a prototypical example, Wang et al. (2020) formulated a combination therapy for simultaneous delivery of a siRNA (against siRlk1) and miRNA (miR-200c) in mesoporous silica nanoparticles, in addition to the photosensitizer indocyanine green (ICG) to enable endosomal escape as well as surface conjugation of iRGD peptide to permit enhanced tumour penetration. This formulation gave increased cellular uptake in both in vitro 3D tumour spheroids and in vivo orthotopic MDA-MB-231 tumours. In addition, intravenous treatment of metastatic breast cancer with this formulation reduced primary tumour growth (Wang et al. 2020). With a different perspective, Zhang et al. (2019) had combined PDA@DOX nanoparticles with near-infrared (NIR) to enhance treatment efficacy. PDA@DOX nanoparticles are formulated by using one-pot synthesis of hollow nanoparticles encapsulated with DOX and modified with polydopamine (PDA) for breast cancer treatment. This formulation had demonstrated 53.16% DOX loading capacity. NIR light was absorbed by PDA outer layer which resulted in simultaneous heat energy generation killing the tumour cells for drug release upon NIR irradiation. This results in the suppression of complete tumour growth in vivo. The nanoparticles also enabled a remarkable ultrasound performance to enable monitoring during the treatment (Zhang et al. 2019). Taken together, simultaneous targeting and combination of nanoparticles with medical device capabilities such as NIR arise as a substantial research focus in breast cancer.

Given the recent exciting developments on nanoparticle technology, and better understanding of breast cancer-associated receptors and signalling mechanisms, strongly performing therapeutic agents are expected to emanate in the near future. The side effects of chemotherapy drugs and development of drug resistance against to current breast cancer chemotherapeutics are major challenges that could be solved with this new modality of drug delivery. Further identification of effective molecular targets, understanding the underlying resistance mechanism of chemotherapeutics and recent developments in nanomedicine hold a great promise to overcome these limitations. The targeted drug delivery capability of nanoparticle-based platforms that increased the treatment efficacy, reduced toxicity to healthy cells and increased intracellular chemotherapeutic accumulation are the key advantages of this platform. Furthermore, recently combinational effects and simultaneous targeting had become increasingly promising in the design of more effective anti-cancer therapies. More effective therapeutic agents are bound to emerge from the combination of distinct mechanism of actions in the specially designed nanoparticles.

References

Abdullah LN, Chow EK-H (2013) Mechanisms of chemoresistance in cancer stem cells. Clin Transl Med 2:3

Ajabnoor GMA, Crook T, Coley HM (2012) Paclitaxel resistance is associated with switch from apoptotic to autophagic cell death in MCF-7 breast cancer cells. Cell Death Dis 3:e260–e260

Akbal Ö, Erdal E, Vural T, Kavaz D, Denkbaş EB (2017) Comparison of protein- and polysaccharide-based nanoparticles for cancer therapy: synthesis, characterization, drug release, and interaction with a breast cancer cell line. Artif Cells Nanomed Biotechnol 45:193–203

Albrecht MA, Evans CW, Raston CL (2006) Green chemistry and the health implications of nanoparticles. Green Chem 8:417–432

Ali S, Rasool M, Chaoudhry HPNP, Jha P, Hafiz A, Mahfooz M, Abdus Sami G, Azhar Kamal M, Bashir S, Ali A, Sarwar Jamal M (2016) Molecular mechanisms and mode of tamoxifen resistance in breast cancer. Bioinformation 12:135–139

Allikmets R, Gerrard B, Hutchinson A, Dean M (1996) Characterization of the human ABC superfamily: isolation and mapping of 21 new genes using the expressed sequence tags database. Hum Mol Genet 5:1649–1655

Al-Mahmood S, Sapiezynski J, Garbuzenko OB, Minko T (2018) Metastatic and triple-negative breast cancer: challenges and treatment options. Drug Deliv Transl Res 8:1483–1507

Ariazi EA, Ariazi JL, Cordera F, Jordan VC (2006) Estrogen receptors as therapeutic targets in breast cancer. Curr Top Med Chem 6:181–202

Bai J-W, Wei M, Li J-W, Zhang G-J (2020) Notch signaling pathway and endocrine resistance in breast cancer. Front Pharmacol 11:924

Banerjee K, Resat H (2016) Constitutive activation of STAT3 in breast cancer cells: a review. Int J Cancer 138:2570–2578

Bartsch R, Bago-Horvath Z, Berghoff A, Devries C, Pluschnig U, Dubsky P, Rudas M, Mader RM, Rottenfusser A, Fitzal F, Gnant M, Zielinski CC, Steger GG (2012) Ovarian function suppression and fulvestrant as endocrine therapy in premenopausal women with metastatic breast cancer. Eur J Cancer 48:1932–1938

Baserga R, Peruzzi F, Reiss K (2003) The IGF-1 receptor in cancer biology. Int J Cancer 107:873–877

Bernard-Marty C, Cardoso F, Piccart MJ (2004) Facts and controversies in systemic treatment of metastatic breast cancer. Oncologist 9:617–632

Bhateja P, Cherian M, Majumder S, Ramaswamy B (2019) The hedgehog signaling pathway: a viable target in breast cancer? Cancers (Basel) 11:1126

Borah A, Pillai SC, Rochani AK, Palaninathan V, Nakajima Y, Maekawa T, Kumar DS (2020) GANT61 and curcumin-loaded PLGA nanoparticles for GLI1 and PI3K/Akt-mediated inhibition in breast adenocarcinoma. Nanotechnology 31:185102

Boyle P, Levin B (2008) World cancer report 2008. IARC Press, International Agency for Research on Cancer, Lyon

Brasseur K, Gévry N, Asselin E (2017) Chemoresistance and targeted therapies in ovarian and endometrial cancers. Oncotarget 8:4008–4042

Byrne JD, Betancourt T, Brannon-Peppas L (2008) Active targeting schemes for nanoparticle systems in cancer therapeutics. Adv Drug Deliv Rev 60:1615–1626

Carnevale RP, Proietti CJ, Salatino M, Urtreger A, Peluffo G, Edwards DP, Boonyaratanakornkit V, Charreau EH, De Kier Joffé EB, Schillaci R, Elizalde PV (2007) Progestin effects on breast Cancer cell proliferation, proteases activation, and in vivo development of metastatic phenotype all depend on progesterone receptor capacity to activate cytoplasmic signaling pathways. Mol Endocrinol 21:1335–1358

Chabbert-Buffet N, Kolanska K, Daraï E, Bouchard P (2018) Selective progesterone receptor modulators: current applications and perspectives. Climacteric 21:375–379

Chen CJ, Chin JE, Ueda K, Clark DP, Pastan I, Gottesman MM, Roninson IB (1986) Internal duplication and homology with bacterial transport proteins in the mdr1 (P-glycoprotein) gene from multidrug-resistant human cells. Cell 47:381–389

Chen X, Mangala LS, Rodriguez-Aguayo C, Kong X, Lopez-Berestein G, Sood AK (2018) RNA interference-based therapy and its delivery systems. Cancer Metastasis Rev 37:107–124

Chin KV, Ueda K, Pastan I, Gottesman MM (1992) Modulation of activity of the promoter of the human MDR1 gene by Ras and p53. Science 255:459–462

Cho K, Wang X, Nie S, Chen ZG, Shin DM (2008) Therapeutic nanoparticles for drug delivery in cancer. Clin Cancer Res 14:1310–1316

Chumsri S, Howes T, Bao T, Sabnis G, Brodie A (2011) Aromatase, aromatase inhibitors, and breast cancer. J Steroid Biochem Mol Biol 125:13–22

Cohen B, Shimizu M, Izrailit J, Ng NF, Buchman Y, Pan JG, Dering J, Reedijk M (2010) Cyclin D1 is a direct target of JAG1-mediated Notch signaling in breast cancer. Breast Cancer Res Treat 123:113–124

Corkery B, Crown J, Clynes M, O'donovan N (2009) Epidermal growth factor receptor as a potential therapeutic target in triple-negative breast cancer. Ann Oncol 20:862–867

Covarrubias G, He F, Raghunathan S, Turan O, Peiris PM, Schiemann WP, Karathanasis E (2019) Effective treatment of cancer metastasis using a dual-ligand nanoparticle. PLoS One 14:e0220474

Crown J, O'Shaughnessy J, Gullo G (2012) Emerging targeted therapies in triple-negative breast cancer. Ann Oncol 23:vi56–vi65

Dana H, Chalbatani GM, Mahmoodzadeh H, Karimloo R, Rezaiean O, Moradzadeh A, Mehmandoost N, Moazzen F, Mazraeh A, Marmari V, Ebrahimi M, Rashno MM, Abadi SJ, Gharagouzlo E (2017) Molecular mechanisms and biological functions of siRNA. Int J Biomed Sci 13:48–57

Das M, Mohanty C, Sahoo SK (2009) Ligand-based targeted therapy for cancer tissue. Expert Opin Drug Deliv 6:285–304

Datta SR, Dudek H, Tao X, Masters S, Fu H, Gotoh Y, Greenberg ME (1997) Akt phosphorylation of BAD couples survival signals to the cell-intrinsic death machinery. Cell 91:231–241

Dawidczyk CM, Kim C, Park JH, Russell LM, Lee KH, Pomper MG, Searson PC (2014) State-of-the-art in design rules for drug delivery platforms: lessons learned from FDA-approved nanomedicines. J Control Release 187:133–144

Di C, Yang L, Zhang H, Ma X, Zhang X, Sun C, Li H, Xu S, An L, Li X, Bai Z (2013) Mechanisms, function and clinical applications of DNp73. Cell Cycle (Georgetown, Tex) 12:1861–1867

Diao Y, Azatyan A, Rahman MF, Zhao C, Zhu J, Dahlman-Wright K, Zaphiropoulos PG (2016) Blockade of the Hedgehog pathway downregulates estrogen receptor alpha signaling in breast cancer cells. Oncotarget 7:71580–71593

Digiovanna MP, Stern DF, Edgerton SM, Whalen SG, Moore D 2nd, Thor AD (2005) Relationship of epidermal growth factor receptor expression to ErbB-2-signaling activity and prognosis in breast cancer patients. J Clin Oncol 23:1152–1160

Duncan R (2006) Polymer conjugates as anticancer nanomedicines. Nat Rev Cancer 6:688–701

Elster N, Collins DM, Toomey S, Crown J, Eustace AJ, Hennessy BT (2015) HER2-family signalling mechanisms, clinical implications and targeting in breast cancer. Breast Cancer Res Treat 149:5–15

Endoh T, Ohtsuki T (2009) Cellular siRNA delivery using cell-penetrating peptides modified for endosomal escape. Adv Drug Deliv Rev 61:704–709

Fang J, Nakamura H, Maeda H (2011) The EPR effect: unique features of tumor blood vessels for drug delivery, factors involved, and limitations and augmentation of the effect. Adv Drug Deliv Rev 63:136–151

Ferrari M (2005) Cancer nanotechnology: opportunities and challenges. Nat Rev Cancer 5:161–171

Feys L, Descamps B, Vanhove C, Vral A, Veldeman L, Vermeulen S, DE Wagter C, Bracke M, DE Wever O (2015) Radiation-induced lung damage promotes breast cancer lung-metastasis through CXCR4 signaling. Oncotarget 6:26615–26632

Fonseca C, Simões S, Gaspar R (2002) Paclitaxel-loaded PLGA nanoparticles: preparation, physicochemical characterization and in vitro anti-tumoral activity. J Control Release 83:273–286

Frank NY, Pendse SS, Lapchak PH, Margaryan A, Shlain D, Doeing C, Sayegh MH, Frank MH (2003) Regulation of progenitor cell fusion by ABCB5 P-glycoprotein, a novel human ATP-binding cassette transporter. J Biol Chem 278:47156–47165

Gallardo A, Lerma E, Escuin D, Tibau A, Muñoz J, Ojeda B, Barnadas A, Adrover E, Sánchez-Tejada L, Giner D, Ortiz-Martínez F, Peiró G (2012) Increased signalling of EGFR and IGF1R, and deregulation of PTEN/PI3K/Akt pathway are related with trastuzumab resistance in HER2 breast carcinomas. Br J Cancer 106:1367–1373

Garcia-Saenz JA, Bermejo B, Estevez LG, Palomo AG, Gonzalez-Farre X, Margeli M, Pernas S, Servitja S, Rodriguez CA, Ciruelos E (2015) SEOM clinical guidelines in early-stage breast cancer 2015. Clin Transl Oncol 17:939–945

Gelbert LM, Cai S, Lin X, Sanchez-Martinez C, DEL Prado M, Lallena MJ, Torres R, Ajamie RT, Wishart GN, Flack RS, Neubauer BL, Young J, Chan EM, Iversen P, Cronier D, Kreklau E, DE Dios A (2014) Preclinical characterization of the CDK4/6 inhibitor LY2835219: in-vivo cell cycle-dependent/independent anti-tumor activities alone/in combination with gemcitabine. Investig New Drugs 32:825–837

Gelderblom H, Verweij J, Nooter K, Sparreboom A (2001) Cremophor EL: the drawbacks and advantages of vehicle selection for drug formulation. Eur J Cancer 37:1590–1598

Giaccone G, Pinedo HM (1996) Drug resistance. Oncologist 1:82–87

Goel S, Decristo MJ, Watt AC, Brinjones H, Sceneay J, Li BB, Khan N, Ubellacker JM, Xie S, Metzger-Filho O, Hoog J, Ellis MJ, Ma CX, Ramm S, Krop IE, Winer EP, Roberts TM, Kim HJ, Mcallister SS, Zhao JJ (2017) CDK4/6 inhibition triggers anti-tumour immunity. Nature 548:471–475

Goepfert TM, Mccarthy M, Kittrell FS, Stephens C, Ullrich RL, Brinkley BR, Medina D (2000) Progesterone facilitates chromosome instability (aneuploidy) in p53 null normal mammary epithelial cells. FASEB J 14:2221–2229

Goldie JH (1983) Drug resistance and cancer chemotherapy strategy in breast cancer. Breast Cancer Res Treat 3:129–136

Graus-Porta D, Beerli RR, Daly JM, Hynes NE (1997) ErbB-2, the preferred heterodimerization partner of all ErbB receptors, is a mediator of lateral signaling. EMBO J 16:1647–1655

Greish K (2007) Enhanced permeability and retention of macromolecular drugs in solid tumors: a royal gate for targeted anticancer nanomedicines. J Drug Target 15:457–464

Gu S, Ngamcherdtrakul W, Reda M, Hu Z, Gray JW, Yantasee W (2018) Lack of acquired resistance in HER2-positive breast cancer cells after long-term HER2 siRNA nanoparticle treatment. PLoS One 13:e0198141

Gupta PB, Onder TT, Jiang G, Tao K, Kuperwasser C, Weinberg RA, Lander ES (2009) Identification of selective inhibitors of cancer stem cells by high-throughput screening. Cell 138:645–659

Hassanpour SH, Dehghani M (2017) Review of cancer from perspective of molecular. J Cancer Res Pract 4:127–129

Hertz DL, Henry NL, Rae JM (2017) Germline genetic predictors of aromatase inhibitor concentrations, estrogen suppression and drug efficacy and toxicity in breast cancer patients. Pharmacogenomics 18:481–499

Hewitt SC, Korach KS (2000) Progesterone action and responses in the alphaERKO mouse. Steroids 65:551–557

Houdaihed L, Evans JC, Allen C (2020) Dual-targeted delivery of nanoparticles encapsulating paclitaxel and everolimus: a novel strategy to overcome breast cancer receptor heterogeneity. Pharm Res 37:39

Huang B, Warner M, Gustafsson J (2015) Estrogen receptors in breast carcinogenesis and endocrine therapy. Mol Cell Endocrinol 418(Pt 3):240–244

Huang Y, Cole SPC, Cai T, Cai YU (2016) Applications of nanoparticle drug delivery systems for the reversal of multidrug resistance in cancer. Oncol Lett 12:11–15

Huang W, Lang Y, Hakeem A, Lei Y, Gan L, Yang X (2018) Surfactin-based nanoparticles loaded with doxorubicin to overcome multidrug resistance in cancers. Int J Nanomedicine 13:1723–1736

Hussain Z, Khan JA, Murtaza S (2018) Nanotechnology: an emerging therapeutic option for breast cancer. Crit Rev Eukaryot Gene Expr 28:163–175

Isakoff SJ (2010) Triple-negative breast cancer: role of specific chemotherapy agents. Cancer J 16:53–61

Ismail-Khan R, Bui MM (2010) A review of triple-negative breast cancer. Cancer Control 17:173–176

Iyer AK, Khaled G, Fang J, Maeda H (2006) Exploiting the enhanced permeability and retention effect for tumor targeting. Drug Discov Today 11:812–818

Jafari R, Majidi Zolbanin N, Majidi J, Atyabi F, Yousefi M, Jadidi-Niaragh F, Aghebati-Maleki L, Shanehbandi D, Soltani Zangbar MS, Rafatpanah H (2019) Anti-Mucin1 aptamer-conjugated chitosan nanoparticles for targeted co-delivery of docetaxel and IGF-1R siRNA to SKBR3 metastatic breast cancer cells. Iran Biomed J 23:21–33

Kipp JE (2004) The role of solid nanoparticle technology in the parenteral delivery of poorly water-soluble drugs. Int J Pharm 284:109–122

Kizilboga T, Baskale EA, Yildiz J, Akcay IM, Zemheri E, Can ND, Ozden C, Demir S, Ezberci F, Dinler-Doganay G (2019) Bag-1 stimulates Bad phosphorylation through activation of Akt and Raf kinases to mediate cell survival in breast cancer. BMC Cancer 19:1254

Koike Y, Ohta Y, Saitoh W, Yamashita T, Kanomata N, Moriya T, Kurebayashi J (2017) Anti-cell growth and anti-cancer stem cell activities of the non-canonical hedgehog inhibitor GANT61 in triple-negative breast cancer cells. Breast Cancer 24:683–693

Koziara JM, Lockman PR, Allen DD, Mumper RJ (2004) Paclitaxel nanoparticles for the potential treatment of brain tumors. J Control Release 99:259–269

Koziara JM, Whisman TR, Tseng MT, Mumper RJ (2006) In-vivo efficacy of novel paclitaxel nanoparticles in paclitaxel-resistant human colorectal tumors. J Control Release 112:312–319

Kubota T, Kuroda S, Kanaya N, Morihiro T, Aoyama K, Kakiuchi Y, Kikuchi S, Nishizaki M, Kagawa S, Tazawa H, Fujiwara T (2018) HER2-targeted gold nanoparticles potentially overcome resistance to trastuzumab in gastric cancer. Nanomedicine 14:1919–1929

Kuo MT (2007) Roles of multidrug resistance genes in breast cancer chemoresistance. Adv Exp Med Biol 608:23–30

Lange CA, Yee D (2008) Progesterone and breast cancer. Womens Health (Lond) 4:151–162

Leclercq G, Lacroix M, Laïos I, Laurent G (2006) Estrogen receptor alpha: impact of ligands on intracellular shuttling and turnover rate in breast cancer cells. Curr Cancer Drug Targets 6:39–64

Lee O, Choi MR, Christov K, Ivancic D, Khan SA (2016) Progesterone receptor antagonism inhibits progestogen-related carcinogenesis and suppresses tumor cell proliferation. Cancer Lett 376:310–317

Leong KG, Niessen K, Kulic I, Raouf A, Eaves C, Pollet I, Karsan A (2007) Jagged1-mediated Notch activation induces epithelial-to-mesenchymal transition through Slug-induced repression of E-cadherin. J Exp Med 204:2935–2948

Li J, Xu W, Yuan X, Chen H, Song H, Wang B, Han J (2017) Polymer-lipid hybrid anti-HER2 nanoparticles for targeted salinomycin delivery to HER2-positive breast cancer stem cells and cancer cells. Int J Nanomedicine 12:6909–6921

Li H, Xu X, Liu Y, Li S, Zhang D, Meng X, Lu L, Li Y (2018) MMP7 induces T-DM1 resistance and leads to the poor prognosis of gastric adenocarcinoma via a DKK1-dependent manner. Anti Cancer Agents Med Chem 18:2010–2016

Liang Y, Goyette S, Hyder SM (2017) Cholesterol biosynthesis inhibitor RO 48-8071 reduces progesterone receptor expression and inhibits progestin-dependent stem cell-like cell growth in hormone-dependent human breast cancer cells. Breast Cancer (Dove Med Press) 9:487–494

Liao D, Zhang W, Gupta P, Lei Z-N, Wang J-Q, Cai C-Y, Vera AAD, Zhang L, Chen Z-S, Yang D-H (2019) Tetrandrine interaction with ABCB1 reverses multidrug resistance in Cancer cells through competition with anti-Cancer drugs followed by downregulation of ABCB1 expression. Molecules (Basel, Switzerland) 24:4383

Liedtke C, Kolberg HC (2016) Systemic therapy of advanced/metastatic breast cancer – current evidence and future concepts. Breast Care (Basel) 11:275–281

Lippert TH, Ruoff HJ, Volm M (2008) Intrinsic and acquired drug resistance in malignant tumors. The main reason for therapeutic failure. Arzneimittelforschung 58:261–264

Longley DB, Johnston PG (2005) Molecular mechanisms of drug resistance. J Pathol 205:275–292

Lu C, Shervington A (2008) Chemoresistance in gliomas. Mol Cell Biochem 312:71–80

Lumachi F, Brunello A, Maruzzo M, Basso U, Basso SM (2013) Treatment of estrogen receptor-positive breast cancer. Curr Med Chem 20:596–604

Lumachi F, Santeufemia DA, Basso SM (2015) Current medical treatment of estrogen receptor-positive breast cancer. World J Biol Chem 6:231–239

Luo Y, Ellis LZ, Dallaglio K, Takeda M, Robinson WA, Robinson SE, Liu W, Lewis KD, Mccarter MD, Gonzalez R, Norris DA, Roop DR, Spritz RA, Ahn NG, Fujita M (2012) Side population cells from human melanoma tumors reveal diverse mechanisms for chemoresistance. J Invest Dermatol 132:2440–2450

Magnifico A, Albano L, Campaner S, Delia D, Castiglioni F, Gasparini P, Sozzi G, Fontanella E, Menard S, Tagliabue E (2009) Tumor-initiating cells of HER2-positive carcinoma cell lines express the highest oncoprotein levels and are sensitive to trastuzumab. Clin Cancer Res 15:2010–2021

Maki S, Konno T, Maeda H (1985) Image enhancement in computerized tomography for sensitive diagnosis of liver cancer and semiquantitation of tumor selective drug targeting with oily contrast medium. Cancer 56:751–757

Mamaeva V, Rosenholm JM, Bate-Eya LT, Bergman L, Peuhu E, Duchanoy A, Fortelius LE, Landor S, Toivola DM, Lindén M, Sahlgren C (2011) Mesoporous silica nanoparticles as drug delivery systems for targeted inhibition of Notch signaling in cancer. Mol Ther 19:1538–1546

Matsumura Y, Maeda H (1986) A new concept for macromolecular therapeutics in cancer chemotherapy: mechanism of tumoritropic accumulation of proteins and the antitumor agent smancs. Cancer Res 46:6387–6392

Matsunaga S, Shuto T, Sato M (2016) Gamma knife surgery for metastatic brain tumors from gynecologic cancer. World Neurosurg 89:455–463

Matter M, Dusmet M, Chevalley F (2000) The place of surgery in the treatment of advanced localized, recurrent and metastatic breast cancer. Rev Med Suisse Romande 120:485–490

Mcarthur HL, Mahoney KM, Morris PG, Patil S, Jacks LM, Howard J, Norton L, Hudis CA (2011) Adjuvant trastuzumab with chemotherapy is effective in women with small, node-negative, HER2-positive breast cancer. Cancer 117:5461–5468

Mcgee S (2010) Understanding metastasis: current paradigms and therapeutic challenges in breast cancer progression. RCSI SMJ Rev 3:56–60

Miele L, Golde T, Osborne B (2006) Notch signaling in cancer. Curr Mol Med 6:905–918

Moasser MM (2007) The oncogene HER2: its signaling and transforming functions and its role in human cancer pathogenesis. Oncogene 26:6469–6487

Mohanty C, Das M, Kanwar JR, Sahoo SK (2011) Receptor mediated tumor targeting: an emerging approach for cancer therapy. Curr Drug Deliv 8:45–58

Mondal L, Mukherjee B, Das K, Bhattacharya S, Dutta D, Chakraborty S, Pal MM, Gaonkar RH, Debnath MC (2019) CD-340 functionalized doxorubicin-loaded nanoparticle induces apoptosis and reduces tumor volume along with drug-related cardiotoxicity in mice. Int J Nanomedicine 14:8073–8094

Muhamad N, Plengsuriyakarn T, Na-Bangchang K (2018) Application of active targeting nanoparticle delivery system for chemotherapeutic drugs and traditional/herbal medicines in cancer therapy: a systematic review. Int J Nanomedicine 13:3921–3935

Murphy CT, Li T, Wang LS, Obeid EI, Bleicher RJ, Eastwick G, Johnson ME, Hayes SB, Weiss SE, Anderson PR (2015) Comparison of adjuvant radiation therapy alone versus radiation therapy and endocrine therapy in elderly women with early-stage, hormone receptor-positive breast cancer treated with breast-conserving surgery. Clin Breast Cancer 15:381–389

Murray S, Briasoulis E, Linardou H, Bafaloukos D, Papadimitriou C (2012) Taxane resistance in breast cancer: mechanisms, predictive biomarkers and circumvention strategies. Cancer Treat Rev 38:890–903

Nagamitsu A, Greish K, Maeda H (2009) Elevating blood pressure as a strategy to increase tumor-targeted delivery of macromolecular drug SMANCS: cases of advanced solid tumors. Jpn J Clin Oncol 39:756–766

Naujokat C, Steinhart R (2012) Salinomycin as a drug for targeting human cancer stem cells. J Biomed Biotechnol 2012:950658

Nayak A, Satapathy SR, Das D, Siddharth S, Tripathi N, Bharatam PV, Kundu C (2016) Nanoquinacrine induced apoptosis in cervical cancer stem cells through the inhibition of hedgehog-GLI1 cascade: role of GLI-1. Sci Rep 6:20600

Neophytou C, Boutsikos P, Papageorgis P (2018) Molecular mechanisms and emerging therapeutic targets of triple-negative breast cancer metastasis. Front Oncol 8:31

Nikolaou M, Pavlopoulou A, Georgakilas AG, Kyrodimos E (2018) The challenge of drug resistance in cancer treatment: a current overview. Clin Exp Metastasis 35:309–318

Normanno N, DI Maio M, De Maio E, De Luca A, De Matteis A, Giordano A, Perrone F (2005) Mechanisms of endocrine resistance and novel therapeutic strategies in breast cancer. Endocr Relat Cancer 12:721–747

O'Shaughnessy J (2005) Extending survival with chemotherapy in metastatic breast cancer. Oncologist 10 (Suppl 3):20–29

O'toole SA, Beith JM, Millar EK, West R, Mclean A, Cazet A, Swarbrick A, Oakes SR (2013) Therapeutic targets in triple negative breast cancer. J Clin Pathol 66:530–542

Oltersdorf T, Elmore SW, Shoemaker AR, Armstrong RC, Augeri DJ, Belli BA, Bruncko M, Deckwerth TL, Dinges J, Hajduk PJ, Joseph MK, Kitada S, Korsmeyer SJ, Kunzer AR, Letai A, Li C, Mitten MJ, Nettesheim DG, Ng S, Nimmer PM, O'connor JM, Oleksijew A, Petros AM, Reed JC, Shen W, Tahir SK, Thompson CB, Tomaselli KJ, Wang B, Wendt MD, Zhang H, Fesik SW, Rosenberg SH (2005) An inhibitor of Bcl-2 family proteins induces regression of solid tumours. Nature 435:677–681

Onyebuchi C, Kavaz D (2019) Chitosan and N, N, N-Trimethyl chitosan nanoparticle encapsulation of Ocimum Gratissimum essential oil: optimised synthesis, in vitro release and bioactivity. Int J Nanomedicine 14:7707–7727

Osborne CK, Wakeling A, Nicholson RI (2004) Fulvestrant: an oestrogen receptor antagonist with a novel mechanism of action. Br J Cancer 90(Suppl 1): S2–S6

Ould-Ouali L, Noppe M, Langlois X, Willems B, Te Riele P, Timmerman P, Brewster ME, Ariën A, Préat V (2005) Self-assembling PEG-p(CL-co-TMC) copolymers for oral delivery of poorly water-soluble drugs: a case study with risperidone. J Control Release 102:657–668

Palmieri C, Patten DK, Januszewski A, Zucchini G, Howell SJ (2014) Breast cancer: current and future endocrine therapies. Mol Cell Endocrinol 382:695–723

Park JH, Gu L, Von Maltzahn G, Ruoslahti E, Bhatia SN, Sailor MJ (2009) Biodegradable luminescent porous silicon nanoparticles for in vivo applications. Nat Mater 8:331–336

Parvani JG, Gujrati MD, Mack MA, Schiemann WP, Lu ZR (2015) Silencing β3 integrin by targeted ECO/siRNA nanoparticles inhibits EMT and metastasis of triple-negative breast cancer. Cancer Res 75:2316–2325

Peer D, Karp JM, Hong S, Farokhzad OC, Margalit R, Langer R (2007) Nanocarriers as an emerging platform for cancer therapy. Nat Nanotechnol 2:751–760

Perou CM, Sorlie T, Eisen MB, Van De Rijn M, Jeffrey SS, Rees CA, Pollack JR, Ross DT, Johnsen H, Akslen LA, Fluge O, Pergamenschikov A, Williams C, Zhu SX, Lonning PE, Borresen-Dale AL, Brown PO, Botstein D (2000) Molecular portraits of human breast tumours. Nature 406:747–752

Pohl S-G, Brook N, Agostino M, Arfuso F, Kumar AP, Dharmarajan A (2017) Wnt signaling in triple-negative breast cancer. Oncogenesis 6:e310–e310

Ranganathan P, Weaver KL, Capobianco AJ (2011) Notch signalling in solid tumours: a little bit of everything but not all the time. Nat Rev Cancer 11:338–351

Rosenblum D, Peer D (2014) Omics-based nanomedicine: the future of personalized oncology. Cancer Lett 352:126–136

Rosenblum D, Joshi N, Tao W, Karp JM, Peer D (2018) Progress and challenges towards targeted delivery of cancer therapeutics. Nat Commun 9:1410

Sada R, Kimura H, Fukata Y, Fukata M, Yamamoto H, Kikuchi A (2019) Dynamic palmitoylation controls the microdomain localization of the DKK1 receptors CKAP4 and LRP6. Sci Signal 12:eaat9519

Sakil HAM, Stantic M, Wolfsberger J, Brage SE, Hansson J, Wilhelm MT (2017) ΔNp73 regulates the expression of the multidrug-resistance genes ABCB1 and ABCB5 in breast cancer and melanoma cells – a short report. Cell Oncol (Dordr) 40:631–638

Samadi P, Saki S, Dermani FK, Pourjafar M, Saidijam M (2018) Emerging ways to treat breast cancer: will promises be met? Cell Oncol (Dordr) 41:605–621

Saraswathy M, Gong S (2014) Recent developments in the co-delivery of siRNA and small molecule anticancer drugs for cancer treatment. Mater Today 17:298–306

Schroeder RL, Stevens CL, Sridhar J (2014) Small molecule tyrosine kinase inhibitors of ErbB2/HER2/Neu in the treatment of aggressive breast cancer. Molecules 19:15196–15212

Shao S, Zhao X, Zhang X, Luo M, Zuo X, Huang S, Wang Y, Gu S, Zhao X (2015) Notch1 signaling regulates the epithelial-mesenchymal transition and invasion of breast cancer in a Slug-dependent manner. Mol Cancer 14:28

Sharma R, Sharma R, Khaket TP, Dutta C, Chakraborty B, Mukherjee TK (2017) Breast cancer metastasis: putative therapeutic role of vascular cell adhesion molecule-1. Cell Oncol (Dordr) 40:199–208

Sharma D, Kumar S, Narasimhan B (2018) Estrogen alpha receptor antagonists for the treatment of breast cancer: a review. Chem Cent J 12:107

Shi J, Kantoff PW, Wooster R, Farokhzad OC (2017) Cancer nanomedicine: progress, challenges and opportunities. Nat Rev Cancer 17:20–37

Shimamura H, Terada Y, Okado T, Tanaka H, Inoshita S, Sasaki S (2003) The PI3-kinase-Akt pathway promotes mesangial cell survival and inhibits apoptosis in vitro via NF-kappa B and Bad. J Am Soc Nephrol 14:1427–1434

Shioi Y, Kashiwaba M, Inaba T, Komatsu H, Sugai T, Wakabayashi G (2014) Long-term complete remission of metastatic breast cancer, induced by a steroidal aromatase inhibitor after failure of a non-steroidal aromatase inhibitor. Am J Case Rep 15:85–89

Singh R, Vyas SP (1996) Topical liposomal system for localized and controlled drug delivery. J Dermatol Sci 13:107–111

Skoda AM, Simovic D, Karin V, Kardum V, Vranic S, Serman L (2018) The role of the hedgehog signaling pathway in cancer: a comprehensive review. Bosn J Basic Med Sci 18:8–20

Sotiriou C, Neo SY, Mcshane LM, Korn EL, Long PM, Jazaeri A, Martiat P, Fox SB, Harris AL, Liu ET (2003) Breast cancer classification and prognosis based on gene expression profiles from a population-based study. Proc Natl Acad Sci U S A 100:10393–10398

Spector NL, Blackwell KL (2009) Understanding the mechanisms behind trastuzumab therapy for human epidermal growth factor receptor 2-positive breast cancer. J Clin Oncol 27:5838–5847

Sun W, Shang J, Zhang J, Chen S, Hao M (2019) Correlations of DKK1 with incidence and prognosis of breast cancer. J BUON 24:26–32

Surapaneni SK, Bashir S, Tikoo K (2018) Gold nanoparticles-induced cytotoxicity in triple negative breast cancer involves different epigenetic alterations depending upon the surface charge. Sci Rep 8:12295

Tang Y, Wang Y, Kiani MF, Wang B (2016) Classification, treatment strategy, and associated drug resistance in breast cancer. Clin Breast Cancer 16:335–343

Tang X, Loc WS, Dong C, Matters GL, Butler PJ, Kester M, Meyers C, Jiang Y, Adair JH (2017) The use of nanoparticulates to treat breast cancer. Nanomedicine 12:2367–2388

Thottassery JV, Zambetti GP, Arimori K, Schuetz EG, Schuetz JD (1997) p53-dependent regulation of MDR1 gene expression causes selective resistance to chemotherapeutic agents. Proc Natl Acad Sci U S A 94:11037–11042

Tiash S, Chowdhury EH (2019) siRNAs targeting multidrug transporter genes sensitise breast tumour to doxorubicin in a syngeneic mouse model. J Drug Target 27:325–337

Tiwari G, Tiwari R, Sriwastawa B, Bhati L, Pandey S, Pandey P, Bannerjee SK (2012) Drug delivery systems: an updated review. Int J Pharm Invest 2:2–11

Umar H, Kavaz D, Rizaner N (2018) Biosynthesis of zinc oxide nanoparticles using Albizia lebbeck stem bark, and evaluation of its antimicrobial, antioxidant, and cytotoxic activities on human breast cancer cell lines. Int J Nanomedicine 14:87–100

Valcourt DM, Dang MN, Scully MA, Day ES (2020) Nanoparticle-mediated co-delivery of Notch-1 antibodies and ABT-737 as a potent treatment strategy for triple-negative breast cancer. ACS Nano 14:3378–3388

Verderio P, Pandolfi L, Mazzucchelli S, Marinozzi MR, Vanna R, Gramatica F, Corsi F, Colombo M, Morasso C, Prosperi D (2014) Antiproliferative effect of ASC-J9 delivered by PLGA nanoparticles against estrogen-dependent breast Cancer cells. Mol Pharm 11:2864–2875

Wagenfeld A, Saunders PT, Whitaker L, Critchley HO (2016) Selective progesterone receptor modulators (SPRMs): progesterone receptor action, mode of action on the endometrium and treatment options in gynecological therapies. Expert Opin Ther Targets 20:1045–1054

Wang J, Xu B (2019) Targeted therapeutic options and future perspectives for HER2-positive breast cancer. Signal Transduct Target Ther 4:34

Wang Y, Xie Y, Kilchrist KV, Li J, Duvall CL, Oupický D (2020) Endosomolytic and tumor-penetrating mesoporous silica nanoparticles for siRNA/miRNA combination cancer therapy. ACS Appl Mater Interfaces 12:4308–4322

Weigelt B, Peterse JL, van't Veer LJ (2005) Breast cancer metastasis: markers and models. Nat Rev Cancer 5:591–602

Wilhelm MT, Rufini A, Wetzel MK, Tsuchihara K, Inoue S, Tomasini R, Itie-Youten A, Wakeham A, Arsenian-Henriksson M, Melino G, Kaplan DR, Miller FD, Mak TW (2010) Isoform-specific p73 knockout mice reveal a novel role for delta Np73 in the DNA damage response pathway. Genes Dev 24:549–560

Xiao YS, Zeng D, Liang YK, Wu Y, Li MF, Qi YZ, Wei XL, Huang WH, Chen M, Zhang GJ (2019) Major vault protein is a direct target of Notch1 signaling and contributes to chemoresistance in triple-negative breast cancer cells. Cancer Lett 440–441:156–167

Yang L, Shi P, Zhao G, Xu J, Peng W, Zhang J, Zhang G, Wang X, Dong Z, Chen F, Cui H (2020) Targeting cancer stem cell pathways for cancer therapy. Signal Transduct Target Ther 5:8

Yao J, Yao X, Tian T, Fu X, Wang W, Li S, Shi T, Suo A, Ruan Z, Guo H, Nan K, Huo X (2017) ABCB5-ZEB1 Axis promotes invasion and metastasis in breast cancer cells. Oncol Res 25:305–316

Yin P, Wang W, Zhang Z, Bai Y, Gao J, Zhao C (2018) Wnt signaling in human and mouse breast cancer: focusing on Wnt ligands, receptors and antagonists. Cancer Sci 109:3368–3375

Yousefi M, Nosrati R, Salmaninejad A, Dehghani S, Shahryari A, Saberi A (2018) Organ-specific metastasis of breast cancer: molecular and cellular mechanisms underlying lung metastasis. Cell Oncol (Dordr) 41:123–140

Yu L, Yang Y, Hou J, Zhai C, Song Y, Zhang Z, Qiu L, Jia X (2015) MicroRNA-144 affects radiotherapy sensitivity by promoting proliferation, migration and invasion of breast cancer cells. Oncol Rep 34:1845–1852

Zhang N, Yin Y, Xu SJ, Chen WS (2008) 5-Fluorouracil: mechanisms of resistance and reversal strategies. Molecules 13:1551–1569

Zhang X, Li X, You Q, Zhang X (2017) Prodrug strategy for cancer cell-specific targeting: a recent overview. Eur J Med Chem 139:542–563

Zhang T, Jiang Z, Xve T, Sun S, Li J, Ren W, Wu A, Huang P (2019) One-pot synthesis of hollow PDA@DOX nanoparticles for ultrasound imaging and chemo-thermal therapy in breast cancer. Nanoscale 11:21759–21766

Zhang L, Jing D, Jiang N, Rojalin T, Baehr CM, Zhang D, Xiao W, Wu Y, Cong Z, Li JJ, Li Y, Wang L, Lam KS (2020) Transformable peptide nanoparticles arrest HER2 signalling and cause cancer cell death in vivo. Nat Nanotechnol 15:145–153

Zuo ZQ, Chen KG, Yu XY, Zhao G, Shen S, Cao ZT, Luo YL, Wang YC, Wang J (2016) Promoting tumor penetration of nanoparticles for cancer stem cell therapy by TGF-β signaling pathway inhibition. Biomaterials 82:48–59

Adv Exp Med Biol - Cell Biology and Translational Medicine (2021) 14: 183–195
https://doi.org/10.1007/5584_2021_638

Published online: 8 May 2021

Telomere Length and Oxidative Stress in Patients with ST-Segment Elevation and Non-ST-Segment Elevation Myocardial Infarction

Nihal Inandiklioğlu, Vahit Demir, and Müjgan Ercan

Abstract

Purpose The telomere length is shown to act as a biomarker, especially for biological aging and cardiovascular diseases, and it is also suggested that with this correlation, increased exposure to the oxidative stress accelerates the vascular aging process. Therefore, this study aims to understand the correlation between the plasma oxidative stress index (OSI) status and leukocyte telomere length (LTL) and cardiologic parameters between the ST-segment elevation myocardial infarction (STEMI) and non-ST-segment elevation myocardial infarction (NSTEMI) groups.

Method One hundred one newly diagnosed patients with STEMI (n = 55) and NSTEMI (n = 46) were included in the study, along with 100 healthy controls who matched the patients in terms of age and gender. Plasma total antioxidant status (TAS), total oxidant status (TOS), and LTL were measured.

Results When LTL, TAS, TOS, and OSI values were evaluated between the patient and control group, OSI ($p = 0.000$) and LTL ($p = 0.05$) values were statistically significant in the patient group compared to the control group. Evaluation was conducted to understand whether there is a difference between the STEMI and NSTEMI groups. The plasma OSI ($p = 0.007$) and LTL ($p = 0.05$) were found to be significantly lower in STEMI patients. However, LTL and OSI results were not statistically significant in NSTEMI patients.

Conclusion This is the first study evaluating telomere length and oxidative stress in STEMI and NSTEMI patients in Turkey. Our results support the existence of short telomere length in STEMI patients. Future studies on telomere length and oxidative stress will support the importance of our findings.

Keywords

Acute myocardial infarction · Non-ST-segment elevation myocardial infarction · Oxidative stress · ST-segment elevation myocardial infarction · Telomere length

N. Inandiklioğlu (✉)
Department of Medical Biology, Faculty of Medicine, Yozgat Bozok University, Yozgat, Turkey
e-mail: nihal.inandiklioglu@yobu.edu.tr

V. Demir
Department of Cardiology, Faculty of Medicine, Yozgat Bozok University, Yozgat, Turkey

M. Ercan
Department of Medical Biochemistry, Faculty of Medicine, Yozgat Bozok University, Yozgat, Turkey

1 Introduction

Acute myocardial infarction (AMI) is one of the leading causes of death in the world. The majority of deaths are associated with acute coronary syndromes (ACS) and their complications. In particular, circulatory system diseases account for 31.9% of total death cases (Turkish Statistical Institute 2018). Patients with myocardial infarction due to ACS are classified according to their electrocardiographic data as ST-segment elevation myocardial infarction or non-ST-segment elevation myocardial infarction. STEMI is a clinical condition that develops as a result of total occlusion of the coronary artery. If early revascularization cannot be achieved after the total occlusion of the thrombus and the ischemia resulting from it, a myocardial necrosis occurs, which spreads from the endocardium to the epicardial area, starting from the 20th min. NSTEMI is a condition in which myocardial ischemia occurs, but myocardial damage and necrosis are limited due to temporary or partial occlusion (Hamm et al. 2011).

Various factors such as age, gender, hypercholesterolemia, hypertension, diabetes mellitus (DM), family history, and smoking are known to increase the risk of ACS. Furthermore, studies have demonstrated that oxidative stress and telomere length are associated with increased cardiovascular risk and mortality (Yeh and Wang 2016; Siti et al. 2015). Telomeres are noncoding tandem repeat DNA sequences (TTAGGG), and their associated proteins are found at the ends of chromosomes. They protect chromosomes against degradation and against the loss of genetic material during cell division (Riethman 2008). Shorter telomere length is linked to age-related diseases such as cardiovascular disease (CVD), DM, hypertension, and cancer. Thus, it has been suggested that telomere length contributes to mortality in many age-related diseases and is a predictor of mortality (Yeh and Wang 2016; Haycock et al. 2014). Oxidative stress causes telomere loss in each cell division, and oxidative damage is less likely to be repaired in telomere DNA than in any part of the chromosome (Von Zglinicki 2002). Some systemic markers of oxidative stress predict the clinical outcome in coronary artery disease. Systemic oxidative stress and inflammation associated with cardiovascular risk can accelerate telomere shortening. Experimental studies suggest that oxidative stress shortens TL in vitro and some studies support the use of TL as an independent marker of MI, but it is not clear whether the blood TL is tissue-specific or whether it is a global biomarker of oxidative stress in acute myocardial infarction (Siti et al. 2015; Von Zglinicki 2002; Margaritis et al. 2017). Although various studies have been carried on telomere length in CVD, only a few studies evaluated the relationship between the effect of leukocyte telomere length (LTL), plasma total antioxidant status (TAS), and total oxidant status (TOS) on clinical data in STEMI and NSTEMI patient groups (Haycock et al. 2014; Margaritis et al. 2017). In this study, we aimed to evaluate the results of telomere length along with the oxidative stress index (OSI) and clinical findings in STEMI and NSTEMI patient groups compared to healthy controls by testing the clinical predictive value of short telomere length in AMI risk groups.

2 Method

2.1 Ethical and Legal Aspects of the Research

The study was performed according to the principles of the Declaration of Helsinki. The ethical trial of this study was approved by the local ethics committee (document no: 2017-06/07). Informed consent was obtained from the patients and controls taking part in our research.

2.2 Study Population

In this case-control study, 101 patients were included, who were diagnosed with STEMI ($n = 55$) and NSTEMI ($n = 46$) at ages ranging from 43 to 67 years old. The control group was composed of 100 healthy volunteers who were age- and sex-matched to chosen patients. Special attention was given to choose controls that did not have any previous CVD history. The STEMI and NSTEMI were diagnosed according to clinical guidelines previously reported (Thygesen et al. 2018). STEMI was defined in the new presence of ST elevation at the J-point with at least two contiguous leads (2.5 mm or more in men younger than 40 years of age, of 2 mm or more in men aged 40 years or older, or of 1.5 mm or more in women in leads V2 to V3 and/or of 1 mm) or in the new left bundle branch block seen in an electrocardiogram (ECG). NSTEMI was diagnosed by a rise or fall of troponin I and absence of persistent ST elevation seen via an ECG. All patients received standard care and underwent primary percutaneous coronary intervention within 2 h after being admitted to the hospital. Thus, use of dual therapy aspirin and a P2Y12 receptor inhibitor (clopidogrel, prasugrel, or ticagrelor) were available for all patients. We conducted clear exclusion criteria for patients by considering the following cases: active infection or chronic inflammatory disease, received fibrinolytic therapy, previous myocardial infarction, malignancy, major surgery required due to coronary artery bypass, chronic obstructive pulmonary disease, renal dysfunction, or hepatic disease. We performed transthoracic echocardiographic examination using an echocardiography device (Philips Affiniti 50, Philips Healthcare, the Netherlands) according to the recommendations of the American Society of Echocardiography (Baumgartner et al. 2017). We determined the left ventricular ejection fraction (LVEF) using the modified Simpson method (Horwitz 2001). BMI was calculated by dividing the body weight by the square of the neck (kg/m^2). All blood values of the participants were examined using the blood taken at the initial diagnosis period.

2.3 Coronary Angiography

Coronary angiography was performed through the right femoral arterial route using the standard Judkins method (Judkins 1968). We visualized coronary arteries using cranial and caudal angles in the right and left oblique plane. Iopromide (Ultravist 370, Schering AG, Berlin, Germany) was used as the contrast agent for coronary angiography. Stenosis and lesion length in the coronary arteries were evaluated using the Digital Imaging and Communications in Medicine (DICOM) program. The SYNTAX scores were used to estimate extent and complexity of coronary lesions and were provided for all patients. The web-based program (www.syntaxscore.com) was used to calculate the SYNTAX score. The severity of coronary lesions was calculated using the Gensini score as a second method (Gensini 1983).

2.4 Measurement of TAS and TOS

The blood samples were collected from the femoral artery at initial diagnosis period before the coronary angiography. Plasma samples were collected during the study and stored at −80 °C. Plasma TAS and TOS levels were measured by Erel's modification methods using commercially available kits (Rel Assay, Turkey). In serum samples, the normal value range for TAS was 1.20–1.50 mmol/L, and the normal value range for TOS was 4.00–6.00 μmol/L. The ratio of TOS to TAS was accepted as the oxidative stress index (OSI). For calculation, the resulting unit of TAS was converted to μmol/L, and the OSI value was calculated according to the following formula, OSI (arbitrary unit-AU) = TOS (μmol H_2O_2 equivalent/L)/TAS (μmol Trolox equivalent/L) (Erel 2004, 2005).

2.5 Measurement of Telomere Length

The blood samples were collected from the femoral artery at the time of initial diagnosis before coronary angiography for an accurate evaluation of biological variables and stored at −80 °C. Genomic DNA was isolated from peripheral blood using the QIAamp DNA kit (QIAGEN, Germany). LTL was measured via the real-time quantitative polymerase chain reaction (RT-PCR) method based on the original study (Cawthon 2002). DNA concentrations of the samples were measured with a fluorometer (QFX, DeNovix, USA). DNA concentrations equaled 5 ng/μl in all samples. The relative telomere length was calculated as the ratio of telomere repeat signal (T) to single-copy gene, 36B4 (acidic ribosomal phosphoprotein P0, S), and copies (T/S ratio). The number of telomere repeats and the quantity of single-copy gene copies for each sample were defined in comparison to a reference sample in a telomere and a single-copy gene quantitative PCR, respectively. The resulting T/S ratio was proportional to the mean LTL. The detailed method is described in another study where T/S ratios were converted into base pairs (bp) with the following formula: bp = 3,274 + 24,133([T/S-0.0545]/1.16) (Verhoeven et al. 2014).

2.6 Statistical Analysis

The statistical analysis was performed using SPSS version 21.0 software (SPSS Inc., ILL Company, USA). Continuous variables were tested for normality using the Kolmogorov-Smirnov test. Continuous data were presented as mean ± standard deviation (SD) or mean (% 95 confidence interval). The Student's t-test was used to compare continuous variables with normal distributions. The Kruskal-Wallis test was used to compare variables with abnormal distributions, and the Pearson (for data with normal distribution) and Spearman correlation analyses (for data not showing normal distribution) were used for correlation analysis. Changes of LTL values, T/S ratio, and TAS, TOS, and OSI levels between groups were tested with one-way ANOVA and post hoc (Tukey) tests for multiple comparisons. Multiple linear regression analysis was conducted with the LTL and T/S ratio as a dependent variable baseline characteristics, and outcome parameters as independent variables. P-values of <0.05 were considered statistically significant.

3 Results

3.1 Characteristics of the Study Population and Telomere Length

A total number of 101 patients who were diagnosed with STEMI and NSTEMI and a total of 100 controls were enrolled in the study. The demographic, biochemical, and clinical data of the patient groups and control group were statistically compared (Table 1). Active smoking ($p = 0.006$), glucose ($p = 0.000$), urea ($p = 0.000$), creatinine ($p = 0.009$), triglyceride ($p = 0.001$), alanine aminotransferase (ALT) ($p = 0.003$), aspartate aminotransferase (AST) ($p = 0.009$), CRP ($p = 0.009$), white blood cell (WBC) ($p = 0.002$), left ventricular ejection fraction (LVEF) ($p = 0.001$), and aorta diameter ($p = 0.016$) values were statistically significant in the patient group compared to the control group. SYNTAX and Gensini scores were also evaluated within the patient groups. A SYNTAX score of 0–22 is considered to be low, 23–32 as medium, and 33 and above as high (Neumann et al. 2019). According to the SYNTAX score, both groups were in the low-risk group. The Gensini score of the patient groups was calculated as 48.2 in NSTEMI and 65.6 in STEMI. Gensini score was found to be higher in STEMI. When the two groups were compared, both SYNTAX ($p = 0.016$) and Gensini ($p = 0.004$) scores were found to be significantly higher in the STEMI group. When TAS, TOS, OSI, and LTL values were compared between the patient and control groups, LTL ($p = 0.05$) and OSI ($p = 0.000$) values were found to be significant in the patient group. When the correlation data was examined in the patient group, a strong and positive correlation was detected between LTL and T/S with TOS ($p = 0.008$, $r = 0.212$; $p = 0.008$, $r = 0.213$), OSI ($p = 0.045$, $r = 0.161$; $p = 0.047$, $r = 0.160$), and troponin I ($p = 0.039$, $r = 0.206$; $p = 0.039$, $r = 0.205$), respectively (Table 2). For the NSTEMI and STEMI groups, the results of the regression analysis regarding the comparison of the TAS, TOS, OSI, LTL, and T/S ratio values to demographic findings of the groups are presented in Table 3.

Table 1 Participants' demographic characteristics

	NSTEMI patients	STEMI patients	Control group	p value
Participants (n)	46	55	100	
Risk factors				
Age (years)	58.7 ± 8.9	57.3 ± 9.3	57.07 ± 9.5	0.185
Male gender (%)	58.7	69.1	46.3	0.054
BMI (kg/m^2)	24.6 ± 2.9	23.4 ± 2.9	23.1 ± 2.6	0.191
Diabetes mellitus (%)	37.0	41.8	33.3	0.656
Hypertension (%)	69.6	54.5	50	0.124
Hyperlipidemia (%)	41.3	49.1	33.3	0.248
Active smoking (%)	43.5	52.7	24.1	0.006[a]
Family history (%)	50.0	56.4	33.3	0.022[a]
Systolic blood pressure (mmHg)	131.3 ± 15.6	129.8 ± 18.5	127.6 ± 12.4	0.510
Diastolic blood pressure (mmHg)	79.2 ± 9.8	80.7 ± 15.5	79.0 ± 5.6	0.285
Heart rate (beats per minute)	80.2 ± 10.5	77.2 ± 6.6	78.7 ± 7.4	0.324
Biological data on admission				
Glucose (mg/dl)	143.2 ± 64.1	151.8 ± 66.3	110.1 ± 35.6	0.000[a]
Urea (mg/dL)	23.6 ± 7.9	31.2 ± 41.1	16.7 ± 6.1	0.000[a]
Creatinine (mg/dL)	1.3 ± 2.8	1.2 ± 1.5	0.80 ± 0.1	0.009[a]
Total cholesterol (mg/dL)	201.2 ± 41.9	200.7 ± 90.1	199.8 ± 43.5	0.427
Triglycerides (mg/dL)	170.5 ± 93.9	125.6 ± 80.0	165.0 ± 72.2	0.001[a]
HDL cholesterol (mg/dL)	47.1 ± 24.6	47.4 ± 22.8	47.7 ± 10.1	0.103
LDL cholesterol (mg/dL)	117.4 ± 38.7	138.7 ± 116.4	115.6 ± 33.2	0.387
ALT (U/l)	26.9 ± 26.8	27.5 ± 15.1	19.9 ± 10.5	0.003[a]
AST (U/l)	38.5 ± 29.1	58.6 ± 55.8	18.7 ± 7.2	0.000[a]
CRP (mg/L)	11.5 ± 24.4	9.9 ± 18.2	3.7 ± 3.6	0.000[a]
WBC (10^3/mm^3)	9.0 ± 2.4	9.6 ± 2.7	7.9 ± 2.3	0.002[a]
Lymphocytes (10^3/mm^3)	2.4 ± 0.8	2.5 ± 1.8	2.3 ± 0.9	0.632
Hemoglobin (g/dL)	14.0 ± 1.8	14.6 ± 4.8	14.3 ± 1.6	0.661
Hematocrit (%)	41.2 ± 7.8	48.2 ± 46.5	42.7 ± 4.6	0.837
Platelets (10^3/mm^3)	237.9 ± 63.2	250.6 ± 68.1	264.3 ± 58.2	0.131
Troponin I (ng/L)	1775.8 ± 2889.3	5865.7 ± 10730.9	–	
CK (U/L)	136.8 ± 149.5	293.4 ± 439.6	–	
CK-MB (U/L)	41.8 ± 54.3	85.5 ± 109.9	–	
Transthoracic echocardiography values				
LVEF (%)	53.1 ± 6.6	51.2 ± 8.3	56.2 ± 5.2	0.001[a]
Aorta diameter (cm)	2.5 ± 0.5	2.6 ± 0.6	2.3 ± 0.4	0.016[a]
Left atrium diameter (cm)	3.6 ± 0.3	3.6 ± 0.4	3.6 ± 0.3	0.966
LVEDD (cm)	4.6 ± 0.5	4.6 ± 0.7	4.4 ± 0.7	0.064
LVESD (cm)	2.8 ± 0.6	2.9 ± 0.7	3.1 ± 0.7	0.233
IVSd (cm)	1.1 ± 0.1	1.2 ± 0.2	1.1 ± 0.2	0.787
Coronary angiography data				
Number of affected vessels			–	
1 (%)	48.6	30.9		
2 (%)	37.1	40.0		
3 (%)	14.3	29.1		
Culprit vessel			–	
No (%)	4.8	0		
LAD (%)	35.7	47.3		
LCX (%)	31.0	16.4		
RCA (%)	28.6	36.4		

(continued)

Table 1 (continued)

	NSTEMI patients	STEMI patients	Control group	*p* value
Lesion length (mm)	19.8 ± 5.7	23.5 ± 7.0	–	
Reference vessel diameter (mm)	2.98 ± 0.4	2.92 ± 0.45	–	
Lesion location			–	
Proximal (%)	21.7	32.7		
Mid (%)	60.9	49.1		
Distal (%)	17.4	18.2		
Percentage of lesion vessel stenosis (%)	95.76 ± 4.33	99.21 ± 3.05	–	
Gensini score	48.2 ± 32.6	65.6 ± 29.6	–	
SYNTAX score	15.7 ± 6.9	19.6 ± 8.7	–	
Analysis data				
TAS (mmolTroloxEquiv./L)	1.47 ± 0.29	1.51 ± 0.20	1.45 ± 0.26	0.339
TOS (μmol H_2O_2 equiv./L)	5.48 ± 5.17	5.27 ± 4.51	5.34 ± 10.6	0.097
OSI (AU)	0.18 ± 0.39	0.05 ± 0.03	0.39 ± 0.93	0.000[a]
LTL (bp)	5013.30 ± 531.65	4817.72 ± 296.01	5049.92 ± 662.17	0.05[a]
T/S ratio	0.41 ± 0.25	0.32 ± 0.14	0.43 ± 0.32	0.079

Values presented as mean ± standard deviation

ALT alanine aminotransferase, *AST* aspartate aminotransferase, *CK* creatinine kinase, *CK-MB* creatinine kinase MB, *CRP* C-reactive protein, *HCT* hematocrit, *HDL* high-density lipoprotein, *IVSd* interventricular septal thickness, *LAD* left anterior descending artery, *LCX* left circumflex artery, *LDL* low-density lipoprotein, *LTL* leukocyte telomere length, *LVDD* left ventricular end-diastolic volume, *LVEF* left ventricular ejection fraction, *LVSD* left ventricular end-systolic volume, *OSI* oxidative stress index, *PLT* platelets, *RCA* right coronary artery, *TAS* total antioxidant status, *TOS* total oxidant status, *WBC* white blood cell

[a]Demographic, clinical, and laboratory data of the study groups

As seen in Table 3, in the NSTEMI group, there was a statistically significant relationship between both LTL and T/S ratio in heart rate (p = 0.018), urea (p = 0.003), aortic diameter (p = 0.05), Gensini score (p = 0.043), TOS (p = 0.018), and OSI values (p = 0.000). On the other hand, results of the regression analysis of TAS, TOS, and OSI levels clearly stated that there was a statistically significant relationship between certain demographic data. TAS with triglyceride and CRP, TOS with hypertension and glucose, and OSI with gender, heart rate, and hemoglobin were found to be statistically significant (Table 3). After a detailed analysis of regression results relevant to LTL, T/S ratio, and demographic data of the STEMI group, it was found that the correlation between glucose (p = 0.05) and TOS (p = 0.017) values was statistically significant. In addition, TAS, TOS, OSI, and demographic data were evaluated based on regression analysis, and it was observed that the relationship between TAS with triglyceride (p = 0.002) and urea (p = 0.009) and OSI with urea (p = 0.037) were also statistically significant.

3.2 Oxidative Stress and Telomere Length

The STEMI and NSTEMI patients and control groups were compared in terms of TAS, TOS, and OSI values. In ANOVA analysis, a significant difference was found among groups at p = 0.09 confidence level. Multiple comparison tables were utilized to see which groups were statistically different from the others (Tukey Test). A comparison of the groups regarding the OSI value acknowledges that the figure for the STEMI group was statistically significant (p = 0.007) and is given in Table 4. The STEMI and NSTEMI patient groups and the control group were also assessed in terms of T/S ratio and LTL values, and a significant difference was found in the confidence level of p = 0.047 as a result of ANOVA analysis. Multiple comparison tables defined the level of differences across groups. Regarding analysis results (Tukey Test), it was noticed that there was a statistically significant difference between the control group and STEMI group (p = 0.05). When ANOVA

Table 2 Data showing significant correlation with TAS, TOS, OSI, LTL, and T/S ratio in patient group

		TAS	TOS	OSI	LTL	T/S ratio
Systolic blood pressure	r	0.192[a]	0.003	−0.001	−0.014	–,015
	p	0.017	0.966	0.986	0.859	,853
Glucose	r	0.141	0.325[b]	−0.106	0.082	0.85
	p	0.080	0.000	0.191	0.308	0.292
Urea	r	0.205[a]	0.186[a]	−0.079	0.148	0.149
	p	0.010	0.021	0.330	0.065	0.064
Creatinine	r	0.218[b]	0.105	−0.023	0.013	0.013
	p	0.007	0.196	0.773	0.873	0.873
Triglycerides	r	0.277[b]	0.051	0.130	0.107	0.108
	p	0.000	0.531	0.108	0.187	0.181
HDL cholesterol	r	−0.180[a]	−0.144	0.048	−0.035	−0.033
	p	0.025	0.073	0.555	0.667	0.681
AST	r	0.192[a]	0.096	−0.238[b]	0.053	0.052
	p	0.017	0.237	0.003	0.512	0.519
CRP	r	0.196[a]	0.207[b]	−0.047	0.122	0.124
	p	0.015	0.010	0.557	0.131	0.125
Lymphocytes	r	−0.074	0.095	−0.170[a]	−0.028	−0.029
	p	0.362	0.240	0.035	0.726	0.722
Troponin I	r	−0.004	0.110	−0.062	0.206[a]	0.205[a]
	p	0.971	0.275	0.539	0.039	0.039
CK	r	0.060	−0.083	−0.290[b]	0.004	0.004
	p	0.558	0.413	0.004	0.969	0.971
CK-MB	r	0.158	−0.033	−0.298[b]	0.063	0.065
	p	0.116	0.746	0.003	0.536	0.520
TOS	r	0.146	–	−0.116	0.212[b]	0.213[b]
	p	0.071	–	0.149	0.008	0.008
OSİ	r	−0.150	−0.116	–	0.161[a]	0.160[a]
	p	0.062	0.149	–	0.045	0.047
LTL	r	0.098	0.212[b]	0.161[a]	–	1.000[b]
	p	0.225	0.008	0.045	–	0.000
T/S ratio	r	0.098	0.213[b]	0.160[a]	1.000[b]	–
	p	0.224	0.008	0.047	0.000	–

[a]Correlation is significant at the 0.05 level
[b]Correlation is significant at the 0.01 level

analysis was used to determine whether there were differences regarding LTL values between the groups, a significant difference was also especially observed in the p = 0.047 confidence level. Furthermore, the Tukey Test indicated a statistically significant difference between the control group and STEMI group (p = 0.05). As a result, in the STEMI group, approximately 232 base pairs of decreased telomere length were noted as a statistical difference, yet there seemed to be no meaningful difference considering the NSTEMI group. The regression analysis identified a statistically significant relationship between LTL and T/S ratio with TOS in both NSTEMI and STEMI groups. OSI values showed significant regression with LTL and the T/S ratio only in the NSTEMI group (Table 3, Figs. 1 and 2).

4 Discussion

After measuring telomere length in 201 individuals from the patients with STEMI and NSTEMI and healthy controls, the study found that telomere length was shorter in STEMI patients than in the control group. As a

Table 3 P-values results of regression analysis in STEMI and NSTEMI groups

	NSTEMI					STEMI				
Risk factors	LTL	T/S ratio	TAS	TOS	OSI	LTL	T/S ratio	TAS	TOS	OSI
Male gender	>0.05	>0.05	>0.05	>0.05	0.008[a]	>0.05	>0.05	>0.05	>0.05	>0.05
Hypertension	>0.05	>0.05	>0.05	0.022[a]	>0.05	>0.05	>0.05	>0.05	>0.05	>0.05
Diastolic blood pressure	>0.05	>0.05	>0.05	>0.05	>0.05	>0.05	>0.05	>0.05	>0.05	>0.05
Heart rate	0.018[a]	0.018[a]	>0.05	>0.05	0.012[a]	>0.05	>0.05	>0.05	>0.05	>0.05
Glucose	>0.05	>0.05	>0.05	0.014[a]	>0.05	0.05[a]	0.05[a]	>0.05	>0.05	>0.05
Hemoglobin	>0.05	>0.05	>0.05	>0.05	0.038[a]	>0.05	>0.05	>0.05	>0.05	>0.05
Triglycerides	>0.05	>0.05	0.01[a]	>0.05	>0.05	>0.05	>0.05	0.002[a]	>0.05	>0.05
Urea	0.003[a]	0.003[a]	>0.05	>0.05	>0.05	>0.05	>0.05	0.009[a]	>0.05	0.037[a]
CRP	>0.05	>0.05	0.003	>0.05	>0.05	>0.05	>0.05	>0.05	>0.05	>0.05
Aorta diameter	0.05[a]	0.05[a]	>0.05	>0.05	>0.05	>0.05	>0.05	>0.05	>0.05	>0.05
Gensini	0.043[a]	0.043[a]	>0.05	>0.05	>0.05	>0.05	>0.05	>0.05	>0.05	>0.05
LTL	–	0.000[a]	>0.05	0.018[a]	0.000[a]	–	0.000[a]	>0.05	0.017[a]	>0.05
T/S ratio	0.000[a]	–	>0.05	0.018[a]	0.000[a]	0.000[a]	–	>0.05	0.017[a]	>0.05
TAS	>0.05	>0.05	–	>0.05	>0.05	>0.05	>0.05	–	>0.05	>0.05
TOS	0.018[a]	0.018[a]	>0.05	–	>0.05	0.017[a]	0.017[a]	>0.05	–	0.000[a]
OSI	0.000[a]	0.000[a]	>0.05	>0.05	–	>0.05	>0.05	>0.05	0.000[a]	–

[a]The mean difference is significant at the 0.05 level

Table 4 Results of TAS, TOS, OSI, LTL, and T/S ratio values compared with control group

	Control	NSTEMI		STEMI	
	Mean difference ± std. deviation	Mean difference ± std. deviation	*p* value	Mean difference ± std. deviation	*p* value
TAS	1.45 ± 0.26	1.47 ± 0.29	0.92	1.51 ± 0.2	0.50
TOS	5.34 ± 10.6	5.48 ± 5.17	0.99	5.27 ± 4.51	0.99
OSI	0.39 ± 0.93	0.18 ± 0.39	0.17	0.05 ± 0.03	0.007[a]
LTL	5049.92 ± 662.17	5013.30 ± 531.65	0.93	4817.72 ± 296.01	0.05[a]
T/S ratio	0.43 ± 0.32	0.41 ± 0.25	0.93	0.32 ± 0.14	0.05[a]

Values presented mean difference (standard deviation)
[a]The mean difference is significant at the 0.05 level

result of the length calculation, the difference was about 232 base pairs. However, no correlation was noticed between NSTEMI and LTL.

There are various studies conducted on the telomere including analysis of the codes of human biological lifetime. It was suggested that telomere length may be a predictor of age-related diseases such as ACS (Zee et al. 2009). Previous reports described that shorter LTL was associated with increased cardiovascular risk and mortality, and it was also predicted to increase proinflammatory activity and high-risk plaque morphology (Zee et al. 2009; Calvert et al. 2011). Studies showing the relationship between telomere shortness and coronary artery diseases in the literature were reported that aged endothelial cells were present in atherosclerosis lesions, and atherosclerotic coronary endothelial cells with coronary artery disease have shorter telomeres than patients without coronary artery disease, and thus, the term "short telomere-mediated aging" in atherosclerotic lesions was devised (Minamino et al. 2002; Ogami et al. 2004). This definition supported a causal link between telomere shortening and coronary artery diseases (Minamino et al. 2002). Shorter telomeres were

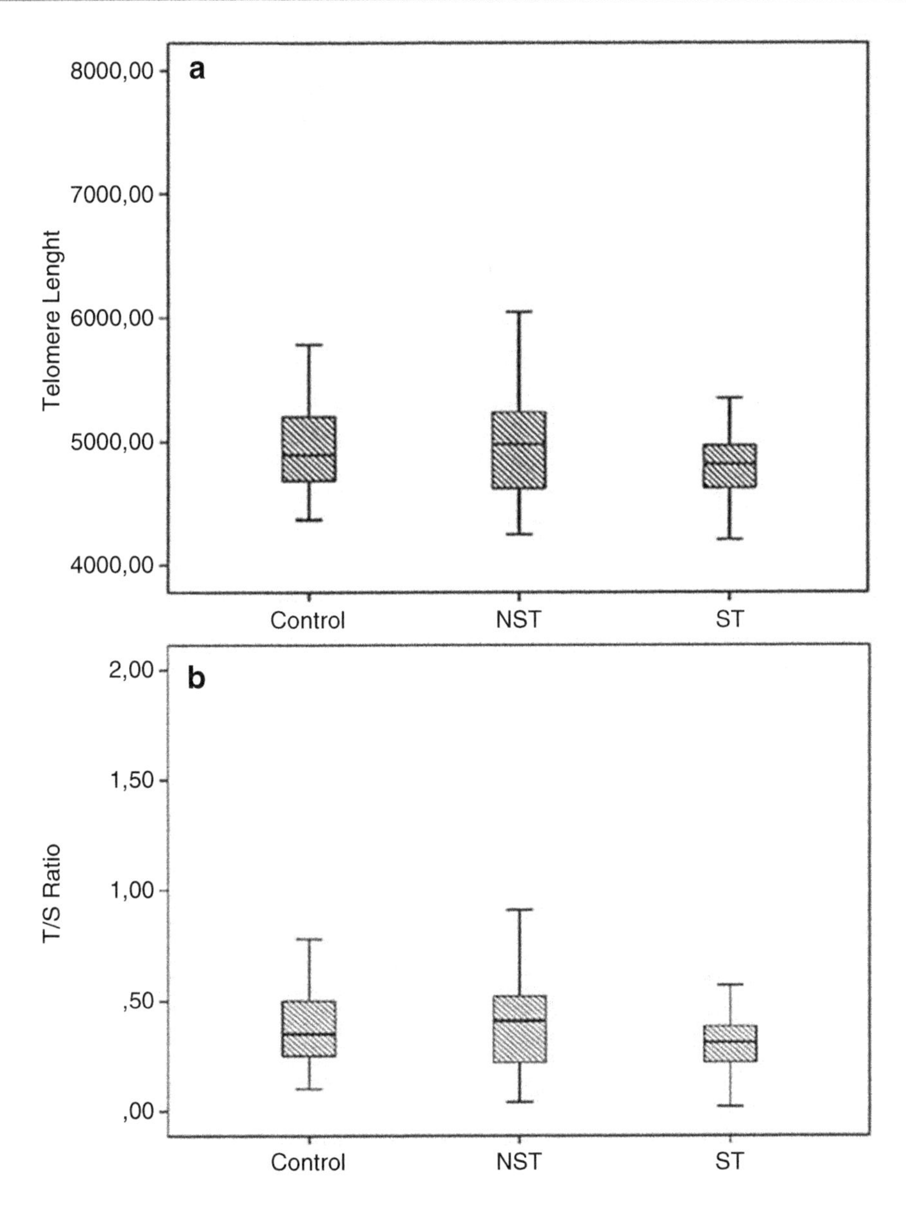

Fig. 1 Box plot of telomere length (**a**) and T/S ratio (**b**) in comparing groups of control, STEMI, and NSTEMI. The thick black line represents median values

being detected in patients with coronary artery diseases (Opstad et al. 2019), and there was an association between shorter telomeres and higher cardiovascular mortality (Cawthon et al. 2003). The study conducted by Cawthon et al., which measured telomere length in DNA samples of 143 people, found that those people with shorter telomeres had a threefold higher mortality rate from heart disease and an eightfold higher mortality rate from infectious diseases (Cawthon et al. 2003). In a meta-analysis study, those with short telomeres had an increased risk of coronary artery disease when compared to those with a long telomere (Haycock et al. 2014). The association of short leukocyte telomere with increased stroke and myocardial infarction (D'Mello et al. 2015) was supported by the demonstration of increased cardiovascular risks in individuals

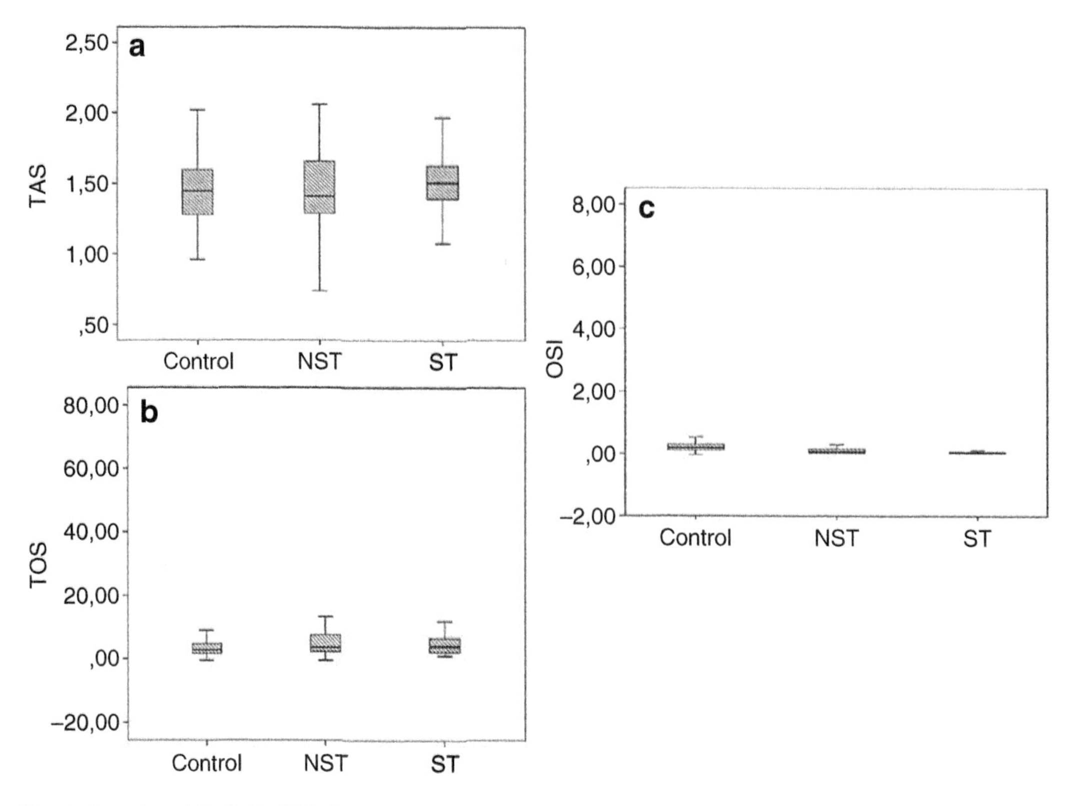

Fig. 2 Box plot of TAS (**a**), TOS (**b**), and OSI (**c**) levels in comparing groups of control, STEMI, and NSTEMI. The thick black line represents median values

with inherited short telomeres (Codd et al. 2013; Zhan et al. 2017). When telomere length was evaluated in both skeletal muscle and leukocyte, a relationship was found between LTL and atherosclerosis, but no relationship was found with muscle telomere length. The difference between leukocyte and muscle telomere length was due to the faster shortening of the leukocyte telomere in early life. It was observed that leukocyte telomere shortening in early life was a risk factor for coronary artery diseases and inherited short telomeres were not risk factors (Sabharwal et al. 2018).

A limited number of LTL studies have so far been reported in STEMI and/or NSTEMI patients. In a study with 353 participants, it was emphasized that LTL was not related after STEMI (Haver et al. 2015). In an incomplete but pre-published study including the NSTEMI group, shorter telomere length was evaluated to determine whether a biomarker predicts adverse incidents in elderly patients undergoing percutaneous coronary intervention (Kunadian et al. 2016). Studies with larger numbers of STEMI and NSTEMI patients will further clarify the relationship between LTL and AMI.

Oxidative stress is caused by increased production of oxidizing products or a decrease in the effectiveness of antioxidant defenses. Various studies have confirmed that heart tolerance to oxidative stress decreases with age due to the reduction in the concentration of antioxidant enzymes. Furthermore, the development of oxidative stress contributes to vascular endothelial dysfunction with aging (Siti et al. 2015; Von Zglinicki 2002). It was shown that aging of endothelial cells of CVD patients with a heavy burden of risk factors starts even before telomeres are shortened to their threshold length (Voghel et al. 2007). Aging, hereditary genomic features, inflammation, and cumulative exposure of oxidative stress refer to the shortened presence of LTL (Siti et al. 2015; Von Zglinicki 2002; Margaritis et al. 2017). Leukocyte telomere length is also associated with oxidative stress and inflammation (Bekaert et al. 2007). Increased oxidative stress and inflammation increase telomere shortening in somatic cells (Zhan et al. 2017; Daniali et al. 2013). Systemic oxidative stress and inflammation associated with cardiovascular risk can accelerate telomere shortening. Thus, the erosion rate hypothesis indicates that telomere shortening in adults is more important than hereditary telomere length (De Meyer et al. 2009). CVD, including STEMI and NSTEMI, are associated with increased oxidative stress as a result of impaired oxidant-antioxidant balance. As Gökdemir et al. showed, plasma TOS and OSI levels were significantly higher in NSTEMI patients, and it was thought to play a role along with the inflammatory process in the pathogenesis of NSTEMI (Gökdemir et al. 2013). Recent studies emphasize that oxidative stress may have an essential role in the pathogenesis of spontaneous reperfusion and development of atrial fibrillation in STEMI patients (Bas et al. 2017; Börekçi et al. 2016). Although there are no studies which show the direct relationship between oxidative stress and telomere length in STEMI and NSTEMI patients, there are certain studies emphasizing the effect of oxidative stress on short telomere length in CVD (Yeh and Wang 2016; Masi et al. 2016).

Since risk factors for CVD are associated with an increase in oxidative stress, in this study, we evaluated TAS, TOS, and OSI values in plasma of participants and found the OSI value to be lower in the STEMI group. OSI value was higher in the NSTEMI group compared to the STEMI group, but OSI was not found to be significant in NSTEMI group when compared with the control group. Studies have shown that oxidative stress indices increase after MI. NSTEMI alters biomarker levels, including oxidative stress indices (Kasap et al. 2007; Kundi et al. 2015). However, the low number of patients may have affected the results. Also interestingly, oxidative stress was found to be low in the NSTEMI group with a high SYNTAX score ($\geq$23), and oxidative stress was found to be high in the NSTEMI group with low SYNTAX score (<23) (Kundi et al. 2015). We found the SYNTAX score was lower in the NSTEMI group compared to the STEMI group, and higher OSI scores were seen in the NSTEMI group than in the STEMI group.

According to the regression analysis, there was a significant correlation between LTL with OSI and TOS values in patients with STEMI. The reason why we found low OSI values in STEMI patients compared to control group was that one of the most critical factors determining the rupture sensitivity of atherosclerotic plaque is the inflammation that develops within the plaque. Depending on the severity of the injury in the atheroma plate, different ACS tables appear. Unless the injury in the atheroma plate is relatively large, the thrombus formed is predominantly thrombocyte, while fibrin content is low. This white thrombus usually does not completely occlude the coronary artery (NSTEMI). However, when thrombus caused by plaque rupture makes total occlusion of the coronary artery, ST elevations occur in the ECG, and all, or almost all, of the affected ventricular wall remains within the necrosis field. This condition is also referred to as transmural AMI (STEMI). The fibrin content of this red thrombus is high. Oxidative stress is closely related to stabilization of atherosclerotic plaque and ACS (Nguyen et al. 2019). Additionally, Lavall et al. suggested that the oxidative profile created by STEMI and NSTEMI was similar regardless of the size of the arterial occlusion created by the thrombus (Lavall et al. 2016).

In conclusion, previous studies dealing with cardiovascular diseases provide strong evidence for the link between the LTL ratio and oxidative stress. However, there are a limited number of studies that provide evidence on the relationship between LTL values and oxidative stress in STEMI and NSTEMI, or state whether they are risk factors. In our study, we demonstrated that OSI levels and LTL values were more statistically significant in STEMI than in NSTEMI, although there were similar risk factors among the groups. Further studies are needed to confirm these findings and explore the potential association between LTL, STEMI, and NSTEMI.

5 Study Limitations

Firstly, we conducted a local study where the number of patients was kept relatively low. However, we carefully selected STEMI and NSTEMI patients based on various demographic and clinical features that may adversely affect the interpretation of our results. Secondly, these results based on the telomere length were measured in blood leukocytes, and the telomere length to be measured in other tissues, such as vascular tissue, may show a stronger correlation with STEMI and NSTEMI. Our analyses reflected the measurement of telomere length at a single time point. Long-term changes in telomere length at different time intervals as a result of years of monitoring of patients may show stronger associations with disease risk. The rate of active smoking and family history was higher in the STEMI group compared to the NSTEMI group. This may explain the significant linkage of oxidative stress and LTL with STEMI, but further studies are needed to validate the findings. The final limitation was that we did not take into account the leukocyte subpopulation distributions, a factor that can determine LTL measurement and the relationship of this parameter to cardiac dysfunction (Blackburn et al. 2015).

Acknowledgments This study was supported by Yozgat Bozok University, Scientific Research Projects Department (Project Number: 6602c-TF/17-91). We wish to thank Yozgat Bozok University, Scientific Research Projects Department, for their supports and Alperen Timucin SÖNMEZ and Cagla SARITÜRK for their statistical analysis assistance.

Conflict of Interest The authors report no relationships that could be construed as a conflict of interest.

Authors' Contributions All authors read and approved the final version of the manuscript.

References

Bas HA, Aksoy F, Icli A et al (2017) The association of plasma oxidative status and inflammation with the development of atrial fibrillation in patients presenting with ST elevation myocardial infarction. Scand J Clin Lab Invest 77(2):77–82

Baumgartner H, Hung J, Bermejo J et al (2017) Recommendations on the echocardiographic assessment of aortic valve stenosis: a focused update from the European association of cardiovascular imaging and the American society of echocardiography. J Am Soc Echocardiogr 30(4):372–392

Bekaert S, De Meyer T, Rietzschel ER, De Buyzere ML, De Bacquer D, Langlois M et al (2007) Telomere length and cardiovascular risk factors in a middle-aged population free of overt cardiovascular disease. Aging Cell 6(5):639–647

Blackburn EH, Epel ES, Lin J (2015) Human telomere biology: a contributory and interactive factor in aging, disease risks, and protection. Science 350 (6265):1193–1198

Börekçi A, Gür M, Türkoğlu C et al (2016) Oxidative stress and spontaneous reperfusion of infarct-related artery in patients with st-segment elevation myocardial infarction. Clin Appl Thromb Hemost 22(2):171–177

Calvert PA, Liew TV, Gorenne I et al (2011) Leukocyte telomere length is associated with high-risk plaques on virtual histology intravascular ultrasound and increased proinflammatory activity. Arterioscler Thromb Vasc Biol 31(9):2157–2164

Cawthon RM (2002) Telomere measurement by quantitative PCR. Nucleic Acids Res 30(10):e47

Cawthon RM, Smith KR, O'Brien E, Sivatchenko A, Kerber RA (2003) Association between telomere length in blood and mortality in people aged 60 years or older. Lancet 361(9355):393–395

Codd V, Nelson CP, Albrecht E, Mangino M, Deelen J, Buxton JL et al (2013) Identification of seven loci affecting mean telomere length and their association with disease. Nat Genet 45(4):422–427

D'Mello MJ, Ross SA, Briel M, Anand SS, Gerstein H, Paré G (2015) Association between shortened leukocyte telomere length and cardiometabolic outcomes: systematic review and meta-analysis. Circ Cardiovasc Genet 8(1):82–90

Daniali L, Benetos A, Susser E, Kark JD, Labat C, Kimura M et al (2013) Telomeres shorten at equivalent rates in somatic tissues of adults. Nat Commun 4(1):1597

De Meyer T, Rietzschel ER, De Buyzere ML, Langlois MR, De Bacquer D, Segers P et al (2009) Systemic telomere length and preclinical atherosclerosis: the Asklepios Study. Eur Heart J 30(24):3074–3081

Erel O (2004) A novel automated direct measurement method for total antioxidant capacity using a new generation, more stable ABTS radical cation. Clin Biochem 37(4):277–285

Erel O (2005) A new automated colorimetric method for measuring total oxidant status. Clin Biochem 38 (12):1103–1111

Gensini GG (1983) A more meaningful scoring system for determining the severity of coronary heart disease. Am J Cardiol 51(3):606

Gökdemir MT, Kaya H, Söğüt O, Kaya Z, Albayrak L, Taşkın A (2013) The role of oxidative stress and inflammation in the early evaluation of acute non-ST-

elevation myocardial infarction: an observational study. Anadolu Kardiyol Derg 13(2):131–136
Hamm CW, Bassand JP, Agewall S et al (2011) ESC Guidelines for the management of acute coronary syndromes in patients presenting without persistent ST-segment elevation: The Task Force for the management of acute coronary syndromes (ACS) in patients presenting without persistent ST segment elevation of the European Society of Cardiology (ESC). Eur Heart J 32(23):2999–3054
Haver VG, Hartman MH, Mateo Leach I et al (2015) Leukocyte telomere length and left ventricular function after acute ST-elevation myocardial infarction: data from the glycometabolic intervention as adjunct to primary coronary intervention in ST elevation myocardial infarction (GIPS-III) trial. Clin Res Cardiol 104(10):812–821
Haycock PC, Heydon EE, Kaptoge S, Butterworth AS, Thompson A, Willeit P (2014) Leucocyte telomere length and risk of cardiovascular disease: systematic review and meta-analysis. BMJ 349:g4227
Horwitz A (2001) A version of Simpson's rule for multiple integrals. J Comput Appl Math 134:1–11
Judkins MP (1968) Percutaneous transfemoral selective coronary arteriography. Radiol Clin N Am 6 (3):467–492
Kasap S, Gönenç A, Sener DE, Hisar I (2007) Serum cardiac markers in patients with acute myocardial infarction: oxidative stress, C-reactive protein and N-terminal probrain natriuretic peptide. J Clin Biochem Nutr 41:50–57
Kunadian V, Neely RD, Sinclair H et al (2016) Study to improve cardiovascular outcomes in high-risk older patients (ICON1) with acute coronary syndrome: study design and protocol of a prospective observational study. BMJ Open 6(8):e012091
Kundi H, Erel Ö, Balun A, Çiçekçioğlu H, Cetin M, Kiziltunç E, Neşelioğlu S, Topçuoğlu C, Örnek E (2015) Association of thiol/disulfide ratio with syntax score in patients with NSTEMI. Scand Cardiovasc J 49 (2):95–100
Lavall MC, Bonfanti G, Ceolin RB, Schott KL, Gonçalves Tde L, Moresco RN, Brucker N, Morsch VM, Bagatini MD, Schetinger MR (2016) Oxidative profile of patients with ST segment elevation myocardial infarction. Clin Lab 62(5):971–973
Margaritis M, Sanna F, Lazaros G et al (2017) Predictive value of telomere length on outcome following acute myocardial infarction: evidence for contrasting effects of vascular vs. blood oxidative stress. Eur Heart J 38 (41):3094–3104
Masi S, D'Aiuto F, Cooper J et al (2016) Telomere length, antioxidant status and incidence of ischaemic heart disease in type 2 diabetes. Int J Cardiol 216:159–164
Minamino T, Miyauchi H, Yoshida T, Ishida Y, Yoshida H, Komuro I (2002) Endothelial cell senescence in human atherosclerosis: role of telomere in endothelial dysfunction. Circulation 105 (13):1541–1544
Neumann FJ, Sousa-Uva M, Ahlsson A, Alfonso F, Banning AP, Benedetto U, Byrne RA, Collet JP, Falk V, Head SJ, Jüni P, Kastrati A, Koller A, Kristensen SD, Niebauer J, Richter DJ, Seferovic PM, Sibbing D, Stefanini GG, Windecker S, Yadav R, Zembala MO, ESC Scientific Document Group (2019) 2018 ESC/EACTS Guidelines on myocardial revascularization. Eur Heart J 40(2):87–165
Nguyen MT, Fernando S, Schwarz N, Tan JT, Bursill CA, Psaltis PJ (2019) Inflammation as a therapeutic target in atherosclerosis. J Clin Med 8(8):1109
Ogami M, Ikura Y, Ohsawa M, Matsuo T, Kayo S, Yoshimi N et al (2004) Telomere shortening in human coronary artery diseases. Arterioscler Thromb Vasc Biol 24(3):546–550
Opstad TB, Kalstad AA, Pettersen AÅ, Arnesen H, Seljeflot I (2019) Novel biomolecules of ageing, sex differences and potential underlying mechanisms of telomere shortening in coronary artery disease. Exp Gerontol 119:53–60
Riethman H (2008) Human telomere structure and biology. Annu Rev Genomics Hum Genet 9:1–19
Sabharwal S, Verhulst S, Guirguis G, Kark JD, Labat C, Roche NE, Martimucci K, Patel K, Heller DS, Kimura M, Chuang D, Chuang A, Benetos A, Aviv A (2018) Telomere length dynamics in early life: the blood-and-muscle model. FASEB J 32(1):529–534
Siti HN, Kamisah Y, Kamsiah J (2015) The role of oxidative stress, antioxidants and vascular inflammation in cardiovascular disease (a review). Vasc Pharmacol 71:40–56
Thygesen K, Alpert JS, Jaffe AS et al (2018) Fourth universal definition of myocardial infarction. J Am Coll Cardiol 72(18):2231–2264
Turkish Statistical Institute. Causes of Death Statistics, 2018. http://www.tuik.gov.tr/PreHaberBultenleri.do?id=30626. Access Sept 2020
Verhoeven JE, Révész D, Epel ES, Lin J, Wolkowitz OM, Penninx BW (2014) Major depressive disorder and accelerated cellular aging: results from a large psychiatric cohort study. Mol Psychiatry 19(8):895–901
Voghel G, Thorin-Trescases N, Farhat N et al (2007) Cellular senescence in endothelial cells from atherosclerotic patients is accelerated by oxidative stress associated with cardiovascular risk factors. Mech Ageing Dev 128(11–12):662–671
Von Zglinicki T (2002) Oxidative stress shortens telomeres. Trends Biochem Sci 27(7):339–344
Yeh JK, Wang CY (2016) Telomeres and telomerase in cardiovascular diseases. Genes (Basel) 7(9):58
Zee RY, Michaud SE, Germer S, Ridker PM (2009) Association of shorter mean telomere length with risk of incident myocardial infarction: a prospective, nested case-control approach. Clin Chim Acta 403 (1–2):139–141
Zhan Y, Karlsson IK, Karlsson R et al (2017) Exploring the causal pathway from telomere length to coronary heart disease: a network mendelian randomization study. Circ Res 121(3):214–219

Adv Exp Med Biol - Cell Biology and Translational Medicine (2021) 14: 197–219
https://doi.org/10.1007/5584_2021_640

Published online: 26 May 2021

Eosinophils as Major Player in Type 2 Inflammation: Autoimmunity and Beyond

Marco Folci, Giacomo Ramponi, Ivan Arcari, Aurora Zumbo, and Enrico Brunetta

Abstract

Eosinophils are a subset of differentiated granulocytes which circulate in peripheral blood and home in several body tissues. Along with their traditional relevance in helminth immunity and allergy, eosinophils have been progressively attributed important roles in a number of homeostatic and pathologic situations. This review aims at summarizing available evidence about eosinophils functions in homeostasis, infections, allergic and autoimmune disorders, and solid and hematological cancers.

Their structural and biological features have been described, along with their physiological behavior. This includes their chemokines, cytokines, granular contents, and extracellular traps. Besides, pathogenic- and eosinophilic-mediated disorders have also been addressed, with the aim of highlighting their role in Th2-driven inflammation. In allergy, eosinophils are implicated in the pathogenesis of atopic dermatitis, allergic rhinitis, and asthma. They are also fundamentally involved in autoimmune disorders such as eosinophilic esophagitis, eosinophilic gastroenteritis, acute and chronic eosinophilic pneumonia, and eosinophilic granulomatosis with polyangiitis. In infections, eosinophils are involved in protection not only from parasites but also from fungi, viruses, and bacteria. In solid cancers, local eosinophilic infiltration is variably associated with an improved or worsened prognosis, depending on the histotype. In hematologic neoplasms, eosinophilia can be the consequence of a dysregulated cytokine production or the result of mutations affecting the myeloid lineage.

Recent experimental evidence was thoroughly reviewed, with findings which elicit a complex role for eosinophils, in a tight balance between host defense and tissue damage. Eventually, emerging evidence about eosinophils in COVID-19 infection was also discussed.

Keywords

Asthma · Autoimmunity · COVID-19 · EGPA · Eosinophils · Inflammation · Neoplasia · Th2

M. Folci (✉) and E. Brunetta
Humanitas Clinical and Research Center – IRCCS, Milan, Italy

Department of Biomedical Sciences, Humanitas University, Milan, Italy
e-mail: marco.folci@humanitas.it; mdfolci@gmail.com

G. Ramponi, I. Arcari, and A. Zumbo
Humanitas Clinical and Research Center – IRCCS, Milan, Italy

Abbreviations

CCL17	C-C motif chemokine ligand 17
CCL22	C-C motif chemokine ligand 22
CCR3	C-C chemokine receptor type 3
FGF-2	fibroblast growth factors 2
GM-CSF	granulocyte-macrophage colony-stimulating factor
IFNγ	interferon gamma
IL-10	interleukin 10
IL-12	interleukin 12
IL-17	interleukin 17
IL-23	interleukin 23
IL-25	interleukin 25
IL-3	interleukin 3
IL-4	interleukin 4
IL-33	interleukin 33
IL-5	interleukin 5
IL-6	interleukin 6
PTX3	pentraxin 3
TGF-β	transforming growth factor β
TNF-α	tumor necrosis factor alpha α
TSLP	thymic stromal lymphopoietin
Th1	T helper 1
Th2	T helper 2
NK cell	natural killer cell
FcεRI	high-affinity IgE receptor
FcεRIα	high-affinity IgE receptor alpha chain
MBP	major basic protein
ECP	eosinophilic cationic protein
EPO	eosinophil peroxidase
ROS	reactive oxygen species
EDN	eosinophil-derived neurotoxin
PMD	piecemeal degranulation
EoSVs	eosinophil sombrero vesicles
EET	eosinophil extracellular traps
ILC2	innate lymphoid cell type 2
PRRs	pattern recognition receptors
PAMPs	pathogen-associated molecular patterns
LPS	lipopolysaccharide
DAMPs	damage-associated molecular patterns
TLRs	toll-like receptors
APCs	antigen-presenting cells
MHC-II	major histocompatibility complex class II
DCs	dendritic cells
GI tract	gastrointestinal tract
RVS	respiratory syncytial virus
HIV	human immunodeficiency virus
AIDS	acquired immunodeficiency syndrome
SARS-CoV-2	SARS-2 coronavirus
COVID-19	coronavirus disease 19
BAL	bronchoalveolar lavage
ADCC	antibody-dependent cellular cytotoxicity
AD	atopic dermatitis
AR	allergic rhinitis
HES	hypereosinophilic syndrome
AEP	acute eosinophilic pneumonia
CEP	chronic eosinophilic pneumonia
EGPA	eosinophilic granulomatosis with polyangiitis
ANCA	antineutrophil cytoplasmic antibodies
TME	tumor microenvironment
TATE	tumor-associated tissue eosinophilia
CTLs	cytotoxic T-lymphocytes
OSCC	oral squamous cell carcinoma
DFS	disease-free survival
MN-eos	eosinophilia-associated myeloid neoplasms
FGFR1	fibroblast growth factor receptor 1
PDGFRα	platelet-derived growth factor receptor a
PDGFRβ	platelet-derived growth factor receptor b
MPN	myeloproliferative neoplasms
CEL-NOS	chronic eosinophilic leukemia-not otherwise specified
CML	chronic myeloid leukemia
Eos-CML	hypereosinophilic variant
HL	Hodgkin's lymphoma

1 Introduction

Eosinophils are a terminally differentiated subset of granulocytes which are developed in the bone marrow and circulate in peripheral blood, from where they are distributed to several organs (Rothenberg and Hogan 2006; Hogan et al. 2008a; Wen and Rothenberg 2016). For a long time, the complex physiological functions of eosinophils resident in human tissues have been underestimated, along with their role in a disease. Eosinophils were relevant only in the realms of parasitic infection (helminths) and allergic disease, with little to no importance in other disorders. In the last few years, however, mounting evidence has emerged that they are involved in a variety of mechanisms, among which regulation of innate and adaptive immunity (Rothenberg and Hogan 2006; Wen and Rothenberg 2016).

2 Eosinophils Biology

Eosinophil development in the bone marrow is driven by IL-3, GM-CSF, and IL-5 effect on their myeloid precursors (Ramirez et al. 2018). IL-5 is the key driver of the latter stages of eosinophilic development and release into peripheral blood. Besides, it stimulates their survival and prevents apoptosis (Ramirez et al. 2018). It is produced by CD34+ hematopoietic cells, Th2 lymphocytes, NK cells, and mast cells (Ramirez et al. 2018). Along with IL-5, the eotaxin family of chemokines has been attributed the role of main eosinophil recruiter in peripheral tissues (Weller and Spencer 2017). Eotaxins (eotaxin-1, eotaxin-2, and eotaxin-3) are produced by local tissue cells and attract eosinophils by acting onto the CCR3 receptor (Wen and Rothenberg 2016; Conroy and Williams 2001; White et al. 1997; Provost et al. 2013).

2.1 Granules

Like other granulocytes, eosinophils contain different granules in their cytoplasm. Primary granules are mainly composed of a hydrophobic protein called galectin-10, which drives the formation of Charcot-Leyden crystals in tissues affected from eosinophilic inflammation (Ramirez et al. 2018; McBrien and Menzies-Gow 2017). Crystalloid or specific granules, containing a crystalline core surrounded by a matrix, are characterized by the presence of high amount of proteins. Among these, preformed cytokines, chemokines, enzymes, growth factors, and several basic proteins lead to the eosinophilic staining pattern of these cells. Major basic protein (MBP), eosinophilic cationic protein (ECP), and eosinophil peroxidase (EPO) are probably the most relevant. MBP is able to exert a cytotoxic role on target cells by interfering with membrane permeability. ECP is a ribonuclease possibly involved in viral infections, while EPO is involved in the production of reactive oxygen species (ROS) targeting extracellular organisms (mirroring the role of myeloperoxidase in neutrophils) (Ramirez et al. 2018). The third most represented granule type is characterized by lipid bodies. Those are lipid-rich cytoplasmic inclusions which characteristically develop in vivo in cells associated with inflammation. The presence of high amount of enzymes such as cyclooxygenases, 5-lipoxygenase, and leukotriene C4 synthase makes these lipid-rich organelles a key site of arachidonic acid esterification and eicosanoid production (Shamri et al. 2011). Furthermore, pleomorphic vesicle-tubular carriers, which were identified and termed eosinophil sombrero vesicles, are actively formed and direct differential and rapid release of eosinophil proteins (Fig. 1; Spencer et al. 2014).

2.2 Activation

The activation of eosinophils occurs under several conditions. Depending on the inciting *noxa,* these cells can elicit several mechanisms of defense ranging from variable levels of enzyme release to exocytosis and extracellular traps. Phagocytosis represents one of the best known mechanisms by means of which the majority of granulocytes and macrophages act against pathogens. Some in vitro

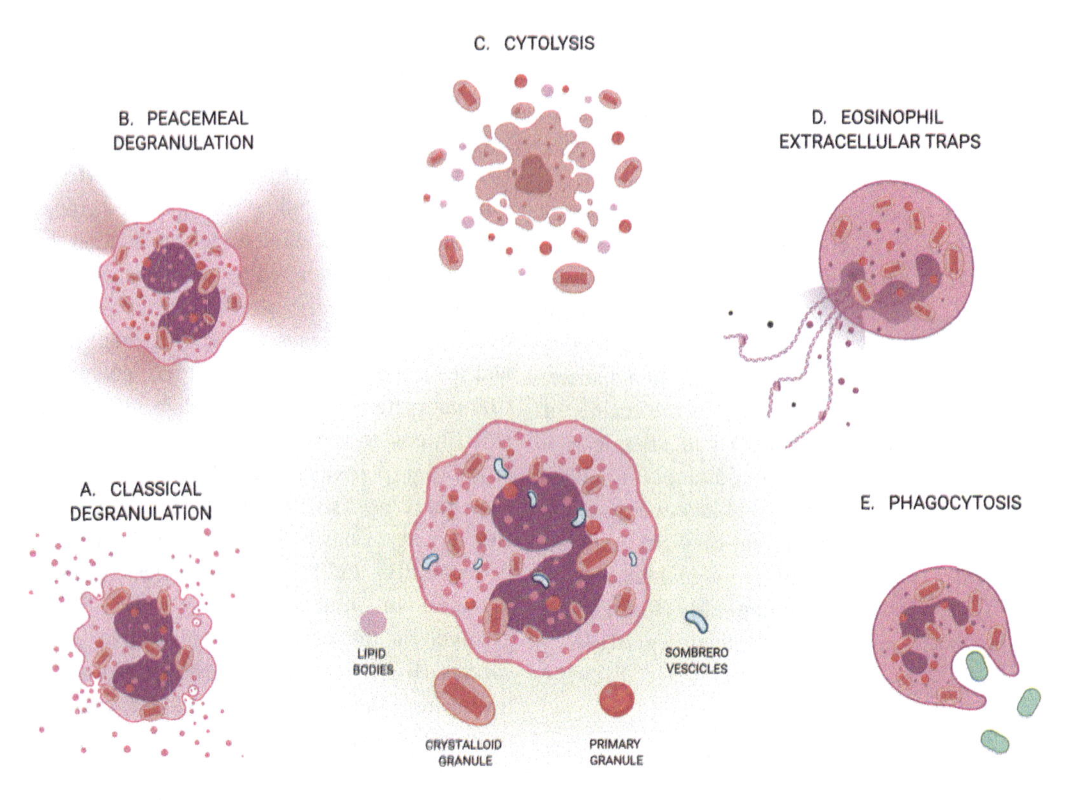

Fig. 1 Eosinophil granules and activation processes **Eosinophils granules** (center) are characterized by four different types. **Crystalloid granules**: unique granules also called specific granules which present a crystalline core surrounded by a fluid matrix; they differentiate from others by the high amount of preformed cytokines, chemokines, and several basic proteins such as MBP; **Primary granules**: composed of a hydrophobic protein called galectin-10; **Lipid bodies**: lipid-rich cytoplasmic inclusions which are characterized by high amounts of enzymes involved in arachidonic acid esterification and eicosanoid production; **Sombrero vesicles**: elongated tubules which appear to be responsible for moving proteins between granules and the plasma membrane. (**a**) **Classical degranulation**: granule content is completely released by the fusion with cytoplasmic membrane through classical exocytosis; (**b**) **Piecemeal degranulation**: defined by the formation of secretory vesicles containing granular proteins which are released into the extracellular space without the direct fusion of granules with cellular membrane; (**c**) **Cytolysis with granule release**: granules are released whole into the extracellular environment during the process of cytolysis; (**d**) **EET**: webs of DNA and granular proteins are ejected into the extracellular environment when organisms are too big to be phagocytized and therefore require extracellular mechanisms of defense; (**e**) **Phagocytosis**: eosinophils are able to engulf materials such as immune complexes, foreign cells, pathogens, or parts of necrotic cells by extruding plasma membrane till enveloping the full particle

studies demonstrated phagocytic capabilities by eosinophils under specific situations;(ARCHER and HIRSCH 1963) indeed, these cells are attracted and are capable of engulfing materials such as immune complexes, foreign cells, or parts of necrotic cells. Evidence proved in vitro phagocytosis of Gram-positive *Staphylococcus aureus* and Gram-negative *Escherichia coli* bacteria, as well as living and dead *Candida albicans* (Cline and Lehrer 1968). Besides bacteria, eosinophils seem to use phagocytosis to destroy parasites such as *Trypanosoma dionisii* (Thorne et al. 1979). Although some controversy exists about the effectiveness of eosinophil phagocytotic mechanisms, they have a variety of mechanisms through which they are able to kill target extracellular organisms (Fig. 1; Wen and Rothenberg 2016; Ramirez et al. 2018).

2.2.1 Classical Degranulation

Degranulation has long been known as the main effector mechanism of eosinophilic activation. Degranulation is the process by means of which eosinophils allow fusion of intracellular granules with the plasma membrane through classical exocytosis (Spencer et al. 2014). Eosinophils contain four different types of granules: crystalloid granules, primary granules, small granules, and secretory vesicles (Hogan et al. 2008a). Within crystalloid granules, basic proteins are stored. These are MBP (in the center of the granule), EPO, ECP, and eosinophil-derived neurotoxin (EDN, within the granule matrix). MBP is a highly cationic protein which exerts a toxic effect against parasites, such as *Schistosoma mansoni*. Furthermore, it has antibacterial properties, and it is able to activate complement through the classical and alternative pathways (Hogan et al. 2008a). For what concerns ECP, it is known to have a bactericidal effect and the capacity to kill parasites. Its mechanism of action seems to be the formation of pores within target membranes, although ECP also possesses RNAse activity (Hogan et al. 2008a). EDN is another RNAse protein which is expressed by eosinophils, monocytes, and polymorphonuclear cells. EDN appears to have antiviral activity in respiratory infections. Interestingly, the EDN family of proteins has one of the highest rates of mutations in the genome of primates (Hogan et al. 2008a). EPO is a haloperoxidase which is involved in bacterial killing. EPO catalyzes the peroxidative oxidation of halides (such as bromide, chloride, and iodide) present in the plasma together with hydrogen peroxide generated by NADPH oxidase (Hogan et al. 2008a).

Classical exocytic degranulation of eosinophils has mostly been observed in vitro (Weller and Spencer 2017). On the contrary, in vivo, it is believed that eosinophils release their granules through cytolysis and piecemeal degranulation in most situations (Weller and Spencer 2017). While this does not apply to interactions with parasites, where classical exocytosis seems to play a major role, it is relevant for other scenarios in which piecemeal degranulation and cytolysis are predominant (Fig. 1; Spencer et al. 2014).

2.2.2 Piecemeal Degranulation

Piecemeal degranulation (PMD) involves the formation of secretory vesicles containing granular proteins which are brought onto the plasma membrane and extruded into the extracellular space (Weller and Spencer 2017). When PMD occurs, not all granules are secreted. Interestingly, some of these remain intact within eosinophils. This may contribute to the ability of eosinophils to release inflammatory mediators in response to repetitive triggers (Melo and Weller 2010). Apparently, eosinophil granules contain a complex network of vesicles and membranes which are an integral component of PMD. Electron microscopy allowed direct visualization of this system (Weller and Spencer 2017). It has also been observed that different stimuli elicit secretion of different proteins, with a trend toward type 1 or type 2 cytokines (Weller and Spencer 2017). Interestingly, IL-4 receptor subunit α (IL-4Rα) is abundant within eosinophil granules at rest. When eosinophils are exposed to eotaxin-1, IL-4Rα is involved in the translocation of IL-4 toward the plasma membrane (Weller and Spencer 2017). This mechanism is selective upon the cytokine transported, with different receptors mediating the transport of different cytokines (Melo and Weller 2010). Eotaxins and other chemokines induce morphological changes on eosinophilic granules (such as the formation of protrusions) which are believed to be necessary for PMD (Fig. 1; Melo and Weller 2010). Moreover, in eosinophils undergoing activation and PMD, a population of vesicles defined eosinophil sombrero vesicles (EoSVs) are frequently represented (Spencer et al. 2014). These are elongated tubules which appear to be mostly involved in the cytolytic release of granules and are thought to be responsible for moving proteins between granules and the plasma membrane (Fig. 1; Melo et al. 2005a, b; Spencer et al. 2006).

2.2.3 Cytolysis with Granule Release

Eventually, eosinophils are able to undergo cytolysis and release their granular contents in vivo (Weller and Spencer 2017). This is also the time when eosinophils release extracellular DNA traps, known as eosinophil extracellular traps. During the process of cytolysis, eosinophil granules seem to be released whole into the extracellular environment. These retain their content of cytokines and cationic proteins and may exert damage onto invading organisms or the host tissue (Weller and Spencer 2017). These granules were found to express chemokine receptors (e.g., CCR3) and secrete cationic proteins (ECP and EPO). Consequently, it is believed that eosinophilic granules released into the extracellular space may function as a fully independent unit capable of secretion and regulation by stimuli (Weller and Spencer 2017). Indeed, during cytolysis, they are released within the affected tissue along with the other granules (Fig. 1; Spencer et al. 2014).

2.2.4 Eosinophil Extracellular Traps (EET)

Together with the previously described granular proteins, it was recently discovered that eosinophils are able to release webs of DNA and granular proteins into the extracellular environment (Yousefi et al. 2008; Mukherjee et al. 2018). These were deemed eosinophil extracellular traps (EETs) as they closely resembled neutrophil extracellular traps, a mechanism of pathogen defense employed by neutrophils (Brinkmann et al. 2004). EETs seem to have a function similar to that of NETs. Although it is still unclear whether the DNA released by eosinophils is mitochondrial or nuclear in origin and whether eosinophils can survive the release of EETs, it appears that they are extremely effective (at least in vitro) in killing bacteria such as *E. coli* (Mukherjee et al. 2018). This finding challenges the dogma by which neutrophils are involved defending the host from bacteria, whereas eosinophils face helminth invasion of the body. Interestingly enough, EETs may also be involved in parasite defense, where organisms are too big to be phagocytosed and therefore require extracellular mechanisms of defense (Mukherjee et al. 2018). As it is described later on in this review, EETs may also have some relevant roles in pathogenesis of autoimmune disorders, just like NETs (Fig. 1; Khandpur et al. 2013).

2.3 Interplay with Innate and Adaptive Immunity in Type 2 Reactions

Eosinophils are mainly recruited and activated by epithelial-derived innate cytokines such as interleukin 33 (IL-33), and thymic stromal lymphopoietin (TSLP). These act in enhancing Th2 responses through dendritic cells, innate lymphoid cell type 2 (ILC2), and basophils (Siracusa et al. 2011; Ziegler et al. 2013). The result of these interactions is both increased eosinophil survival (Wong et al. 2010) and activity with the effect of modulating innate and adaptive immunity. Eosinophils express pattern recognition receptors (PRRs) which allow them to be directly activated by recognizing specific pathogen-associated molecular patterns (PAMPs) of bacteria, such as lipopolysaccharide (LPS), and fungi, including beta-glucans (Shamri et al. 2011).

Among PAMPs, several toll-like receptors (TLRs) are expressed, among which the most abundant is TLR7, which is activated by single-stranded RNA (Phipps et al. 2007). Eosinophils also respond to damage-associated molecular patterns (DAMPs), and for this reason, they are attracted to damaged tissues and necrotic cells (Shamri et al. 2011). Through these mechanisms, eosinophils play a direct role in innate immune response to a wide variety of pathogens such as helminths, viruses, bacteria, and fungi contributing to tissue homeostasis.

Moreover, eosinophils can respond to the complement cascade, i.e., granule proteins, specifically MBP, ECP, and EPO, regulating both classical and alternative complement pathways (Bass 1975). It has been demonstrated how activated murine and human eosinophils express major histocompatibility complex class II

(MHC-II) and co-stimulatory molecules which allow them to function as antigen-presenting cells (APCs) for viral antigens (Vrtis et al. 2020; Del Pozo et al. 1992). By acting as APCs, eosinophils have the ability to activate adaptive immunity and polarize the immune reaction through the release of soluble mediators. Eosinophil activation and release of granule proteins seem to be pivotal in activating and enhancing Th2 responses. Indeed, two of the most representative granule molecules, EDN and EPO, are proved to be strong chemoattractant (Yang et al. 2003) for dendritic cells (DCs) inducing their maturation and activation (Yang et al. 2004) as well as migration to draining lymph nodes (Chu et al. 2014).

The importance of eosinophils for Th2 priming is also demonstrated by weak or even absent reactions in those sites usually devoid of eosinophils such as the peritoneum, skin, or rectum. The ability to induce Th2 cell recruitment is proved by the induction of specific chemokines such as CCL22 and CCL17 (Jacobsen et al. 2008) but also by the polarization of naive T helper cells via IL-4, IL-25, and indoleamine 2,3-dioxygenase, which is selectively pro-apoptotic toward Th1 cells (Odemuyiwa et al. 2004).

Eosinophils have also shown to be important in cytokine production, especially IL-4, IL-13, and IL-25 (Fulkerson and Rothenberg 2018). IL-4 induces immunoglobulin switching toward local production of IgEs and also induces B-cell and IgM-producing plasma cell differentiation (Jordan et al. 2004). Moreover, it determines leukocyte transmigration into inflamed tissues, along with B-cell proliferation, survival, and antibody production upon co-culture in vitro (Wong et al. 2014). On the contrary, IL-13 is linked to mucus secretion by goblet cells in lungs (Fulkerson and Rothenberg 2018). IL-25 has proved to increase Th2 polarization and proliferation and cytokine production by T cells which, in turn, are stimulated to release IL-5 supporting eosinophilic differentiation and recruitment (Wang et al. 2007).

Eventually, this cell lineage has a variety of immunoregulatory functions (Wen and Rothenberg 2016; Weller and Spencer 2017). Eosinophils interact with B cells leading to their survival and activation (Jordan et al. 2004; Cambier and Morrison 1991). Besides, through antigen processing, they can function as antigen-presenting cells inducing immunoglobulin production. The release of cytokines can provide Th2-polarizing signals during the interaction with T cells. Besides, they seem to promote maturation of bone marrow plasma cells and to regulate mucosal IgA secretion (Weller and Spencer 2017).

According to experiments in mice, eosinophils may also be involved in central T-cell tolerance, as they interact with the negative selection process of T cells (Throsby et al. 2000). Mast cells are not only in close proximity but also functionally linked to eosinophils; in fact, these cells are found in strict relation both in physiological conditions and in pathologic disorders such as allergic lung and inflamed gut (Elishmereni et al. 2011). Studies show how MBP, EPO, and ECP are able to trigger production of cytokines and increased release of histamine by mast cells, in the form of IgE-independent activation (Hogan et al. 2008a). Moreover, mast cell-specific proteases were not only observed to be able to recruit eosinophils into tissues but also to promote their lifetime and secretion of a variety of molecules, such as GM-CSF, IL-3, IL-5, and TNF-α (Wong et al. 2009; Shakoory et al. 2004).

3 Tissue Homeostasis and Eosinophil Physiological Role

Tissue eosinophils have some steady-state functions which appear to be mostly independent from both allergy and infection. For instance, eosinophils and M2 macrophages are required for the constitution of the mammary gland stroma and are involved in the development of ductal tree (Weller and Spencer 2017; Gouon-Evans et al. 2002; Aupperlee et al. 2014). Although the precise mechanism has not been fully elucidated, murine knockout models for eosinophils show a diminished ductal branching (Gouon-Evans et al. 2000). Intestinal eosinophils are necessary for mucosal homeostasis and allow production of sufficient amounts of IgA and prevent alterations of the microbiome and disruption of the mucosal integrity (Mantis et al. 2011). Moreover, damages involving epithelial cells determine the productions of alarmins which result in a potent chemoattractant stimulus to recruit and activate eosinophils. In this condition, the releasing of eosinophil-derived FGF-2 and transforming growth factor beta (TGF-β) is linked to epithelial repair and remodeling (Stenfeldt and Wennerås 2004). By observing the pattern of eotaxin expression, researchers were able to improve their understanding about eosinophil target organs, among which the lungs and the gastrointestinal tract (GI tract) are clearly the most well-known. Eosinophils also home in the thymus, the spleen, and lymph nodes. Furthermore, their presence was actually observed in most human organs, including the uterus, the mammary gland, the liver, the heart, and the peripheral nervous system (Ramirez et al. 2018). In damaged tissues, eosinophils are associated with regenerative responses. This was observed in the skeletal muscle and in the liver, where the interplay of IL-4 and eosinophil presence seems to regulate the function of resident cells and the recruitment of nonimmune cells (Fig. 2; Weller and Spencer 2017; Goh et al. 2013).

3.1 Infections

Eosinophils are known to be fundamental in helminth immunity. Along with their role in defense from parasites, eosinophils are also pivotal in bacterial, fungal, and viral infections (Fig. 2).

3.1.1 Eosinophils in Parasitic Infections

The role of eosinophils in parasitic infections has been widely accepted. Those cells represent not only one of the most powerful defensive end effectors involved against helminths but also a key regulator of the immune reaction that develops during worm infestation by obtaining the characteristics of antigen-presenting cells (Butterworth 1985). Indeed, helminth antigens were found to induce on eosinophil co-stimulatory molecules such as CD80, CD86, and CD40 which led them to actively interact with other immune cells (Mawhorter et al. 1993). Moreover, eosinophils from murine models, infected with filarial parasite, upregulate surface MHC-II (Mawhorter et al. 1993). Antigen presentation and T-cell response initiations were also observed in studies with *Strongyloides stercoralis* both in vitro and in vivo (Padigel et al. 2007). The role of eosinophils as mediators of Th2 inflammation was demonstrated with their capacity to home and stimulate dendritic cells by releasing granule contents (Jacobsen et al. 2008). Activated eosinophils are able to strongly influence the phenotype of recruited immune cells, directing their actions toward the expulsion of parasite. The enhanced Th2 responses are represented by the production of IL-4, IL-5, and IL-13 and have been observed with *Strongyloides stercoralis*, *Trichinella spiralis*, and in the mouse model of *Trichuris trichiura* (Rosenberg et al. 2013). Even if the clearance of worms from tissues is the result of a concert of strategies which involve displacement, deprivation of nutrients, and disruption of habitat, a strong Th2 inflammation represents the central player of host reaction, which is displayed through the recruitment and activation of eosinophils at the site of infection (Rosenberg et al. 2013; Kopf et al. 1996). Studies on eosinophils in infectious diseases are demonstrating how the classical role of eosinophils as end effector cells involved mainly in parasitic infestations is obsolete. More and more evidence highlights how these cells are dynamic elements that could even be recruited by the pathogens to exert a protection against the host immune system. The concept of eosinophils as mediators in killing parasites is based on histopathologic evidence of these cells surrounding dying parasites as well as in vitro experiments demonstrating eosinophil degranulation with helminth clearance (Klion and Nutman 2004). Recently, this paradigm has been challenged and a new role for eosinophils in parasitic infections has been proposed. Some studies suggest that eosinophils could be recruited by helminths during different phases of the infection with the aim of obtaining protection from the immune system

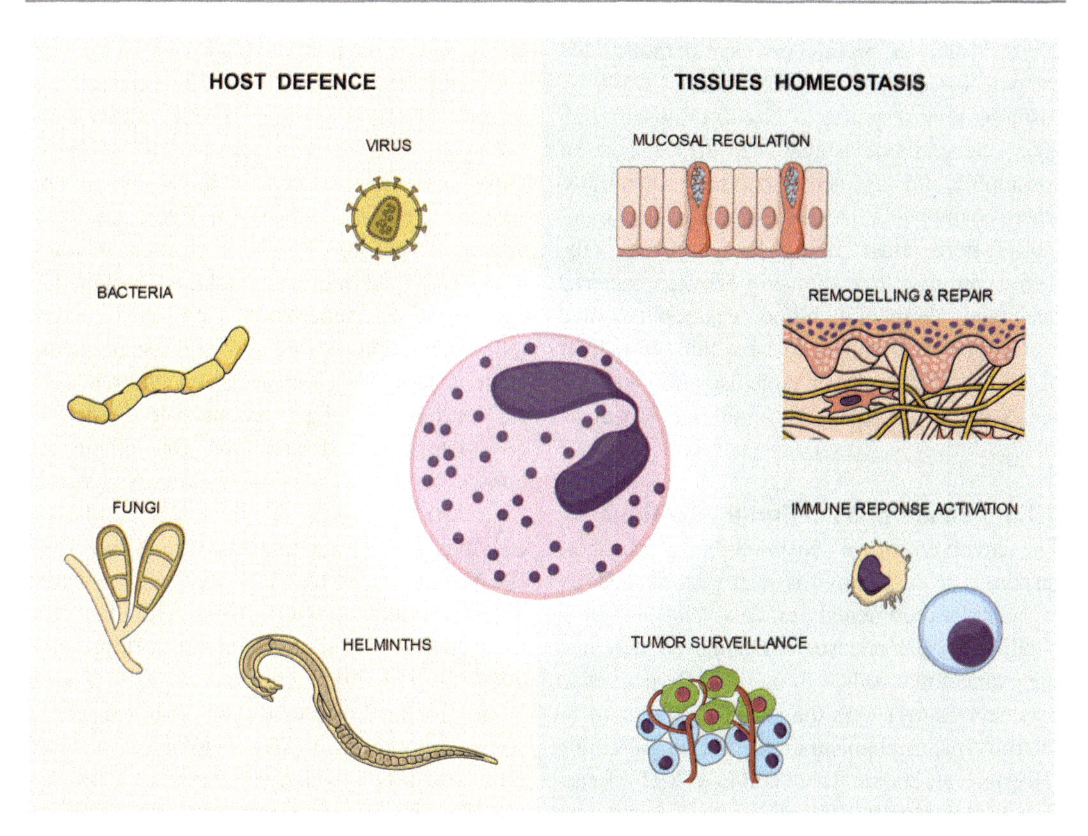

Fig. 2 Eosinophils' role in health
Eosinophils' roles in health are focused not only on host defense (yellow square) but also on tissue homeostasis (green square). The yellow square depicts actions against pathogens such as helminths, fungi, bacteria, and viruses, while the green square highlights the mucosal regulation and tissue remodeling and repair besides immune response activation and tumor surveillance

reaction and repair tissues damaged by parasite invasion (Huang and Appleton 2016). In a mouse model of *T. spiralis* infection, eosinophils promote the expansion of dendritic cells and CD4+ T-lymphocytes by the production of IL-10, causing tissue larvae survival (Huang et al. 2014).

3.1.2 Eosinophils in Bacterial Infections

Eosinophils have been shown to interact with bacterial pathogens in a variety of ways depending on the bacterial species and the microenvironment condition (Svensson and Wennerås 2005; Ravin and Loy 2015). Engulfing bacterial organisms by phagocytosis (Weller and Goetzl 1980) was observed in vitro with *Staphylococcus aureus*, *Escherichia coli*, and *Listeria monocytogenes* even if this mechanism seems to be not so effective. Degranulation with the release of enzymatic proteins, such as ECP, EPO, and MBP, showed strong bactericidal activities against *S. aureus* as well as with *Escherichia coli* (Hogan et al. 2008a; Bystrom et al. 2011; Persson et al. 2001). Studies demonstrated how eosinophil granule proteins have also been implicated in the killing of another Gram-negative bacterium, *Pseudomonas aeruginosa*, and how ECP has affinity for bacterial lipopolysaccharide and peptidoglycan (Hogan et al. 2008a; Torrent et al. 2008). Another effective mechanism is represented by the production of extracellular mitochondrial DNA traps in response to bacteria. Both ECP and MBP were found to be localized within the extracellular DNA suggesting that bacterial killing was mediated by eosinophil granule proteins (Yousefi et al. 2008; Von Köckritz-Blickwede and Nizet

2009). Moreover, researchers have demonstrated catapult-like ejection of mitochondrial DNA by eosinophils in response to *E. coli* (Yousefi et al. 2008). Several studies evaluated the behavior of eosinophils during acute bacterial infections, where eosinopenia has been shown to be a common feature (Bass 1975). During bacteremia, there is an inverse relationship between bacterial load and peripheral blood eosinophils, and eosinopenia was shown to be a reliable marker of bacterial etiology in patients admitted with sepsis to the intensive care unit (Davido et al. 2017; Abidi et al. 2008).

3.1.3 Eosinophils in Fungi Infections

The involvement of eosinophils in the host response against fungi is well established by in vitro experimental models and in vivo conditions. The release of granule content into the extracellular milieu through TLRs activation by direct contact with the pathogen seems to be the principal mechanism implied in fungal killing (Hogan et al. 2008a; Ravin and Loy 2015). Reaction to *Alternaria alternata* or to *Coccidioides immitis* is documented in human infections (Garro et al. 2011). Activated eosinophils are demonstrated in lesions caused by *Paracoccidioides brasiliensis* (Garro et al. 2011). Besides, it has been proven how *Cryptococcus neoformans* can be phagocytosed by eosinophils, which are able to release IL-12, IFN-gamma, and TNF (Garro et al. 2011). A recent study provides evidence of a dual behavior of eosinophils after a challenge with *Aspergillus fumigatus* (Guerra et al. 2017). While the conidial killing ability of eosinophils and the increased susceptibility to *Aspergillus* infection of eosinophil-ablated mice were confirmed, eosinophils were also shown to be a prominent source of IL-23 and IL-17, which might play a crucial, detrimental role in the induction and maintenance of inflammation in allergic aspergillosis (Pégorier et al. 2006).

3.1.4 Eosinophils in Viral Infections

The role of eosinophils in mucosal immune responses against viruses, particularly viral respiratory infections, is fostered by evidence of their direct involvement in vivo as well as in vitro findings (Piehler et al. 2011). The expression of several TLRs, including TLR3, TLR7, and endosomal TLR9, leads eosinophils to detect viral- or microbe-associated molecular patterns such as genetic fragments of RNA and DNA (Wong et al. 2007). The activation of these receptors determines cytokine production, degranulation, superoxide and nitric oxide (NO) generation, as well as the release of chemotactic factors (Mantis et al. 2011). Different clinical settings shed light on the role exerted by eosinophils in viral infections. One of the best known instances refers to respiratory syncytial virus (RSV) disease where eosinophil-derived enzymes are proven to reduce virus infectivity (Domachowske et al. 1998) and the production of azote monoxide (NO). Along with this, EETs have direct antiviral effects in vitro (Phipps et al. 2007; Su et al. 2015; Silveira et al. 2019; Yousefi et al. 2018). Eosinophils are also capable of secreting preformed TH1 cytokines, including IL-12 and IFN-γ, which are important for mounting effective antiviral responses (Davoine and Lacy 2014). Furthermore, gene and cytokine depletion studies highlight the role of Th2 spectrum cytokines as critical in response to formalin-inactivated RSV (Hogan et al. 2008a). Another example refers to the host response against influenza viruses. Recent studies showed that, after a challenge with influenza A virus, eosinophils are able to undergo piecemeal degranulation, upregulate antigen presentation molecules, and enhance CD8+ T-cell response (Samarasinghe et al. 2017). Retrospective clinical studies may suggest a role of chronic eosinophilic inflammation in the response to viral infections. During the 2009 H1N1 influenza pandemic, patients affected by asthma had a higher risk of being hospitalized; however, they also had a lower risk of complications or death (van Kerkhove et al. 2011; Louie et al. 2009). Other studies show how pediatric patients who developed acute pneumonia had high IL-5 levels and peripheral eosinophilia (Vaillant et al. 2009), suggesting that eosinophil recruitment may be crucial for late-stage anti-influenza defense. Moreover, allergic mice seem to be protected from airway damage induced by influenza A virus (Terai et al. 2011). Besides respiratory viruses, eosinophilia is a frequent finding in patients affected by human immunodeficiency virus (HIV) from the early stage until the onset of acquired immunodeficiency syndrome (AIDS) (Vrtis et al. 2020), even in the absence of other infective triggers or allergic conditions.(Chou and Serpa 2015). The role of eosinophils in HIV is still unclear: although some enzymes released during their activation, such as EDN, have been proved to possess inhibitory activity against the virus (Fulkerson and Rothenberg 2013), the increasing level of eosinophils observed in the late stage of disease does not seem to support a major defensive role (Platt et al. 1998). This peculiar finding could be determined by a progressive shift in cytokine production toward a Th2 pattern (Platt et al. 1998), a phenomenon observed in patients progressing toward AIDS (Taylor et al. 2004) which could be the result of a pathologic mechanism impairing CD8+ T-cell response of the host (Barker et al. 1995).

COVID-19

Coronavirus disease 19 (COVID-19) caused by the SARS-2 coronavirus (SARS-CoV-2) has emerged as the third severe lower respiratory tract infection caused by a coronavirus in the twenty-first century. Nowadays, it represents a great burden for all health systems and the economy of the countries involved. One of the most surprising findings is the presence of severe eosinopenia in a large amount of infected people (Zhang et al. 2020). This biochemical feature seems to be related to the gravity of the disease, and, most importantly, it has been demonstrated to revert in patients who recover (Du et al. 2020). Samples from lung biopsies or bronchoalveolar lavage (BAL) from COVID-19 patients show an important inflammation characterized by the infiltration of lymphocytes and macrophages with hyperexpression of pentraxin 3 (PTX3),(Brunetta et al. 2020) but there were no signs of eosinophils (Hotez et al. 2020; Liao et al. 2020). Even though the pathophysiology of eosinopenia remains unclear, it is hypothesized that it could result from a combination of factors, including the inhibition of bone marrow eosinophilopoiesis, a reduced expression of chemokine receptors/adhesion factors (Bass 1975; Hassani et al. 2020), a direct induction of apoptosis during the acute infection, and even an excessive consumption by eosinophil antiviral actions (Wardlaw 1999) as a marker of host exhaustion (Jesenak et al. 2020). In conclusion, eosinopenia in COVID-19 seems a frequent feature, but more evidence is required to disclose the biological meaning of what has been observed.

4 Detrimental Type 2 Inflammation: Eosinophils as Mediator of Damage

Storing a plethora of cytokines, chemokines, and growth factors in their granules, eosinophils represent one of the most powerful agents involved in augmenting Th2 inflammation. Those molecules could be released in a very rapid way under pro-inflammatory stimuli such as alarmins and ILC2 activation. Moreover, expressing cognate receptors for many of the mediators they secrete (such as IL-5, GM-CSF, and eotaxin-1), eosinophils can be stimulated by autocrine regulation. The uncontrolled release of cationic proteins could lead to tissue damages and fibrosis; indeed, MBP results toxic not only to helminths but also to the airway epithelium by causing disarray within the lipid bilayer and increasing the cell permeability (Hogan et al. 2008b). In patients affected by asthma, some studies reveal a direct correlation between severity of bronchial hyperreactivity and MBP concentrations in BAL (Gleich and Adolphson 1993; Leigh et al. 2000; Gleich 2000). Furthermore, MBP associates with epithelium damage in chronic rhinosinusitis (Ponikau et al. 2005). Bronchoconstriction and responsiveness to inhaled methacholine were both augmented by instillation of human MBP and EPO (Gleich and Adolphson 1993). Neutralization of endogenously secreted MBP seemed to prevent antigen-induced bronchial hyperreactivity in guinea pigs (Costello et al. 1999). On the other hand, both ECP and EDN have been proven to possess neurotoxic activities (Durack et al.

1979; Gleich and Adolphson 1986). Damaged tissues of patients affected by eosinophilic esophagitis were analyzed, and studies demonstrated the presence of EDN in active lesions (Kephart et al. 2010). On the contrary, deposition of this enzyme seems to be efficaciously reduced in patients treated with anti-IL-5 antibody (Straumann et al. 2010).

As they express complement receptors (Fischer et al. 1986), eosinophils are capable of activating their lytic actions as demonstrated by C5a fraction (Zeck-Kapp et al. 1995). Furthermore, these cells could act through antibody-dependent cellular cytotoxicity (ADCC) against parasites and mammalian targets (Hallam et al. 1982) by activation of their Fc receptors (FcαR, FcγRI–III, and FcεRI–II) (Gounni et al. 1994; Decot et al. 2005; Grangette et al. 1989; Hartnell et al. 1992). Eosinophils promote fibroblast proliferation (Minshall et al. 1997), matrix metalloproteinase and TGF-β expression, and extracellular matrix protein synthesis (Pégorier et al. 2006) with active production of TGF-β1. As a matter of fact, degranulating eosinophils are found in areas of fibrogenesis, suggesting a potential profibrotic role (Birring et al. 2005). Locating near nerves in chronic inflammatory conditions (Smyth et al. 2013; Fryer et al. 2006), eosinophils could be also responsible for damaging these structures (Hogan et al. 2001) as well as mediating hyperactivation (Elbon et al. 1995). By means of DNA and granule protein extrusion into the extracellular space, eosinophils give rise to EETs which are well described in allergic asthma (Dworski et al. 2011), drug hypersensitivity reactions, and allergic contact dermatitis (Simon et al. 2011). Another overlooked aspect due to eosinophil activation refers to their prothrombotic activity which was proven in several pathological conditions even if not well characterized (Fig. 3; Kojima and Sasaki 1995).

4.1 When Things Go Wrong: Allergy

While eosinophils are associated with a variety of autoimmune diseases, they have mostly been studied in disorders which have allergy and Th2 response as their main pathogenetic drivers. Recruited by damaged epithelia which lead to ILC2 activation through alarmins, eosinophils are involved in stimulating the Th2 pathway. Indeed, through the secretion of high amounts of preformed cytokines, these cells are directly involved in homing and polarizing of lymphocytes, along with isotype switching toward immunoglobulin E (Fig. 3).

4.1.1 Atopic Dermatitis

In atopic dermatitis (AD), a defective barrier function of the skin allows antigens to be exposed more easily to the skin's immune system, leading in turn to humoral and cellular immune activation (Ramirez et al. 2018; Furue et al. 2017; Werfel et al. 2016). It has been proposed that AD patients may have two subsets of the disease, one mostly driven by Th2 and one by Th1/Th17 immune responses. Eosinophil infiltration of the skin and blood eosinophilia is common in AD, with evidence of blood levels correlating with severity of the disease. Interestingly, the relevance of eosinophils in AD may have been hinted by the efficacy of dupilumab, a monoclonal antibody which targets a receptor chain (IL-4Rα) common to the IL-4 and IL-13 receptors (Ramirez et al. 2018).

4.1.2 Upper Airway Disorders

In allergic rhinitis (AR), along with chronic rhinitis and nasal polyposis, tissue eosinophilia is an almost invariable finding (Ramirez et al. 2018). These infections are commonly associated with asthma and, just like in the latter, exhibit a most likely IL-5-driven Th2 immune response (Bachert et al. 2006). In AR, volatile allergens lead to the recruitment of eosinophils several hours after the encounter. Eosinophils interact with mast cells and Th2 cells leading plasma cells to an IgE class switch and increased IgE production and mast cells to increased degranulation (Ramirez et al. 2018). Besides, eosinophil degranulation leads to direct tissue damage to the nasal epithelium. Chronic rhinosinusitis and nasal polyposis also appear to be strongly influenced by an environment rich in IL-5 (Ramirez et al. 2018).

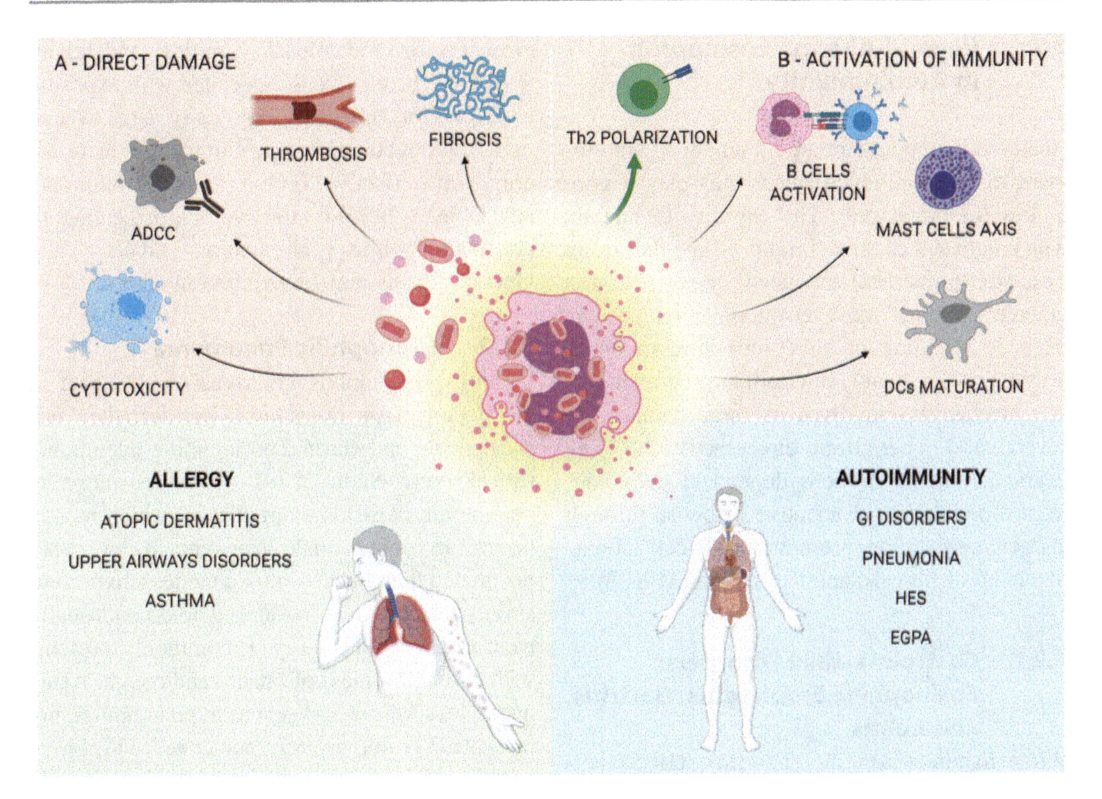

Fig. 3 Eosinophils' role in disease
(**a**) **Direct damage**: eosinophil activation exerts damages to cells and tissues through several mechanisms among which the principal is represented by granule release causing surrounding cytotoxicity. Furthermore, expressing several Fc receptors, eosinophils are able to induce cytotoxicity mediated by antibodies (ADCC). Other pathological consequences are mediated by the high amount of cytokines and growth factors such as TGF-β which link, respectively, to eosinophil activation to thrombosis and fibrosis. (**b**) **Immunity activation**: innate and adaptive immunity can be triggered, ruled, and polarized toward Th2 responses under the plethora of cytokines and chemokines secreted by activated eosinophils. Green (allergy) and blue (autoimmunity) squares represent the principal disease classes in which eosinophils have a pathogenic role (see paragraph IV)

4.1.3 Asthma

For what concerns asthma, eosinophilic involvement defines the subset of "eosinophilic asthma," which has a severe phenotype with increased risk of severe exacerbations and near-fatal events (Fulkerson and Rothenberg 2018; Moore et al. 2010; Yancey et al. 2017). In these patients, drugs targeting the IL-5 pathway proved to be highly effective (Fulkerson and Rothenberg 2013). Mepolizumab and the more recent benralizumab and reslizumab all showed promising results in reducing the rates of asthma exacerbations in patients with eosinophilic asthma (Ramirez et al. 2018; Ortega et al. 2014). Genetic, environmental local host factors may play a determining role in shaping the asthma subset. A likely candidate for local immune regulation is pentraxin 3 (PTX3). On the one hand, PTX3 seems to limit airway remodeling (Zhang et al. 2012). On the other hand, PTX3 stimulates high levels of eosinophil inflammation (Ramirez et al. 2018). Eosinophils were also postulated to be causing irreversible fibrotic remodeling through the release of TGF-β (Ramirez et al. 2018; Ohno et al. 1996).

4.2 Beyond Allergy: Eosinophils in Autoimmunity

Apart from atopy, there are a number of autoimmune diseases in which eosinophils play a more or less important role. The capacity to interact with innate and adaptive immunity and the strong cytotoxic properties, mediated mostly through granule proteins, make eosinophils one of the main actors initiating or contributing to organ destruction in many autoimmunity conditions. The mechanisms involved in tissue damage are several and range from direct activation, i.e., degranulation or antibody-dependent cytotoxicity, to modulation of adaptive response through antigen presentation, promotion of B-cell actions, induction of fibrosis, and tissue repair (Fig. 3).

4.2.1 Gastrointestinal Disorders: Eosinophilic Esophagitis, Gastritis, and Colitis

While hypereosinophilic syndrome (HES) is a systemic syndrome, there are several organ-restricted manifestations of eosinophilia. Eosinophilic gastritis and eosinophilic gastroenteritis are GI tract disorders in which GI mucosa is infiltrated with abnormal numbers of eosinophils (≥20 eosinophils per high-power field) (Rached and Hajj 2016). The histological finding of local eosinophilia is often coupled with elevated levels of eosinophils in blood (Fulkerson and Rothenberg 2013). Symptoms are generally vague and include abdominal pain, nausea, and vomiting (Zhang and Li 2017). The differential diagnosis is extensive and includes infections and inflammatory bowel disorder. Treatment is based upon glucocorticoids or elimination diets, when sensitization to specific allergens is found (Ramirez et al. 2018). Eosinophilic esophagitis is a more common disorder, with a prevalence of about 0,4% in the Western world (Furuta and Katzka 2015). It is characterized by extensive esophageal eosinophilia (≥15 eosinophils per high-power field), and symptoms vary depending on age (Ramirez et al. 2018). In adults, eosinophilic esophagitis is characterized by dysphagia due to bolus impactions. In children, it can present with food refusal, vomiting, and failure to thrive. The disease is associated with exposure to several food allergens (e.g., dairy, eggs, wheat, or soy products) although elimination diets are not always effective. Proton pump inhibitors are advisable as they are effective in a percentage of patients, while elimination diets and glucocorticoids may be required in others.

4.2.2 Eosinophilic Pneumonia

In lungs, eosinophils notoriously participate in the pathogenesis of asthma. Besides, they can cause acute and chronic eosinophilic pneumonia (Suzuki and Suda 2019). Acute eosinophilic pneumonia (AEP) is a rare disease which mostly occurs in young male smokers. It has been postulated that tobacco smoke incites a hypersensitivity reaction in individuals who do not usually have a history of allergy. It commonly presents with fever, shortness of breath, and cough. Acute respiratory failure and severe hypoxemia are not rare. AEP is not strongly associated with blood eosinophilia, while diagnosis is performed with the aid BAL. If alveolar fluid demonstrates a level of eosinophilia major to 25%, along with suggestive symptoms, is considered to be diagnostic for AEP. Treatment is based upon systemic glucocorticoids for 2–4 weeks, starting with high doses when clinically needed. Chronic eosinophilic pneumonia (CEP) is a form of interstitial lung disease which can occur at any age, from childhood to old age, and mostly occurs in women. It is more strongly associated with allergy while smokers are uncommon. Cough and shortness of breath with a subtle onset are the most common presentation, while acute deterioration is rare. Blood eosinophilia is observed in most patients, together with elevated IgE levels. In this case, BAL eosinophilia (40–60%) allows to establish the diagnosis. Treatment includes prolonged treatment with systemic glucocorticoids, and, later during the course of the disease, inhaled steroids may be used to reduce the burden of side effects. Omalizumab (anti-IgE monoclonal antibody) and mepolizumab also showed some efficacy in reducing the need for steroids in these patients (Suzuki and Suda 2019).

4.2.3 Hypereosinophilic Syndrome

Firstly, HES should be mentioned. HES is a group of diseases where hypereosinophilia (peripheral blood eosinophils >1.5 × 10 (McBrien and Menzies-Gow 2017) eosinophils per liter) leads to extensive tissue infiltration of eosinophils (Fulkerson and Rothenberg 2018; Simon et al. 2010). Tissue damage ensues and organs targeted comprise the lungs, the skin, the GI tract, and the cardiovascular system. Manifestations can be various and include asthma, pleural effusions, thromboembolic disease, peripheral neuropathy, vasculitis, hematologic abnormalities, myocarditis, and others (Williams et al. 2016). Management depends on the underlying cause, which can be myeloproliferative, lymphocytic (in which eosinophil production is secondary to T-cells' excessive IL-5 production), associated with other conditions, or unknown. While steroids are the mainstay of treatment, at least during the acute phase of the disease, they are less effective in the subset of myeloproliferative HES and anyhow bear considerable side effects. Hydroxyurea and tyrosine kinase inhibitors have been used to treat HES chronically, with good results upon selected subsets of patients. Recently, mepolizumab (anti-IL-5 antibody) was shown to be highly effective in patients without mutations of the FIP1L1 and PDGFRα genes. In this subset of myeloproliferative HES patients, imatinib leads to the best outcome (Curtis and Ogbogu 2016).

4.2.4 EGPA

Eventually, eosinophils are fundamental in the pathogenesis of eosinophilic granulomatosis with polyangiitis (EGPA, previously known as Churg-Strauss syndrome) (Greco et al. 2015). According to some, EGPA may be the crossing bridge between HES and ANCA-associated vasculitides, although formally it belongs to the latter group. Organ damage in EGPA is due to vasculitis (mediated by ANCA in a subset of around 40% of patients) and eosinophil infiltration. It is also hypothesized that some patients may have mostly an ANCA-driven and others an eosinophil-driven pathogenesis, which would be reflected in the different pattern of organ involvement of ANCA+ and ANCA- patients (Schroeder et al. 2019). Eosinophils are observed at high levels both in blood and within lesions, where a Th2 response (along with Th1 and Th17) is present. High levels of eotaxin seem to drive eosinophil infiltration within lesions (Greco et al. 2015). In EGPA, manifestations vary depending upon the phase of the disease. Initially, patients may have suffered from asthma and nasal polyposis for several years. Later, hypereosinophilia and organ infiltration occur. However, it is only during the third stage of the disease that vasculitic lesions appear. These can affect most organs, although the upper airways, the peripheral nervous system, the kidneys (especially in the ANCA+ subset), the myocardium (especially in ANCA- patients), the lungs, the GI tract, and the skin are the most common targets of the disease (Sablé-Fourtassou et al. 2005; McKinney et al. 2014). Diagnosis is based upon the presence of suggestive manifestations and peripheral eosinophilia. Treatment consists of steroids and immunosuppressants, among which cyclophosphamide has historically been the most well studied. Recently, mepolizumab has been introduced in the management of EGPA with excellent results (Wechsler et al. 2017; McBrien and Menzies-Gow 2018).

5 The Emerging Role of Eosinophils in Malignancy

The tumor microenvironment (TME) plays a significant role in cancer growth, and numerous studies have shown the positive correlation between lymphocyte infiltrates in tumors and survival outcomes (Gajewski et al. 2013). Eosinophils, in the form of tumor-associated tissue eosinophilia (TATE), constitute a significant fraction of the leukocyte infiltrate surrounding different cancer histotypes, both solid and hematological. However, their role is more ambiguous than other inflammatory cells: the exact relationship between eosinophils and cancer outcomes is still unclear. One possible explanation is the fact that the definition and measurement of tissue eosinophilia are difficult and not uniform among authors, precluding a comparison of the results obtained from different studies (Alkhabuli and High 2006). Moreover, different cancer histotypes have different responses to TATE, and even when there's a positive correlation, some authors support the hypothesis that eosinophils are crucial in the host defense against cancer while others suggest that the antitumor effect of TATE is barely influential (Sakkal et al. 2016; Samoszuk 1997; Pereira et al. 2011). Limited evidence shows that cancer cells recruit eosinophils by producing chemokines, such as eotaxin-1/CCL11 and eotaxin-2/CCL24 in oral carcinoma as well as Hodgkin's lymphoma, and CCL17 in peripheral T-cell lymphomas (Thielen et al. 2008; Teruya-Feldstein et al. 1999; Lorena et al. 2003). Additionally, necrotic areas of tumors release alarmins or DAMPs that directly recruit various immune cells including eosinophils (Bertheloot and Latz 2017). Lastly, eosinophil infiltration of cancers can also be mediated by the production of chemotactic factors by tumor-infiltrating immune cells. Although there are numerous ways in which eosinophils can suppress tumor growth, none is proven to be the predominant. Eosinophils in the TME are thought to express the same receptors and mediators as cytotoxic T-lymphocytes (CTLs) and to be directly involved in antitumor response by secreting chemokines to promote tumor immune surveillance mediated by CD8+ T cells and M1 macrophages (Gatault et al. 2012; Carretero et al. 2015; Biswas and Mantovani 2010). In some cases, such as in the colo-205 cell line, eosinophils can induce cell death with some selectivity in their tumoricidal properties, which are dependent on the CD11a-/CD18-mediated stable contacts with target cells, while in vitro studies proved that eosinophils can directly kill cancer cells by releasing cytotoxic granules (Legrand et al. 2010; Capron Loiseau et al. 2016; Costain et al. 2001). Some studies also suggested that eosinophils act as antigen-presenting cells. If it is true that the mechanisms by which eosinophils can counteract tumor growth are numerous, it is also true that eosinophils lead to the Th2 response, which does not correlate positively with the peritumoral inflammatory response. Th2 phenotypes are particularly undesirable in early tumor development as they do not promote or improve the killing capacity of cells associated with antitumor responses. They may even favor the growth of cancer with enhanced angiogenesis and release of growth factors depending on the surrounding stimuli (Esposito et al. 2004; Puxeddu et al. 2010; Murdoch et al. 2008).

5.1 Solid Tumors

Some solid tumors may induce an increase in peripheral eosinophil count, but this paraneoplastic eosinophilia is for the most part not clinically relevant and is usually rare, fluctuating between 0.5% and 7% (Falchi and Verstovsek 2015). Previous studies demonstrated that different types of tumor correlate differently with TATE: in most cancer types (skin melanoma and colorectal, breast, gastric, oral, larynx, and nasopharyngeal cancers), eosinophils have an antitumoral role, whereas in other tumors, eosinophils play a pro-tumoral action (lung and uterine cervix cancer) or no effect at all (brain and bladder cancer) (Subeikshanan et al. 2016; Varricchi et al. 2018). A positive prognostic impact of TATE for oral squamous cell carcinoma (OSCC) patients has been described in the previous literature. An association between the intensity of TATE and an increased disease-free survival (DFS) as well as overall survival was proved (Rakesh et al. 2015). However, some publications actually described eosinophilic tumor infiltration as a marker of unfavorable prognosis in OSCC patients (Rakesh et al. 2015; Yellapurkar et al. 2016). For what concerns colorectal cancer, an association between decreasing eosinophilic infiltration and increasing aggressiveness was confirmed by some researchers (Kiziltaş et al. 2008). Moreover, the occurrence of colorectal cancer was found to be inversely correlated with circulating eosinophils, suggesting a protective role for eosinophils in cancer development (Prizment et al. 2011;

Saraiva and Carneiro 2018). In gastric cancer, high levels of TATE correlated with an improved survival rate (Cuschieri et al. 2002). A similar positive association of TATE with improved prognosis was observed in esophageal cancer (Zhang et al. 2014) as well as in breast cancer where was described a reduced risk of disease recurrence in patients with higher peripheral eosinophil count (Ownby et al. 1983).

A recent meta-analysis showed that TATE was notably associated with improved overall survival in patients with solid tumors of the GI tract (especially colorectal cancer and esophageal cancer), whereas no difference in overall survival existed in oral cancer, laryngeal cancer, or cervical cancer (Saraiva and Carneiro 2018; Reichman et al. 2019). There appears to be no correlation between DFS and TATE, even if TATE may be inversely correlated with lymphoid invasion and lymph vessel metastasis (Hu et al. 2020; Kurose et al. 2019). These apparently conflicting results suggest that the role of eosinophils and their mediators in human tumors could be specific to different types of cancer.

5.2 Eosinophilia and Hematologic Malignancies

Different from solid tumors, eosinophilia is a recurrent phenomenon in both myeloid and lymphoid disorders. By using the International Classification of Disease for Oncology (version 3), the Surveillance, Epidemiology and End Results (SEER) database, it is shown that the incidence rate of eosinophilia due to onco-hematological causes was approximately 0.36 per million from 2001 to 2005 (Johnson and George 2013). In myeloid malignancies, eosinophilia is a direct product of clonal expansion of a mutated hematopoietic stem cell, whereas in lymphoid malignancies, eosinophils are usually part of the TME and are recruited by cytokines such as IL-5 and IL-13 (Schrezenmeier et al. 1993). Three major eosinophilia-associated myeloid neoplasms (MN-eos) are defined. The first category is the most common and is driven by constitutively activated tyrosine kinase fusion genes, such as platelet-derived growth factor receptor a (PDGFRα), platelet-derived growth factor receptor b (PDGFRβ), or fibroblast growth factor receptor 1 (FGFR1) (Noel 2012; Reiter and Gotlib 2017; Shomali and Gotlib 2019). The second WHO-defined MN-eos is chronic eosinophilic leukemia-not otherwise specified (CEL-NOS), which is included among the myeloproliferative neoplasms (MPN). This definition is operational and requires the absence of the Philadelphia chromosome or rearrangements of PDGFRα, PDGFRβ, and FGFR1, demonstration of more than 2% of blast cells in the peripheral blood or more than 5% in the bone marrow, and evidence of clonality of the eosinophil population (Shomali and Gotlib 2019). The third condition is idiopathic HES, a diagnosis of exclusion when other benign and malign cause of hypereosinophilia can't be diagnosed (Shomali and Gotlib 2019). Other myeloid neoplasms that may present themselves with an increased eosinophil count are Philadelphia-positive BCR-ABL+ chronic myeloid leukemia (CML), especially in the hypereosinophilic variant (eos-CML) and acute leukemia, in particular acute myelomonocytic leukemia (Valent et al. 2004). With the exception of CML, in which the eosinophil count is traditionally considered an unfavorable prognostic factor (Hasford et al. 1998), the prognostic importance of eosinophilia was studied only in limited cases: in a large study of 1008 patients, eosinophilia predicted a significantly reduced overall survival without influencing DFS (Wimazal et al. 2010). Another study revealed that eosinophilia was linked with poorer prognosis due to reduction in both overall survival and DFS (Matsushima et al. 2003). Lymphoid malignancies are usually associated with eosinophils in a similar way of solid tumors, interacting with cancer cells in the TME and influencing cancer growth. Peripheral blood increases in eosinophil count (although mild) and eosinophil infiltration in the tumor are commonly described in Hodgkin's lymphoma (HL), especially the mixed cellularity or nodular sclerosis types, since Reed-Sternberg cells interact directly with eosinophils by secreting numerous cytokines and chemokines such as GM-CSF and IL-5. Tumor eosinophilia indicates poor prognosis, probably caused by eosinophil-induced stimulation of tumor cells, but some studies have shown that eosinophils may have a cytotoxic effect to cancer cells through the release of ECP (Subeikshanan et al. 2016; Varricchi et al. 2018; Cyriac et al. 2008; Glimelius et al. 2011). TATE is also present in other lymphoid malignancies, in particular precursor B-cell and T-cell lymphoblastic lymphoma/leukemia and cutaneous T-cell lymphoma. In these disorders, TATE is generally associated with a worse outcome (Suchin et al. 2001; Roufosse et al. 2012; Jin et al. 2015). The "lymphocytic variant of HES" (L-HES) represents a clonal disorder in its own right: mature T cells with Th2 differentiation produce IL-5 and consequently cause peripheral blood eosinophilia (Simon et al. 1999). The prevalence of this disorder in patients with unexplained HES is estimated to be up to 27%. The diagnosis is frequently difficult and requires the demonstration of a clonally rearranged T-cell receptor on aberrant T cell with peculiar immunophenotype (CD3+, CD4−, CD8− cells or CD3−, CD4+ cells) (Simon et al. 1996). Together with this, elevated TARC (a chemokine implicated in Th2-mediated diseases) may be helpful in supporting the diagnosis. Fortunately, it is a pathology with a slow rate of progression, which resembles an inflammatory disease (Roufosse et al. 2007).

6 Conclusion

Eosinophils appear to be more and more fundamental in Th2 inflammation. The paradigm of end effector cells involved in helminth immunity and allergy is slowly fading in favor of a highly active inflammatory cell type able to enhance the Th2 response in different physiological and pathological settings. More studies are required to dissect the intrinsic ability eosinophils show to interact avidly with epithelia, tissue-resident immune cells, and adaptive immunity, in the context of autoimmune diseases.

Acknowledgments Figures 1 and 3, created with BioRender.com

Figure 2 created with MindGraph.

Conflict of Interest All authors declare they have no conflict of interest and they have not received any financial support for this work.

Ethical Approval The authors declare that this article does not contain any studies with human participants or animals.

References

Abidi K et al (2008) Eosinopenia is a reliable marker of sepsis on admission to medical intensive care units. Crit Care 12

Alkhabuli JO, High AS (2006) Significance of eosinophil counting in tumor associated tissue eosinophilia (TATE). Oral Oncol 42:849–850

ARCHER GT, HIRSCH JG (1963) Motion picture studies on degranulation of horse eosinophils during phagocytosis. J Exp Med 118:287–294

Aupperlee MD et al (2014) Epidermal growth factor receptor (EGFR) signaling is a key mediator of hormone-induced leukocyte infiltration in the pubertal female mammary gland. Endocrinology 155:2301–2313

Bachert C, Patou J, Van Cauwenberge P (2006) The role of sinus disease in asthma. Curr Opin Allergy Clin Immunol 6:29–36

Barker E, Mackewicz CE, Levy JA (1995) Effects of TH1 and TH2 cytokines on CD8+ cell response against human immunodeficiency virus: implications for long-term survival. Proc Natl Acad Sci U S A 92:11135–11139

Bass DA (1975) Behavior of eosinophil leukocytes in acute inflammation. II Eosinophil dynamics during acute inflammation. J Clin Invest 56:870–879

Bertheloot D, Latz E (2017) HMGB1, IL-1α, IL-33 and S100 proteins: dual-function alarmins. Cell Mol Immunol

Birring SS et al (2005) Sputum eosinophilia in idiopathic pulmonary fibrosis. Inflamm Res 54:51–56

Biswas SK, Mantovani A (2010) Macrophage plasticity and interaction with lymphocyte subsets: Cancer as a paradigm. Nat Immunol

Brinkmann V et al (2004) Neutrophil extracellular traps kill bacteria. Science 303(80):1532–1535

Brunetta E et al (2020) Macrophage expression and prognostic significance of the long pentraxin PTX3 in COVID-19. Nat Immunol 22

Butterworth AE (1985) Cell-mediated damage to helminths. Adv Parasitol 23:143–235

Bystrom J, Amin K, Bishop-Bailey D (2011) Analysing the eosinophil cationic protein – a clue to the function of the eosinophil granulocyte. Respir Res 12

Cambier J, Morrison D (1991) Modeling of T cell contact-dependent B cell activation. IL-4 and antigen receptor

ligation primes quiescent B cells to mobilize calcium in response to Ia cross-linking. J Immunol:2075–2082
Capron Loiseau M et al (2016) Human eosinophils exert TNF- human eosinophils exert TNF-a and granzyme A-mediated tumoricidal activity toward colon carcinoma cells. J Immunol
Carretero R et al (2015) Eosinophils orchestrate cancer rejection by normalizing tumor vessels and enhancing infiltration of CD8 + T cells. Nat Immunol
Chou A, Serpa JA (2015) Eosinophilia in patients infected with human immunodeficiency virus. Curr HIV/AIDS Rep 12:313–316
Chu DK et al (2014) Indigenous enteric eosinophils control DCs to initiate a primary Th2 immune response in vivo. J Exp Med 211:1657–1672
Cline MJ, Lehrer RI (1968) Phagocytosis by human monocytes. Blood 32:423–435
Conroy DM, Williams TJ (2001) Eotaxin and the attraction of eosinophils to the asthmatic lung. Respir Res 2:150–156
Costain DJ, Guha AK, Liwski RS, Lee TDG (2001) Murine hypodense eosinophils induce tumour cell apoptosis by a granzyme B-dependent mechanism. Cancer Immunol Immunother
Costello RW et al (1999) Antigen-induced hyperreactivity to histamine: Role of the vagus nerves and eosinophils. Am J Physiol 276
Curtis C, Ogbogu P (2016) Hypereosinophilic syndrome. Clin Rev Allergy Immunol 50:240–251
Cuschieri A et al (2002) Influence of pathological tumour variables on long-term survival in resectable gastric cancer. Br J Cancer
Cyriac S, Sagar TG, Rajendranath R, Rathnam K (2008) Hypereosinophilia in hodgkin lymphoma. Indian J Hematol Blood Transfus
Davido B et al (2017) Changes in eosinophil count during bacterial infection: revisiting an old marker to assess the efficacy of antimicrobial therapy. Int J Infect Dis 61:62–66
Davoine F, Lacy P (2014) Eosinophil cytokines, chemokines, and growth factors: emerging roles in immunity. Front Immunol 5
Decot V et al (2005) Heterogeneity of expression of IgA receptors by human, mouse, and rat eosinophils. J Immunol 174:628–635
Del Pozo V et al (1992) Eosinophil as antigen-presenting cell: activation of T cell clones and T cell hybridoma by eosinophils after antigen processing. Eur J Immunol 22:1919–1925
Domachowske JB, Dyer KD, Adams AG, Leto TL, Rosenberg HF (1998) Eosinophil cationic protein/ RNase 3 is another RNase A-family ribonuclease with direct antiviral activity. Nucleic Acids Res 26
Du Y et al (2020) Clinical features of 85 fatal cases of COVID-19 from Wuhan. A retrospective observational study. Am J Respir Crit Care Med 201:1372–1379
Durack DT, Sumi SM, Klebanoff SJ (1979) Neurotoxicity of human eosinophils. Proc Natl Acad Sci U S A 76:1443–1447
Dworski R, Simon HU, Hoskins A, Yousefi S (2011) Eosinophil and neutrophil extracellular DNA traps in human allergic asthmatic airways. J Allergy Clin Immunol 127:1260–1266
Elbon CL, Jacoby DB, Fryer AD (1995) Pretreatment with an antibody to interleukin-5 prevents loss of pulmonary M2 muscarinic receptor function in antigen-challenged Guinea pigs. Am J Respir Cell Mol Biol 12:320–328
Elishmereni M et al (2011) Physical interactions between mast cells and eosinophils: a novel mechanism enhancing eosinophil survival in vitro. Allergy Eur J Allergy Clin Immunol 66:376–385
Esposito I et al (2004) Inflammatory cells contribute to the generation of an angiogenic phenotype in pancreatic ductal adenocarcinoma. J Clin Pathol
Falchi L, Verstovsek S (2015) Eosinophilia in hematologic disorders. Immunol Allergy Clin N Am 35:439–452
Fischer E, Capron M, Prin L, Kusnierz JP, Kazatchkine MD (1986) Human eosinophils express CR1 and CR3 complement receptors for cleavage fragments of C3. Cell Immunol 97:297–306
Fryer AD et al (2006) Neuronal eotaxin and the effects of CCR3 antagonist on airway hyperreactivity and M2 receptor dysfunction. J Clin Invest 116:228–236
Fulkerson PC, Rothenberg ME (2013) Targeting eosinophils in allergy, inflammation and beyond. Nat Rev Drug Discov 12:117–129
Fulkerson PC, Rothenberg ME (2018) Eosinophil development, disease involvement, and therapeutic suppression. Adv Immunol 138:1–34
Furue M et al (2017) Atopic dermatitis: immune deviation, barrier dysfunction, IgE autoreactivity and new therapies. Allergol Int 66:398–403
Furuta GT, Katzka DA (2015) Eosinophilic esophagitis. N Engl J Med 373:1640–1648
Gajewski TF, Schreiber H, Fu YX (2013) Innate and adaptive immune cells in the tumor microenvironment. Nat Immunol
Garro AP, Chiapello LS, Baronetti JL, Masih DT (2011) Rat eosinophils stimulate the expansion of Cryptococcus neoformans-specific CD4+ and CD8+ T cells with a T-helper 1 profile. Immunology 132:174–187
Gatault S, Legrand F, Delbeke M, Loiseau S, Capron M (2012) Involvement of eosinophils in the anti-tumor response. Cancer Immunol Immunother 61:1527–1534
Gleich GJ (2000) Mechanisms of eosinophil-associated inflammation. J Allergy Clin Immunol 105:651–663
Gleich GJ, Adolphson CR (1986) The eosinophilic leukocyte: structure and function. Adv Immunol 39:177–253
Gleich GJ, Adolphson C (1993) Bronchial hyperreactivity and eosinophil granule proteins. Agents Actions 43:223–230
Glimelius I et al (2011) Effect of eosinophil cationic protein (ECP) on Hodgkin lymphoma cell lines. Exp Hematol
Goh YPS et al (2013) Eosinophils secrete IL-4 to facilitate liver regeneration. Proc Natl Acad Sci U S A 110:9914–9919

Gounni AS et al (1994) High-affinity IgE receptor on eosinophils is involved in defence against parasites. Nature 367:183–186
Gouon-Evans V, Rothenberg ME, Pollard JW (2000) Postnatal mammary gland development requires macrophages and eosinophils. Development 127:2269–2282
Gouon-Evans V, Lin EY, Pollard JW (2002) Requirement of macrophages and eosinophils and their cytokines/chemokines for mammary gland development. Breast Cancer Res 4:155–164
Grangette C et al (1989) IgE receptor on human eosinophils (FcERII). Comparison with B cell CD23 and association with an adhesion molecule. J Immunol 143
Greco A et al (2015) Churg-Strauss syndrome. Autoimmun Rev 14:341–348
Guerra ES et al (2017) Central role of IL-23 and IL-17 producing eosinophils as immunomodulatory effector cells in acute pulmonary aspergillosis and allergic asthma. PLoS Pathog 13
Hallam C, Pritchard DI, Trigg S, Eady RP (1982) Rat eosinophil-mediated antibody-dependent cellular cytotoxicity: investigations of the mechanisms of target cell lysis and inhibition by glucocorticoids. Clin Exp Immunol 48:641–648
Hartnell A, Kay AB, Wardlaw AJ (1992) IFN-gamma induces expression of Fc gamma RIII (CD16) on human eosinophils. J Immunol 148
Hasford J et al (1998) A new prognostic score for survival of patients with chronic myeloid leukemia treated with Interferon Alfa Writing Committee for the Collaborative CML Prognostic Factors Project Group. JNCI J Natl Cancer Inst
Hassani M et al (2020) Differentiation and activation of eosinophils in the human bone marrow during experimental human endotoxemia. J Leukoc Biol
Hogan SP et al (2001) A pathological function for eotaxin and eosinophils in eosinophilic gastrointestinal inflammation. Nat Immunol 2:353–360
Hogan SP et al (2008a) Eosinophils: biological properties and role in health and disease. Clin Exp Allergy 38:709–750
Hogan SP et al (2008b) Eosinophils: biological properties and role in health and disease. Clin Exp Allergy 38:709–750
Hotez PJ, Bottazzi ME, Corry DB (2020) The potential role of Th17 immune responses in coronavirus immunopathology and vaccine-induced immune enhancement. Microb Infect 22
Hu G et al (2020) Tumor-associated tissue eosinophilia predicts favorable clinical outcome in solid tumors: a meta-analysis. BMC Cancer 20:1–9
Huang L, Appleton JA (2016) Eosinophils in helminth infection: defenders and dupes. Trends Parasitol 32:798–807
Huang L et al (2014) Eosinophil-derived IL-10 supports chronic nematode infection. J Immunol 193:4178–4187
Jacobsen EA et al (2008) Allergic pulmonary inflammation in mice is dependent on eosinophil-induced recruitment of effector T cells. J Exp Med 205:699–710
Jesenak M, Banovcin P, Diamant Z (2020) COVID-19, chronic inflammatory respiratory diseases and eosinophils – observations from reported clinical case series. Allergy
Jin JJ, Butterfield JH, Weiler CR (2015) Hematologic malignancies identified in patients with hypereosinophilia and hypereosinophilic syndromes. J Allergy Clin Immunol Pract 3:920–925
Johnson RC, George TI (2013) The differential diagnosis of eosinophilia in neoplastic hematopathology. Surg Pathol Clin 6:767–794
Jordan MB, Mills DM, Kappler J, Marrack P, Cambier JC (2004) Promotion of B cell immune responses via an alum-induced myeloid cell population. Science 304 (80):1808–1810
Kephart GM et al (2010) Marked deposition of eosinophil-derived neurotoxin in adult patients with eosinophilic esophagitis. Am J Gastroenterol 105:298–307
Khandpur R et al (2013) NETs are a source of citrullinated autoantigens and stimulate inflammatory responses in rheumatoid arthritis. Sci Transl Med 5:178ra40-178ra40
Kiziltaş Ş, Ramadan SS, Topuzoğlu A, Küllü S (2008) Does the severity of tissue eosinophilia of colonic neoplasms reflect their malignancy potential? Turkish J Gastroenterol
Klion AD, Nutman TB (2004) The role of eosinophils in host defense against helminth parasites. J Allergy Clin Immunol 113:30–37
Kojima K, Sasaki T (1995) Veno-occlusive disease in Hypereosinophilic syndrome. Intern Med 34:1194–1197
Kopf M et al (1996) IL-5-deficient mice have a developmental defect in CD5+ B-1 cells and lack eosinophilia but have normal antibody and cytotoxic T cell responses. Immunity 4:15–24
Kurose N et al (2019) Adenosquamous carcinoma of the uterine cervix displaying tumor-associated tissue eosinophilia. SAGE Open Med Case Rep 7:2050313X1982823
Legrand F et al (2010) Human eosinophils exert TNF-α and granzyme A-mediated tumoricidal activity toward colon carcinoma cells. J Immunol 185:7443–7451
Leigh R et al (2000) Eosinophil cationic protein relates to sputum neutrophil counts in healthy subjects. J Allergy Clin Immunol 106:593–594
Liao M et al (2020) Single-cell landscape of bronchoalveolar immune cells in patients with COVID-19. Nat Med 26:842–844
Lorena SCM, Oliveira DT, Dorta RG, Landman G, Kowalski LP (2003) Eotaxin expression in oral squamous cell carcinomas with and without tumour associated tissue eosinophilia. Oral Dis
Louie JK et al (2009) Factors associated with death or hospitalization due to pandemic 2009 influenza A

(H1N1) infection in California. JAMA – J Am Med Assoc 302:1896–1902
Mantis NJ, Rol N, Corthésy B (2011) Secretory IgA's complex roles in immunity and mucosal homeostasis in the gut. Mucosal Immunol 4:603–611
Matsushima T et al (2003) Prevalence and clinical characteristics of myelodysplastic syndrome with bone marrow eosinophilia or basophilia. Blood
Mawhorter SD, Pearlman E, Kazura JW, Henry Boom W (1993) Class II major histocompatibility complex molecule expression on murine eosinophils activated In Vivo by Brugia Malayi. Infect Immun 61
McBrien CN, Menzies-Gow A (2017) The biology of eosinophils and their role in asthma. Front Med 4:93
McBrien CN, Menzies-Gow A (2018) Mepolizumab for the treatment of eosinophilic granulomatosis with polyangiitis. Drugs Today 54:93
McKinney EF, Willcocks LC, Broecker V, Smith KGC (2014) The immunopathology of ANCA-associated vasculitis. Semin Immunopathol 36:461–478
Melo RCN, Weller PF (2010) Piecemeal degranulation in human eosinophils: a distinct secretion mechanism underlying inflammatory responses. Histol Histopathol 25:1341–1354
Melo RCN, Perez SAC, Spencer LA, Dvorak AM, Weller PF (2005a) Intragranular vesiculotubular compartments are involved in piecemeal degranulation by activated human eosinophils. Traffic 6:866–879
Melo RCN et al (2005b) Human eosinophils secrete preformed, granule-stored interleukin-4 through distinct vesicular compartments. Traffic 6:1047–1057
Minshall EM et al (1997) Eosinophil-associated TGF-β1 mRNA expression and airways fibrosis in bronchial asthma. Am J Respir Cell Mol Biol 17:326–333
Moore WC et al (2010) Identification of asthma phenotypes using cluster analysis in the severe asthma research program. Am J Respir Crit Care Med 181:315–323
Mukherjee M, Lacy P, Ueki S (2018) Eosinophil extracellular traps and inflammatory pathologies-untangling the web! Front Immunol 9
Murdoch C, Muthana M, Coffelt SB, Lewis CE (2008) The role of myeloid cells in the promotion of tumour angiogenesis. Nat Rev Cancer
Noel P (2012) Eosinophilic myeloid disorders. Semin Hematol 49:120–127
Odemuyiwa SO et al (2004) Cutting edge: human eosinophils regulate T cell subset selection through Indoleamine 2,3-dioxygenase. J Immunol 173:5909–5913
Ohno I et al (1996) Transforming growth factor β1 (TGFβ1) gene expression by eosinophils in asthmatic airway inflammation. Am J Respir Cell Mol Biol 15:404–409
Ortega HG et al (2014) Mepolizumab treatment in patients with severe eosinophilic asthma. N Engl J Med 371:1198–1207
Ownby HE, Roi LD, Isenberg RR, Brennan MJ (1983) Peripheral lymphocyte and eosinophil counts as indicators of prognosis in primary breast cancer. Cancer
Padigel UM et al (2007) Eosinophils act as antigen-presenting cells to induce immunity to Strongyloides stercoralis in mice. J Infect Dis 196:1844–1851
Pégorier S, Wagner LA, Gleich GJ, Pretolani M (2006) Eosinophil-derived cationic proteins activate the synthesis of remodeling factors by airway epithelial cells. J Immunol 177:4861–4869
Pereira MC, Oliveira DT, Kowalski LP (2011) The role of eosinophils and eosinophil cationic protein in oral cancer: a review. Arch Oral Biol 56:353–358
Persson T et al (2001) Bactericidal activity of human eosinophilic granulocytes against Escherichia coli. Infect Immun 69:3591–3596
Phipps S et al (2007) Eosinophils contribute to innate antiviral immunity and promote clearance of respiratory syncytial virus. Blood 110:1578–1586
Piehler D et al (2011) Eosinophils contribute to IL-4 production and shape the T-helper cytokine profile and inflammatory response in pulmonary cryptococcosis. Am J Pathol 179:733–744
Platt EJ, Wehrly K, Kuhmann SE, Chesebro B, Kabat D (1998) Effects of CCR5 and CD4 cell surface concentrations on infections by Macrophagetropic isolates of human immunodeficiency virus type 1. J Virol 72:2855–2864
Ponikau JU et al (2005) Striking deposition of toxic eosinophil major basic protein in mucus: implications for chronic rhinosinusitis. J Allergy Clin Immunol 116:362–369
Prizment AE, Anderson KE, Visvanathan K, Folsom AR (2011) Inverse association of eosinophil count with colorectal cancer incidence: atherosclerosis risk in communities study. Cancer Epidemiol Biomark Prev
Provost V et al (2013) CCL26/eotaxin-3 is more effective to induce the migration of eosinophils of asthmatics than CCL11/eotaxin-1 and CCL24/eotaxin-2. J Leukoc Biol 94:213–222
Puxeddu I et al (2010) Osteopontin is expressed and functional in human eosinophils. Allergy Eur J Allergy Clin Immunol
Rached AA, Hajj WE (2016) Eosinophilic gastroenteritis: approach to diagnosis and management. World J Gastrointest Pharmacol Ther 7:513
Rakesh N, Devi Y, Majumdar K, Reddy SS, Agarwal K (2015) Tumour associated tissue eosinophilia as a predictor of locoregional recurrence in oral squamous cell carcinoma. J Clin Exp Dent
Ramirez GA et al (2018) Eosinophils from physiology to disease: a comprehensive review. Biomed Res Int 2018
Ravin KA, Loy M (2015) The eosinophil in infection. Clin Rev Allergy Immunol 50:214–227
Reichman H et al (2019) Activated eosinophils exert antitumorigenic activities in colorectal cancer. Cancer Immunol Res 7:388–400
Reiter A, Gotlib J (2017) Myeloid neoplasms with eosinophilia. Ann Intern Med 129:704–714

Rosenberg HF, Dyer KD, Foster PS (2013) Eosinophils: changing perspectives in health and disease. Nat Rev Immunol 13:9–22

Rothenberg ME, Hogan SP (2006) The Eosinophil. Annu Rev Immunol 24:147–174

Roufosse F, Cogan E, Goldman M (2007) Lymphocytic variant hypereosinophilic syndromes. Immunol Allergy Clin N Am

Roufosse F, Garaud S, De Leval L (2012) Lymphoproliferative disorders associated with hypereosinophilia. Semin Hematol

Sablé-Fourtassou R et al (2005) Antineutrophil cytoplasmic antibodies and the Churg-Strauss syndrome. Ann Intern Med 143:632–638

Sakkal S, Miller S, Apostolopoulos V, Nurgali K (2016) Eosinophils in cancer: favourable or unfavourable? Curr Med Chem

Samarasinghe AE et al (2017) Eosinophils promote antiviral immunity in mice infected with influenza a virus. J Immunol 198:3214–3226

Samoszuk M (1997) Eosinophils and human cancer. Histol Histopathol

Saraiva AL, Carneiro F (2018) New insights into the role of tissue eosinophils in the progression of colorectal cancer: a literature review. Acta Med Portuguesa 31:329–337

Schrezenmeier H, Thome SD, Tewald F, Fleischer B, Raghavachar A (1993) Interleukin-5 is the predominant eosinophilopoietin produced by cloned T lymphocytes in hypereosinophilic syndrome. Exp Hematol

Schroeder JW et al (2019) Anti-neutrophil cytoplasmic antibodies positivity and anti-leukotrienes in eosinophilic granulomatosis with polyangiitis: a retrospective monocentric study on 134 Italian patients. Int Arch Allergy Immunol 180:64–71

Shakoory B, Fitzgerald SM, Lee SA, Chi DS, Krishnaswamy G (2004) The role of human mast cell-derived cytokines in eosinophil biology. J Interferon Cytokine Res 24:271–281

Shamri R, Xenakis JJ, Spencer LA (2011) Eosinophils in innate immunity: an evolving story. Cell Tissue Res 343:57–83

Shomali W, Gotlib J (2019) World Health Organization-defined eosinophilic disorders: 2019 update on diagnosis, risk stratification, and management. Am J Hematol 94:1149–1167

Silveira JS et al (2019) Respiratory syncytial virus increases eosinophil extracellular traps in a murine model of asthma. Asia Pac Allergy 9

Simon HU et al (1996) Expansion of cytokine-producing CD4-CD8- T cells associated with abnormal Fas expression and hypereosinophilia. J Exp Med

Simon HU, Plötz SG, Dummer R, Blaser K (1999) Abnormal clones of T cells producing interleukin-5 in idiopathic eosinophilia. N Engl J Med

Simon HU et al (2010) Refining the definition of hypereosinophilic syndrome. J Allergy Clin Immunol 126:45–49

Simon D et al (2011) Eosinophil extracellular DNA traps in skin diseases. J Allergy Clin Immunol 127:194–199

Siracusa MC et al (2011) TSLP promotes interleukin-3-independent basophil haematopoiesis and type 2 inflammation. Nature 477:229–233

Smyth CM et al (2013) Activated eosinophils in association with enteric nerves in inflammatory bowel disease. PLoS One 8:e64216

Spencer LA et al (2006) Cytokine receptor-mediated trafficking of preformed IL-4 in eosinophils identifies an innate immune mechanism of cytokine secretion. Proc Natl Acad Sci U S A 103:3333–3338

Spencer LA, Bonjour K, Melo RCN, Weller PF (2014) Eosinophil secretion of granule-derived cytokines. Front Immunol 5

Stenfeldt AL, Wennerås C (2004) Danger signals derived from stressed and necrotic epithelial cells activate human eosinophils. Immunology 112:605–614

Straumann A et al (2010) Anti-interleukin-5 antibody treatment (mepolizumab) in active eosinophilic oesophagitis: a randomised, placebo-controlled, double-blind trial. Gut 59:21–30

Su Y-C et al (2015) Dual Proinflammatory and antiviral properties of pulmonary eosinophils in respiratory syncytial virus vaccine-enhanced disease. J Virol 89:1564–1578

Subeikshanan V et al (2016) A prospective comparative clinical study of peripheral blood counts and indices in patients with primary brain tumors. J Postgrad Med

Suchin KR et al (2001) Increased interleukin 5 production in eosinophilic Sézary syndrome: regulation by interferon alfa and interleukin 12. J Am Acad Dermatol

Suzuki Y, Suda T (2019) Eosinophilic pneumonia: a review of the previous literature, causes, diagnosis, and management. Allergol Int 68:413–419

Svensson L, Wennerås C (2005) Human eosinophils selectively recognize and become activated by bacteria belonging to different taxonomic groups. Microbes Infect 7:720–728

Taylor RJ, Schols D, Wooley DP (2004) Restricted entry of R5 HIV type 1 strains into eosinophilic cells. AIDS Res Hum Retrovir 20:1244–1253

Terai M et al (2011) Early induction of interleukin-5 and peripheral eosinophilia in acute pneumonia in Japanese children infected by pandemic 2009 influenza a in the Tokyo area. Microbiol Immunol 55:341–346

Teruya-Feldstein J et al (1999) Differential chemokine expression in tissues involved by Hodgkin's disease: direct correlation of eotaxin expression and tissue eosinophilia. Blood

Thielen C et al (2008) TARC and IL-5 expression correlates with tissue eosinophilia in peripheral T-cell lymphomas. Leuk Res

Thorne KJI, Glauert AM, Svvennsen RJ, Franks D (1979) Phağocytosis and killinġ of Trypanosoma dionisii by human neutrophils, eosinophils and monocytes. Parasitology 79:367–379

Throsby M, Herbelin A, Pléau J-M, Dardenne M (2000) CD11c + eosinophils in the murine thymus:

developmental regulation and recruitment upon MHC class I-restricted thymocyte deletion. J Immunol 165:1965–1975

Torrent M, Navarro S, Moussaoui M, Nogués MV, Boix E (2008) Eosinophil cationic protein high-affinity binding to bacteria-wall lipopolysaccharides and peptidoglycans. Biochemistry 47:3544–3555

Vaillant L, La Ruche G, Tarantola A, Barboza P (2009) Epidemiology of fatal cases associated with pandemic H1N1 influenza 2009. Euro Surveill 14

Valent P, Sperr WR, Schwartz LB, Horny HP (2004) Diagnosis and classification of mast cell proliferative disorders: delineation from immunologic diseases and non-mast cell hematopoietic neoplasms. J Allergy Clin Immunol

van Kerkhove MD et al (2011) Risk factors for severe outcomes following 2009 influenza a (H1N1) infection: a global pooled analysis. PLoS Med 8

Varricchi G et al (2018) Eosinophils: the unsung heroes in cancer? OncoImmunology

Von Köckritz-Blickwede M, Nizet V (2009) Innate immunity turned inside-out: antimicrobial defense by phagocyte extracellular traps. J Mol Med 87:775–783

Vrtis, W. M. et al. Virus-specific T cells eosinophils bind rhinovirus and activate. (2020)

Wang YH et al (2007) IL-25 augments type 2 immune responses by enhancing the expansion and functions of TSLP-DC-activated Th2 memory cells. J Exp Med 204:1837–1847

Wardlaw AL (1999) Molecular basis for selective eosinophil trafficking in asthma: a multistep paradigm. J Allergy Clin Immunol 104:917–926

Wechsler ME et al (2017) Mepolizumab or placebo for eosinophilic granulomatosis with Polyangiitis. N Engl J Med 376:1921–1932

Weller PF, Goetzl EJ (1980) The human eosinophil: roles in host defense and tissue injury. Am J Pathol 100:791–820

Weller PF, Spencer LA (2017) Functions of tissue-resident eosinophils. Nat Rev Immunol 17:746–760

Wen T, Rothenberg ME (2016) The regulatory function of eosinophils. In: Myeloid cells in health and disease, vol 4. American Society of Microbiology, pp 257–269

Werfel T et al (2016) Cellular and molecular immunologic mechanisms in patients with atopic dermatitis. J Allergy Clin Immunol 138:336–349

White JR et al (1997) Cloning and functional characterization of a novel human CC chemokine that binds to the CCR3 receptor and activates human eosinophils. J Leukoc Biol 62:667–675

Williams KW et al (2016) Hypereosinophilia in children and adults: a retrospective comparison. J Allergy Clin Immunol. Pract 4:941–947.e1

Wimazal F et al (2010) Evaluation of the prognostic significance of eosinophilia and basophilia in a larger cohort of patients with myelodysplastic syndromes. Cancer

Wong CK, Cheung PFY, Ip WK, Lam CWK (2007) Intracellular signaling mechanisms regulating toll-like receptor-mediated activation of eosinophils. Am J Respir Cell Mol Biol 37:85–96

Wong CK, Ng SSM, Lun SWM, Cao J, Lam CWK (2009) Signalling mechanisms regulating the activation of human eosinophils by mast-cell-derived chymase: implications for mast cell-eosinophil interaction in allergic inflammation. Immunology 126:579–587

Wong CK, Hu S, Cheung PFY, Lam CWK (2010) Thymic stromal lymphopoietin induces chemotactic and prosurvival effects in eosinophils: implications in allergic inflammation. Am J Respir Cell Mol Biol 43:305–315

Wong TW, Doyle AD, Lee JJ, Jelinek DF (2014) Eosinophils regulate peripheral B cell numbers in both mice and humans. J Immunol 192:3548–3558

Yancey SW et al (2017) Biomarkers for severe eosinophilic asthma. J Allergy Clin Immunol 140:1509–1518

Yang D et al (2003) Eosinophil-derived neurotoxin (EDN), an antimicrobial protein with chemotactic activities for dendritic cells. Blood 102:3396–3403

Yang D et al (2004) Human ribonuclease a superfamily members, eosinophil-derived neurotoxin and pancreatic ribonuclease, induce dendritic cell maturation and activation. J Immunol 173:6134–6142

Yellapurkar S et al (2016) Tumour-associated tissue eosinophilia in oral squamous cell carcinoma- a boon or a bane? J Clin Diagnos Res

Yousefi S et al (2008) Catapult-like release of mitochondrial DNA by eosinophils contributes to antibacterial defense. Nat Med 14:949–953

Yousefi S et al (2018) Oxidative damage of SP-D abolishes control of eosinophil extracellular DNA trap formation. J Leukoc Biol 104:205–214

Zeck-Kapp G, Kroegel C, Riede UN, Kapp A (1995) Mechanisms of human eosinophil activation by complement protein C5a and platelet-activating factor: similar functional responses are accompanied by different morphologic alterations. Allergy 50:34–47

Zhang MM, Li YQ (2017) Eosinophilic gastroenteritis: a state-of-the-art review. J Gastroenterol Hepatol (Australia) 32:64–72

Zhang J et al (2012) Pentraxin 3 (PTX3) expression in allergic asthmatic airways: role in airway smooth muscle migration and chemokine production. PLoS One 7

Zhang Y et al (2014) Clinical impact of tumor-infiltrating inflammatory cells in primary small cell esophageal carcinoma. Int J Mol Sci

Zhang JJ et al (2020) Clinical characteristics of 140 patients infected with SARS-CoV-2 in Wuhan, China. Allergy Eur J Allergy Clin Immunol

Ziegler SF et al (2013) The biology of thymic stromal lymphopoietin (TSLP). In: Advances in pharmacology, vol 66. Academic, pp 129–155

Adv Exp Med Biol - Cell Biology and Translational Medicine (2021) 14: 221–224
https://doi.org/10.1007/978-3-030-80492-3

Index

Printed by Books on Demand, Germany